MW00452637

STARTUP CXO

A Field Guide to Scaling Up Your Company's Critical Functions and Teams

Matt Blumberg with
Peter M. Birkeland

WILEY

Copyright © 2021 by Bolster Network and Matthew Y. Blumberg.
All rights reserved.

Published by John Wiley & Sons, Inc., Hoboken, New Jersey.

Published simultaneously in Canada.

No part of this publication may be reproduced, stored in a retrieval system, or transmitted in any form or by any means, electronic, mechanical, photocopying, recording, scanning, or otherwise, except as permitted under Section 107 or 108 of the 1976 United States Copyright Act, without either the prior written permission of the Publisher, or authorization through payment of the appropriate per-copy fee to the Copyright Clearance Center, Inc., 222 Rosewood Drive, Danvers, MA 01923, (978) 750-8400, fax (978) 646-8600, or on the Web at www.copyright.com. Requests to the Publisher for permission should be addressed to the Permissions Department, John Wiley & Sons, Inc., 111 River Street, Hoboken, NJ 07030, (201) 748-6011, fax (201) 748-6008, or online at www.wiley.com/go/permissions.

Limit of Liability/Disclaimer of Warranty: While the publisher and author have used their best efforts in preparing this book, they make no representations or warranties with respect to the accuracy or completeness of the contents of this book and specifically disclaim any implied warranties of merchantability or fitness for a particular purpose. No warranty may be created or extended by sales representatives or written sales materials. The advice and strategies contained herein may not be suitable for your situation. You should consult with a professional where appropriate. Neither the publisher nor author shall be liable for any loss of profit or any other commercial damages, including but not limited to special, incidental, consequential, or other damages.

For general information on our other products and services or for technical support, please contact our Customer Care Department within the United States at (800) 762-2974, outside the United States at (317) 572-3993, or fax (317) 572-4002.

Wiley publishes in a variety of print and electronic formats and by print-on-demand. Some material included with standard print versions of this book may not be included in e-books or in print-on-demand. If this book refers to media such as a CD or DVD that is not included in the version you purchased, you may download this material at http://booksupport.wiley.com. For more information about Wiley products, visit www.wiley.com.

Library of Congress Cataloging-in-Publication Data is Available:

ISBN 9781119772576 (Hardback)
ISBN 9781119774068 (ePub)
ISBN 9781119774075 (ePDF)

Cover Design: Wiley

SKY10025958_042721

This book is dedicated to the entrepreneurs, founders, and leadership teams in startups and scaleups everywhere who are constantly striving to do better so they can turn ideas into products and companies that transform communities, markets, and the world.

If you want to go fast, go alone. If you want to go far, go together.
—African proverb

I haven't come this far, only to come this far.

—Unknown

Contents

Foreword xvii

PART ONE: Introduction **1**

Matt Blumberg

Introduction **2**

Chapter 1: The Nature of a CXO's Role **9**

Chapter 2: Scaling a CXO **12**

PART TWO: Finance and Administration **15**

Jack Sinclair

Chief Financial Officer **16**

Chapter 3: In the Beginning: Laying the CFO Foundation **19**

Chapter 4: Fundraising **22**

Chapter 5: Size of Opportunity **25**

Chapter 6: Financial Plan **27**

Chapter 7: Unit Economics and KPIs **30**

Chapter 8: Investor Ecosystem Research **32**

Chapter 9: Pricing and Valuation **34**

Chapter 10: Due Diligence and Corporate Documentation **37**

Chapter 11: Using External Counsel 40

Chapter 12: Operational Accounting 42

Chapter 13: Treasury and Cash Management 49

Chapter 14: Building an In-House Accounting Team 52

Chapter 15: International Operations 55

Chapter 16: Strategic Finance 58

Chapter 17: Other Areas to Partner With 67

Chapter 18: High Impact Areas for the Startup CFO as Partner 71

Chapter 19: Board and Shareholder Management 77

Chapter 20: Equity 80

Chapter 21: Mergers and Acquisitions (M&A) 85

Chapter 22: Bonus Section: What We Used for Our Internal Systems When We Started Bolster 91

Chapter 23: CEO-to-CEO Advice About the Finance Role 97

Matt Blumberg

PART THREE: People and Human Resources 101

Cathy Hawley

Chief People Officer 102

Chapter 24: Values and Culture 105

Chapter 25: Diversity, Equity, and Inclusion (DE&I) 111

Chapter 26: Building Your Team 113

Chapter 27: Organizational Design and Operating Systems 118

Chapter 28: Team Development 124

Chapter 29: Leadership Development 127

Chapter 30: Talent and Performance Management 130

Chapter 31: Career Pathing 132

Chapter 32: Role-Specific Learning and Development 134

Chapter 33: Employee Engagement 136

Chapter 34: Rewards and Recognition 138

Chapter 35: Reductions in Force 140

Chapter 36: Recruiting 142

Chapter 37: Onboarding 149

Chapter 38: Compensation 152

Chapter 39: People Operations 154

Chapter 40: Systems 164

Chapter 41: CEO-to-CEO Advice About the People/HR Role 167

Matt Blumberg

PART FOUR: Marketing 173

Nick Badgett and Holly Enneking

Chief Marketing Officer 174

Chapter 42: Where to Start 177

Chapter 43: Generating Demand for Sales 181

Chapter 44: Supporting the Company Culture 186

Chapter 45: Breaking Down Marketing's Functions 191

Chapter 46: Events 204

Chapter 47: Content and Communications 212

Chapter 48: Product Marketing 218

Chapter 49: Marketing Operations 223

Chapter 50: Sales Development 226

Chapter 51: Marketing as a Partner/Collaborating with the Rest of the C-suite 233

Chapter 52: Building a Marketing Machine (Scaleup) 236

Chapter 53: CEO-to-CEO Advice About the Marketing Role 243

Matt Blumberg

PART FIVE: Sales 249

Anita Absey

Chief Revenue Officer 250

Chapter 54: In the Beginning: From Prospect to Customer 251

Chapter 55: Hiring the Right People 254

Chapter 56: Profile of Successful Salespeople 257

Chapter 57: Some Myth Busting 260

Chapter 58: Compensating Sales Team Members 262

Chapter 59: Pipeline 266

Chapter 60: Scaling the Sales Organization 268

Chapter 61: Scaling Your Team Through Culture 271

Chapter 62: Scaling Sales Process and Methodology 276

Chapter 63: Scaling the Operating System 279

Chapter 64: Marketing Alignment 282

Chapter 65: Market Assessment and Alignment 286

Chapter 66: Expanding Distribution Channels 288

Chapter 67: Geographic Expansion 291

Chapter 68: Pricing and Packaging 294

Chapter 69: CEO-to-CEO Advice About the Sales Role 300

Matt Blumberg

PART SIX: Business/Corporate Development 305

Ken Takahashi

Chief Business Development Officer 306

Chapter 70: How to Make the Biggest Impact as a CBDO 311

Chapter 71: Building Your Influence Internally 314

Chapter 72: Building Your Influence Externally 318

Chapter 73: Where Internal and External Meet: Your Relationship with Your CEO 323

Chapter 74: Influence Meets Operating System 325

Chapter 75: Develop External Trust for the Company 327

Chapter 76: Build Your Influence in Strategy 329

Chapter 77: Building Your Influence in Business Development 330

Chapter 78: When Things Go Wrong in a Partnership ... and They Will 335

Chapter 79: Geographic Expansion 338

Chapter 80: M&A: Buy Side 341

Chapter 81: M&A: Sell Side 344

Chapter 82: CEO-to-CEO Advice About the Business/Corporate Development Role 348

Matt Blumberg

PART SEVEN: Customer Success/Account Management 353

George Bilbrey

Chief Customer Officer 354

Chapter 83: Five Misperceptions 357

Chapter 84: Startup Customer Success Organization 360

Chapter 85: Scaling the Service Organization 362

Chapter 86: Timing: When to Hire Your Team 366

Chapter 87: Customer Segmentation and Journey 368

Chapter 88: Understanding Customers 372

Chapter 89: Understanding Customers Through Metrics 374

Chapter 90: Foundations of a Great Customer Service Organization 379

Chapter 91: Building an Effective Team 385

Chapter 92: Partnering with the Organization 387

Chapter 93: Five Eternal Questions 391

Chapter 94: CEO-to-CEO Advice About the Customer Success Role 396

Matt Blumberg

PART EIGHT: Product and Engineering 401

Shawn Nussbaum

Chief Product Officer and Chief Technology Officer 402

Chapter 95: The Product Development Leaders 405

Chapter 96: Product Development Culture 412

Chapter 97: Technical Strategy: Proportional Engineering Investment and Managing Technical Debt 416

Chapter 98: Shifting to a New Development Culture 424

Chapter 99: Starting Things 427

Chapter 100: Hiring Product Development Team Members 434

Chapter 101: Increasing the Funnel and Building Diverse Teams 442

Chapter 102: Retaining and Career Pathing People 446

Chapter 103: Hiring and Growing Leaders 449

Chapter 104: Organizing, Collaborating with, and Motivating Effective Teams 455

Chapter 105: Due Diligence and Lessons Learned from a Sale Process 468

Chapter 106: Selling Your Company: Preparation 475

Chapter 107: Selling Your Company: Telling the Story 479

Chapter 108: CEO-to-CEO Advice About the Product/Engineering Role 482

Matt Blumberg

PART NINE: Privacy 487

Dennis Dayman

Chief Privacy Officer 488

Chapter 109: The Role of Privacy Officer 491

Chapter 110: Privacy Advice for Startups 494

Chapter 111: Legal Documents 500

Chapter 112: The European Union 505

Chapter 113: Data Mapping 507

Chapter 114: Data Breach 510

Chapter 115: Least Privileged Access 515

Chapter 116: Employee ~~Training~~ Engagement 519

Chapter 117: Building Your Privacy Team in a Startup 522

Chapter 118: Building Your Privacy Team as You Scaleup 525

Chapter 119: Certifications 527

Chapter 120: Assessments 529

Chapter 121: CEO-to-CEO Advice About the Privacy Role 536

Matt Blumberg

PART TEN: Operations 541

Jack Sinclair

Chief Operating Officer 542

Chapter 122: CEO-to-CEO Advice About the Operating Role 549

Matt Blumberg

PART ELEVEN: The Future of Executive Work 551

Chapter 123: The Future of Executive Work 553

Matt Blumberg

Chapter 124: Fractional Chief Financial Officer 556

John McCarthy

Chapter 125: Fractional Chief People Officer 562

Courtney Graeber

Chapter 126: Fractional Chief Marketing Officer 567

Scott Kabat

Chapter 127: Fractional Chief Revenue Officer 571

B.J. Bushur

Chapter 128: Fractional Chief Revenue Officer 576

Sherri Sklar

Chapter 129: Fractional Chief Business Development Officer 580

Jon Guttenberg

Chapter 130: Fractional Chief Customer Officer **586**

Amy Mustoe

Chapter 131: Fractional Chief Product/Technology Officer **590**

Drew Dillon

Chapter 132: Fractional Chief Privacy Officer **594**

Teresa Troester-Falk

Conclusion 599

Epilogue 601

Pete Birkeland

References 603

Acknowledgments 604

About the Authors 606

Index 613

Foreword

As a first-time tech founder and CEO at ExactTarget, one phrase kept ringing through my head: "I don't know what I don't know." Even with 10+ years of business experience and a freshly minted MBA degree, I had so many blind spots having never built software, raised venture capital or even led a multi-functional organization. Filling in these gaps took many years and lots of trial and error.

One of my gap-filling strategies was to learn from other CEOs going through similar highs and lows of scaling their company. On this journey, I was fortunate to meet Matt Blumberg. As CEO of Return Path, Matt was building a high growth company in the same digital marketing industry as me. I was impressed by his leadership, strategic thinking, and commitment to helping other entrepreneurs. We became fast friends and we both looked forward to learning from one another.

Matt was always gracious and willing to take my call or have dinner together. Our conversations covered every topic imaginable from leadership to board management to strategic partnerships to international expansion. My advice to entrepreneurs and leaders—build your peer network and spend time developing and nurturing these relationships. While board members and advisors are an important source of knowledge, learning from peers can be invaluable.

One of the highlights of my relationship with Matt was when we were both invited to the White House to witness President Obama signing the Jumpstart our Business Startups Act (or JOBS Act) in April of 2012. With bi-partisan support, the law opened up crowdfunding for startups and streamlined the IPO path. ExactTarget had just gone public two weeks prior so I knew the benefits that the JOBS Act would bring to entrepreneurs. But what I didn't know in April 2012 was that the event would foreshadow my relationship with Matt and how we would work together supporting entrepreneurs and startup ecosystems.

Fast forward to 2021. We are facing unprecedented challenges in the world and the need for innovation and leadership has never been greater. CEOs and functional leaders need tools and resources to accelerate their learning curves and the learning curves of those around them. Speed of learning, thought, and action are more important now than ever. This is why I am so excited about Matt's latest book, *Startup CXO*.

Startup CXO provides a comprehensive field guide to starting and scaling tech companies. Really, the information is super helpful to any company. It provides a "book within a book" framework to enable and empower readers to jump into any section as needed. And it's written by practitioners who provide tons of tangible advice and actionable insights. By reading this book, I believe that leaders will be better equipped to build great companies and anticipate what's around every corner.

In my view, the best CEOs have a grasp of all functions. They can go a mile wide and a couple inches deep. They hire A+ talent and build a culture that brings out the very best in people. They understand how Sales and Marketing fit together, they value HR, Finance, and Legal and understand their interdependencies, they have a clear vision for how Product and Engineering fit together, they know how to be aggressive and how to manage risk, and so much more.

So, *Startup CXO* is an amazing resource for CEOs but also for functional leaders and professionals at any stage of their career. The best functional leaders and professionals understand that cross-functional teamwork is everything. It's so important to have insight and empathy for how other areas of the organization operate. The big picture is needed to see how all of the puzzle pieces fit together.

I feel so lucky that through our venture studio, High Alpha, I have the opportunity to work with Matt and his leadership team as we build Bolster—a talent marketplace for startup and scaleup tech companies. We are living and applying the concepts and lessons contained in this very book!

My wish for you is that reading *Startup CXO* minimizes your "I don't know what I don't know" list; that it accelerates your development, your curiosity, your ability to ask the right questions, and helps you surround yourself with the right talent. My wish is that you dream big, lead with purpose and integrity, and master your craft. I hope—and believe—that *Startup CXO* will be a helpful companion for you on your company-building journey.

Good luck!

Scott Dorsey
Managing Partner, High Alpha
April 2021

Part One

Introduction

Matt Blumberg

Introduction

After selling Return Path, and finishing up the second edition of *Startup CEO: A Field Guide to Scaling Up Your Business*, my colleagues and I at our new startup Bolster started envisioning a new book as a sequel or companion to *Startup CEO*.

Simply put, the first book left me with the nagging feeling that it wasn't enough to only help CEOs excel, because starting and scaling a business is a collective effort. What about the other critical leadership functions that are needed to grow a company? If you're leading HR, or Finance, or Marketing, or any key function inside a startup, what resources are available to you? What should you be thinking about? What does "great" look like for your function? What challenges lurk around the corner as you scale your function that you might not be focused on today? What are your fellow executives focused on in their own departments, and how can you best work together? If you're a CEO who has never managed all these functions before, what should you be looking for when you hire and manage all these people? If you're an aspiring executive, from entry-level to manager to director, what do you need to think about as you grow your career and develop your skills? And if you're a board member or investor, what scorecard or metrics are you using to ensure your companies and investments are achieving greatness?

That was the origin of this new book, *Startup CXO: A Field Guide to Scaling Up Your Company's Critical Functions and Teams*. This book is a "book of books," with eleven separate, detailed Parts, one for each major function inside a company, each composed of several discrete short chapters outlining the key playbooks for each functional leadership role in the company. Because it covers CFOs, CMOs, CPOs, etc.—we landed on "Startup CXO" as the name.

America's "startup revolution" continues to gather steam. There are increasing numbers of venture capital investors, seed funds, and accelerators supporting increasing numbers of entrepreneurial ventures. While there are a number of books in the marketplace about CEOs and leadership, and some about individual functional disciplines (lots of books on the topic of Sales, the topic of Product Development, and the like), there are very few books that are practical how-to guides for any individual function, and none that wrap all these functions into a compendium that can be used by a whole startup executive team. Very simply, each Part in this book will serve as a how-to guide for a given executive, and taken together, the book will be a good how-to guide for startup executive teams in general. The eleven Parts are:

Part 1 Introduction (Chapters 1 and 2)
Part 2 Finance and Administration (Chapters 3–23)
Part 3 People and Human Resources (Chapters 24–41)
Part 4 Marketing (Chapters 42–53)
Part 5 Sales (Chapters 54–69)
Part 6 Business/Corporate Development (Chapters 70–82)
Part 7 Customer Success and Account Management (Chapters 83–94)
Part 8 Product and Engineering (Chapters 95–108)
Part 9 Privacy (Chapters 109–121)
Part 10 Operations (Chapter 122)
Part 11 The Future of Executive Work (Chapters 123–132)

We also have a number of ancillary materials—templates and charts—that would normally be included in an Appendix. Because the book is already over 600 pages we decided to provide that material externally on our Startup Revolution website: "http://www.startuprev.com" www.startuprev.com.

This book carries my name as its principal author, and although I'm writing parts of it and editing it, I'm not THE author, I'm AN author. That's an important point. The book has a very large number of contributors and external reviewers and you'll probably notice a different "voice" for each Part. That's what you should see with multiple authors, but despite the various perspectives, we're all focused on providing a playbook for each functional area including the mistakes we made, what we would do differently, and what worked really well. Each Part has one or two principal authors who have the experience, credibility, and expertise to share something of value with others in their specific functional disciplines—most of my Bolster co-founders are writing Parts, and the others are being written by former Return Path executive colleagues or members of Bolster's network. And that is a good lead-in to a few caveats before you embark on the book.

First, although most of the book is being written by former Return Path executives, it is not meant to be the Return Path story. Every author here has 20–30 years of experience working at multiple companies of different sizes and at different stages and in different sectors on which he or she is drawing. It's also not the story of Bolster, the new company that a number of us started earlier this year, although *Startup CXO* is pretty closely related to Bolster's business of helping assess and place on-demand CXO talent.

Related, this book is based on the experience of the contributors—as with *Startup CEO*, for the most part, that means U.S.-based tech or tech-enabled services businesses. Most of us have more B2B experience than B2C, although a number of us have both. Some of the authors write a lot about people, some write a lot about process, some are philosophical, some are more practical and tactical. They reflect the nature of those functions and the nature of the writers of those Parts. I hope the book proves to be timeless and that it spans cultural and industry boundaries but there will be some inherent limitations based on our own experience.

A few notes on language. We realize that not every leadership role in a startup is actually a "C"-level role. Sometimes the most senior person running a functional department is an SVP, a VP, or even a Director or Manager. But *Startup Functional Leader* is a lousy title for a book. There are also some elements of language worth noting up front. First, we got a lot of feedback from people we trust and respect on how to handle gender pronouns, and unfortunately, the feedback was not consistent—some felt we should alternate masculine and feminine pronouns, and some felt we should go with the plural. Others thought we should use female pronouns to compensate for what has historically been a male-dominated perspective. Given the lack of an obvious standard here, for this book, we chose to use the gender-neutral plural terms (even though it looks a little funny to a grammar stickler like me). Second, we use the words startup and scaleup in the book without precise revenue-based or employee-count-based definitions, but you should assume that startups are smaller companies, whereas scaleups are ones that have already reached some meaningful level of critical mass. Third, we use terms like "executive team," "leadership team," and "executive committee" interchangeably to refer to a company's senior-most group of leaders. Finally, we frequently refer to the concept of an "operating system." I talk about this at length in *Startup CEO*, but basically, it means—whether for a person, a team, or a company—the collection of meeting and communication routines and operating practices that form the cadence of a team's work.

A final note about the roles we selected for this book. We tried to stick with the basics: Heads of Finance, Marketing, Sales, Customer Service, Product/Technology, HR/People, Business Development, and Privacy. Some of these don't necessarily exist in all companies, for example, a lot of B2C companies don't have a dedicated Head of Sales, and instead a lot of the responsibility for revenue generation sits within Marketing or Business Development. And not everyone has a Head of Privacy or data protection (although these days, most companies probably should). These Parts are still worth reading, as someone in your company will be fulfilling those responsibilities. There are certainly other C-level roles you'll find in large companies, and even in smaller ones that we could have added but chose not to—mainstream ones like Chief Information Security Officer and General Counsel, and even newer niche ones like Chief Diversity Officer and Chief Sustainability Officer. And there are plenty of roles in other industries that this book skips entirely for now, like Manufacturing and Logistics. Just because we don't cover a role in this book doesn't mean we don't value it ... we just had to draw the line somewhere. We also had a lengthy discussion before writing the book about what to do with the role of COO (Chief Operating Officer). Because the responsibilities associated with the COO role vary widely company by company, we landed on including a single small chapter to just talk about the different types of COO out there. Most of the functions covered by COOs are covered elsewhere in the book, or in *Startup CEO*.

Because *Startup CXO* is a book of books, and designed as a field guide, it's not necessary to read it from front to back, although you certainly can. The choice of the words "Field Guide" in the subtitle was deliberate because a field guide is a handy resource that you quickly consult. On the continuum of written works, you have a dictionary on one end (understand the meaning of words) and a book on the other end (understand ideas and concepts). A field guide sits right in the middle. A field guide has elements of definitions, ideas, and concepts, but the main purpose is to help the reader identify something, understand it quickly, and be able to apply what they've read to their situation. If you aren't reading cover to cover, read the two general chapters up front and then pick and choose. Pick your own function if you're a CXO and start there. Then move on to the function of one of your colleagues where maybe you're having some kind of friction at the moment, so you can build empathy with that colleague. We've organized the contributions into closely related groups like corporate (Finance, HR), go-to-market (Marketing, Sales, Business Development, Customers), and product (Product, Privacy, Operations) to help you learn about functions that likely interact with each other extensively.

If you're a CEO, you could start with the function you "grew up in" and then move on to whatever function you need to hire, or you're most concerned with, or even the one that's working the best so you can gain some additional insight into why—and how to replicate that success in other places. I also have a "CEO-to-CEO Advice" section at the end of each functional Part and in those I share my thoughts on what "great" looks like for each CXO, signs that your CXO isn't scaling, and how I engage with the CXO. CEOs, Board members, and investors can quickly get an overview and understanding of each function by reading those.

Regardless of what role you play in a company and what experience you bring to the role, I hope this book speaks to you and inspires you in some way—that it's a playbook for something meaningful to you. If you're a CEO, maybe it will help you figure out who to hire or how to more effectively manage a direct report by telling you what "great" looks like for that function. If you're a functional leader in a startup, maybe it will help you focus on some aspect of your role you hadn't thought about yet. If you're an aspiring leader, maybe it will give you some insight into the kinds of steps you need to take in order to grow your career. Whichever persona you are, on behalf of all of the book's contributors, we hope you gain some insight, and we thank you for reading *Startup CXO*.

Chapter 1

The Nature of a CXO's Role

I was struck by something as I read over the nearly complete manuscript of *Startup CXO* for the first time: each CXO believes that their part of the business is the most important part. And they make a compelling set of arguments:

Shawn: If you don't have a good product, you don't have a business.

Anita: If you don't have revenues, you don't have a business.

Ken: If you don't develop the ecosystem, you don't have a business.

Nick: If you don't generate market opportunities, you don't have a business.

George: If you don't create exceptional customer experiences, you don't have a business.

Cathy: If you don't recruit, train, and develop the right people, you don't have a business.

Jack: If you don't have the cash, you don't have a business.

Dennis: If you don't bake privacy in at the beginning, you don't have a business.

We had a debate years ago at a Return Path Board meeting as to whether we were a sales-driven business or a product-driven business—and more important, whether we should be one or the other. Two of our Board members, both of whom I respect tremendously, were anchoring the different points of view, Scott Petry, on the product side, talking about how successful Apple was at getting customers to camp out overnight to be the first ones to buy the newest iThing; and Greg Sands, on the sales side, talking about how successful Oracle was at getting product into the hands of customers. I took a devil's advocate point of view in the conversation, true to our operating philosophy at Return Path, which was that HR/People was the most important function because we were a people-driven business.

So, who is right? Are the best companies sales-driven, product-driven, people-driven, or something else? Which of the CXO's functions is the most important? My answer is—they all are important, just in different ways, at different times, and in different combinations. While it's the CEO's job to balance the functions out—to figure out which lever to pull at which time, it's the CXO's job to be at the ready when their lever is pulled. And that gets to the important question of what the nature of a CXO role is, and why those roles can be tricky. CXOs have three principal jobs that they must keep in balance at all times, although there is a clear priority in my mind of the three jobs.

CXOs are first and foremost members of the company's Executive Team. They must, must, must put that team, understanding of the different functions, and the relationships on it at the top of their agenda. They shouldn't show up on the team only advocating for their own team. CEOs must insist on that behavior and mentality. Without it, a company simply can't function sustainably. This concept is one that we have always called the First Team concept, and it's articulated very eloquently by Patrick Lencioni in a number of his books, particularly in *The Five Dysfunctions of a Team* and *The Advantage*. As members of the Executive Team, all CXOs are accountable to each other for the success of the business as a whole and must partner with each other to achieve that success.

CXOs are also the head of their respective functional departments. They must carry the flag of their team and wave it proudly throughout the organization, especially when working with their teams. They are the functional role model, the functional mentor, and the functional decision-maker for the people on their functional team. To be an effective leader, they must be The Quintessential X (sales professional, engineer, marketer, etc.).

Finally, CXOs are company leaders. They are role models for company values. They should always be on alert for things that are going well or going poorly around them. Things that need attention or recognition. Situations that need calming down. Guests who are sitting unattended in the office lobby. Delivery people who need a check signed and who need to be tipped. Putting the new bottle of water onto the water cooler. You get the idea. Company leaders have the actual and moral authority to step outside of their departments and handle things as they need to be handled, regardless of which employees are involved.

Chapter 2

Scaling a CXO

Congratulations, you just got promoted from Director of X to CXO! You're now in charge of a whole functional department, you now report to the CEO, you're now on the Executive Committee. You have a whole bunch of direct reports that either represent the team you used to lead or yesterday were your peers. You've reached the pinnacle of your career in X function. The only other ways to grow your career vertically are to lead your function at a larger and larger company, or to become a CEO. Wow!

That feeling of euphoria is wonderful. I remember having it when I worked at MovieFone and became the head of marketing and product management instead of just the "Internet guy." It definitely led to a nice celebratory night out in Manhattan with friends.

But then, the reality set in the next morning. Uh oh. I've never done this job before. Maybe I know how to do 25% of it. I'm only 26 years old. Is anyone going to respect me? I have so much to learn. Can I fake it? How on earth did I find myself here? This phenomenon is called the Imposter Syndrome, and it's totally normal. In fact, if you grow your career quickly, it would be weird not to have at least a touch of it.

The good news is, you're not the first person to be promoted to an executive role for the first time (and of course you're not the last, either). Every single executive, at any company, had their first executive role at some point. While there's some credence to the expression "fake it till you make it," there's a more methodical approach you can take to scaling yourself as a CXO—or if you're the CEO, to helping your new CXO scale. Think of the journey in three steps that can be taken in any order.

First, *master the tactics.* You need to understand all of the things that happen in your department. Some, you will know well because they're the ones you've done over time. Some you won't know at all. Make sure you do a complete inventory of the functional competencies for your role and all the roles reporting to you. Depending on how organized your company is with job descriptions and what's often known as a RACI (responsible-accountable-consulted-informed) analysis, this may be as easy as pulling something off the shelf and having a series of meetings with the people on your team to walk you through what they do. If your company isn't that organized, you may want to take the opportunity to proactively build that kind of functional competency/RACI list for everything in your team. That is no small exercise, but it's one that will pay back massive dividends. As one of my long-time colleagues and the writer of the CRO/Sales Part of this book, Anita Absey, says, "What gets measured gets managed." I'd add to that: if you don't know something even exists, you can't begin to measure it, let alone manage it!

Second, *form your strategic approach.* Every single function in a company has tactical and transactional elements to it—and every single function can be ONLY tactical if you let it. That's the lowest common denominator. HR can be about benefits and payroll. Sales can be about pipeline management and closing deals. Marketing can be about blog posts and SEO. A transactional focus is especially true of corporate functions like HR and Finance, but it's true of all functions. But just as every function has its tactical elements that must be attended to, every function CAN be strategic. As you settle into your new role, and as you grow into the role of senior executive and learn the First Team lesson of putting the needs of the business before the needs of your department, you will be able to start thinking more holistically about the business and how your department fits into it, so when your CEO pulls the lever that indicates they need your team to step up and lead, to be strategic on some topic, you are ready. What does it mean to be strategic vs. tactical? It's the difference between eating what's on your plate and planning out next week's menu. What are the ways in which your function can produce competitive differentiation for the business? What are the frameworks that will guide your decision-making about resource allocation or prioritization? How can you best support the other departments in the company? Those are the kinds of things you need to master in step 2. As my colleague Dave Wilby once said about one of the teams he was managing, "We have to figure out how to be the nose, not the tail."

Finally, *look around the corner to see what's next for you and for your team.* Senior executives constantly need to be toggling between different execution and planning horizons. You need to make your goals this quarter, and to make them, you have to hit daily or weekly activity metrics and milestones. But what about next quarter? Or next year? Or what happens if your company doubles in size in the next six months and is set to double again? Start by revisiting that functional competency/RACI list from step 1 and stress-test every element of it. Ask yourself, What must be true of this line item when the company is twice its current size? While you have to develop and scale as a leader—with all that goes into that in terms of soft skills—the only way to scale yourself as a CXO is to understand what great looks like for your role at the next stage of the company's life, and make sure you don't get there after your company needs you to.

All three of these steps—mastering the tactics of your department, forming your strategic approach, and understanding what's next—are things you may be able to do on your own to a point. That said, they will all go more quickly and with a higher probability of success if you engage your CEO, your Head of HR, members of your Board, or outside mentors or coaches to assist you on your journey.

Part Two

Finance and Administration

Jack Sinclair

Chief Financial Officer

There's no one path to the role of "startup CFO" and it's likely that anyone with that title could have taken one of many paths to get there. Some come from an accounting background, they have a CPA and possibly have experience as a controller, mastering accounting systems, GAAP rules, and providing accurate and timely financials. Other startup CFOs have had experience in investment banking, focusing on mergers and acquisitions, initial public offerings, and key performance indicators (KPIs) before jumping over to the finance team of a company. And some, like myself, come from an unrelated field. I came to the role of startup CFO from consulting where I learned strategic analysis, competitive benchmarking and, most importantly, helping clients as a partner to solve problems. I also had a two-year experience starting a company, where a friend and I tried to start a business creating corporate finance software. There were a lot of learning lessons there as well, especially that I should not be a primary coder in the business. Overall an effective CFO has to eventually master all of the basics—the accounting and finance requirements of the role—but often your early days as a startup CFO will focus on where you came from. And more often than not, a first-time "Head of Finance" has to learn the areas they do not have expertise in on the fly.

Coming from the consulting world, I was missing a lot of the standard skills that CFOs need to have. I had never made a journal entry, never closed the accounting books, never formed a corporate entity in a foreign country, and never managed a cross-functional team. I had a lot to learn and luckily was able to do a lot of it on the job. But while I was missing some of the standard CFO skills, I had deep experience and skills in some of the hard-to-teach things like financial analysis, competitive benchmarking, thinking strategically, and managing transactions. And, I was instinctively drawn to what I now believe is the secret to a great startup CFO: the concept of "Finance as a Partner."

Most people, both in and out of Finance, think of the Finance division as either a gatekeeper to financial resources, or they think of us as the people who require budgets and justification for internal funding. If you accept the definition that Finance acts *only* as an approval gate, or a speed bump to getting things done, or a black hole for requests, you're limiting yourself and your team to the bare routine tasks of the role and your organization on a whole will not operate as smoothly as it can. I have always thought that Finance should be a partner to the organization and that mind shift, from making requests to working collaboratively with others both on finance issues but also on general business questions, is critically important for the Finance function and the company.

The other mind shift that is required if you're a startup and you want to help accelerate your company's growth, is that you'll have to constantly look for new innovations, new technologies, new ways of doing things. Too often decisions are made simply because that is the way they were always done and these outdated rules lose their effectiveness. For example, a requirement that people work in the office 40 hours a week was not possible in 2021, given the COVID-19 pandemic. That's an extreme example, of course, but there are a lot of "rules" that made sense when they were created but lose their effectiveness over time. So, try to let go of these perceived guardrails and rules … and help teach others to do so as well.

Organizations can't scale if investments are not clearly understood or if areas of the business, such as Product and Marketing, suffer from lack of analytic support. If you're not a partner to the business, Finance will be seen as a speed bump, as a drag on growth rather than an accelerator of growth. Being seen as a gatekeeper, as a speed bump, can come from the best intentions of people in your company because they often see Finance as the guardian of cash. Also, since startups are often losing money while they get to scale and figure out their product-market fit, careful management of cash is a priority. As the CFO in a startup, it's easy to just say "no" to people who want an exception or stick blindly to your budgeted spend. But if you approach the role differently, if you operate as a partner and ask others what Finance can do to help provide them the right information at the right time to make the best decisions, the organization is able to scale much more effectively.

An important part of collaborating with others as partners is to enable other functions to be more self-sufficient and you do this by teaching them about finance and by helping them learn how to make an effective business case. Most people won't be well-versed in finance. Some people might even have a fear of finance, and they may think that they're not good with numbers. Your number one job with helping your company scale is to help people learn how to best think about the issues of investing, measuring, making data-based decisions, and managing financial resources. By taking on the mindset of enabler, of teacher, or helper, the startup CFO can develop a stronger grasp of which systems are needed, where data has to live, what analytic support is required, what types of training are required, and how investments can be made for a greater impact. You should also embrace being taught. Asking questions of other areas to better understand their function and challenges will help make an effective partnership. An effective startup CFO should keep the collaborative "partner" model in mind as they build their team, their systems, and their processes. And strive to create an environment where there are no surprises.

Chapter 3

In the Beginning: Laying the CFO Foundation

Finance

The earliest days of the startup are a constant battle between doing things the "right way" and being scrappy. The CFO is going to face the same quandary many times, and for many decisions: do you go the fast, cheap, and easy route, or do you spend more money, take more time to weigh options, and socialize your decision with a lot of stakeholders so that you can create scalable data stores and processes?

During these first few months, you will face countless decisions every day and the choices you make will reverberate for years. A small example is keeping excellent Board meeting minutes. If you're a C-corp, or even an S-corp, you'll have to have Board meetings and you'll need to take minutes of the meeting. It's easy when you start out to not take it completely seriously since you might have three people on the Board and you might just log the date, time started and finished, people present, and the results of any votes. Not a big deal when you're starting out, but fast forward to a time when you're considering a financing, merger, acquisition, or being acquired. Now it's a different story because now every document will be scrutinized by lawyers for accuracy, thoroughness, and as a window into how you operate your business. During our early Return Path days, I needed to painfully reconstruct what was discussed at a number of our early Board meetings the week before the next one. And then go through signing each one before due diligence during a transaction.

From then on I always took minutes during the meeting and filed and signed them right after, and kept them in a consistent place. I could have saved myself a lot of time and trouble if I had done it right when we started.

Every business will have unique processes that are critical to a strong foundation but there are a handful of areas that will be important for all businesses, like the Board minutes I mentioned and others such as:

- File storage and management
- Initial interest to closed sale ("interest to order")
- Order to cash
- Chart of accounts
- HR information system (HRIS)
- Employee expense policy
- Option grant policy and budget
- Federal and state registrations
- Sales tax

As a startup you will want to be as scrappy as you can be but you don't want to skimp on these key areas in Table 3.1.

I go into greater detail on each one of these tasks later in the book, but all of them have the same two characteristics: They're easy to do quickly and they're easy to ignore. Don't ignore them or you'll put your company in a difficult situation sometime down the road and create a mountain of work for yourself and your team trying to fix and/or re-create later. One that is particularly important for you, as CFO, is to make sure you have a deep understanding of the entire journey of a customer, from initial contact with your company to closed sale. Of course, your Sales and Marketing teams will have their own ideas and understanding of what that journey looks like, but your understanding—your model—needs to be far more detailed and needs to include costs, timelines, and return on investment.

Table 3.1 Key foundational tasks.

Key foundational tasks	Definition
File storage and management	How you electronically store all of your important corporate documents, signed client documents, non-disclosure agreements (NDAs), etc. It is a good idea to have a different process for your core corporate documents (Articles of Incorporation, Stock Purchase agreements, etc.), your company documents (leases, compliance filings, tax returns, etc.) and sales and business development agreements.
Interest to order	The journey of every contact and sales lead all the way to a closed deal. The most effective way of illustrating this is via a flowchart. The flowchart tracks a lead as it moves through the sales cycle eventually becoming an order. The flowchart will include all automations, like email triggers and alerts, and also include all manual touch points and process owners. This is helpful to understand all of the systems and databases and also highlight areas that are a source of likely data entry errors or bottlenecks in the process.
Order to cash	This process tracks what happens once a sale closes, all the way to collection. It can include how information gets sent to your financial system, customer provisioning, revenue recognition, deferred revenue tracking, and accounts receivable management.
Chart of accounts	Whatever accounting system you choose will have a default set of accounts that you use for bookkeeping. The only certain thing is that the default settings, at least for your Profit and Loss statement, will be wrong for your business. It is worth taking some time in the very beginning to think about what revenue and cost accounts you will want and how you want to categorize them. Keep in mind that your reporting requirements will be a helpful guide to the accounts and their structure.
HRIS (Human Resources information system)	Setting up your employee information, payroll, departments, and benefits correctly from day one. Most startup-focused HRIS do a good job helping ensure you have all of the right information and stay compliant so they are worth the early investment.
Employee expense policy	Documenting allowable expenses and the process to submit, approve, and categorize is best done as early as possible
Option grant policy and budget	You generally want to have a clear and transparent option grant policy before ramping up your hiring. The option plan documentation and details are discussed later in Part 2, and you also want to have almost an algorithm that determines the grant amount depending on role and market compensation.
Federal and state registrations	Once you set up your company, you should make sure to make all the proper filings in any state you have an employee. You should also understand each state's (and some municipalities') requirements for fees, sales, and use taxes.
Sales tax	This is a complicated area that has become increasingly important to start correctly. Most financial systems have easy integrations with sales tax software that should make this relatively painless if you start early. The tough part is to know if and where you have to charge sales tax.

Chapter 4

Fundraising

At some point, unless you're incredibly fortunate and are able to bootstrap or already are very wealthy, you'll need to get a cash infusion into your business and there are several different ways to do that, typically based on your stage of growth.

Seed equity is usually the first funding a company receives from outside investors. You could get seed funding on day one or at a later date if the founders bootstrap the startup. A generation ago, seed capital came almost entirely from friends and family and the amounts were relatively small—whatever they could scrape up. In recent years there are more and more professional seed investors, more sophistication, and higher expectations. The seed round can often have the valuation pushed to the next round, where the seed investors will end up having their shares valued at some discount to the next round.

Venture equity refers to money provided by venture capitalists and typically happens in the next few rounds. Venture capitalists take risks at this stage because companies are usually in a high growth phase and not that close to profitability—they could be losing tons of money. My perspective here is that at these stages getting a simple and straightforward preferred investment is worth giving up some valuation. For example, it is probably better to have a $15 million valuation on a simple preferred round with typical investor rights and protections than an $18 million valuation where the preferred comes with extra rights and participation. In this case, "participation" means that the preferred investor will receive their investment back (or even a multiple of their investment) AND participate as a common shareholder.

There are many, many books and seminars that can help you figure out the right financing path as an early stage startup, including *Venture Deals* by Brad Feld and Jason Mendelson and Matt's chapter in *Startup CEO.* I'd suggest reading and learning as much as possible before you accept venture capital, and get yourself a good lawyer with specific experience in startup financing who will give you some clarity on cost.

Growth equity is an investment in more mature companies with a clear economic model that are looking to make big investments to expand more quickly, go into new markets, or make acquisitions. The businesses pursuing growth equity will typically have clear product-market fit and are looking to make a big change in their growth trajectory. This is also a time where they may be able to purchase existing shares (called "secondary investments") that may enable founders and early investors to achieve some liquidity.

Commercial debt is senior debt, typically from your primary operational banker. The closer to profitability you are, the better terms you can get. A simple term loan or a revolver loan based on some asset of the company provides liquidity and cash without having to negotiate and write up contracts each time you need cash. Commercial debt typically will have covenants that the company will need to meet in order for there to be uninterrupted lending.

Venture debt is a bit of a hybrid of debt and equity that usually has a higher interest rate than simple commercial debt and usually includes an equity component in the form of warrants. It usually has less onerous covenants and longer terms.

Preparing for a Fundraising Event

Now that I've outlined some of the basic fundraising terms, what is the CFO's role in fundraising? How can you be an effective partner to the founders, the CEO, and the company? As CFO, you'll be responsible for collecting a wide range of information—far more than financial—and presenting it in a compelling, engaging, and truthful way. This could vary a lot depending on the CEO and the stage of the company. Early on, a CEO will take the bulk of the responsibilities but later on, and again depending on the CEO's interest/knowledge of the finance world, the CFO will need to be a more active and visible partner for the CEO. All the information you collect will be used to create the investment story, which is collated into the "pitch deck."

The Pitch Deck

Your pitch deck is used mainly as preliminary reading before an investor call or as a takeaway after the call and its purpose is to pique the interest of investors, provide information, or clarify your business—not to "sell" them on your company. Don't waste their time with a dictionary-sized pitch deck. If you're invited to a conversation with investors, the pitch itself is usually done by the founder/CEO and is much more free-flowing than simply following a deck. The best pitch decks usually have at least the following sections in roughly this order:

1. The Opportunity: Clearly illustrate the business idea and the core value propositions.
2. The Business: Description of the approach and the business model.
3. The Product: Description of the product with relevant screenshots and roadmap.
4. Size of Opportunity: Presentation of the Total Addressable Market (TAM).
5. Current Business Status: Current clients and partners, high-level KPIs and any other relevant information.
6. Financials: Overview of current financials with appropriate projections.
7. Senior Team: A slide on the Senior Team with their roles and key experiences.

The pitch deck doesn't have to be a slide presentation or PDF of slides. Quality and clarity of thinking in discussing your opportunity, product, and results are the most important things, not the pitch deck format and not any bells and whistles that you include.

As the company becomes more mature, the overall structure of the pitch deck will remain similar to what I've outlined above but the current business status will include more metrics, a product vision, and a supporting discussion around projections.

Chapter 5

Size of Opportunity

Certainly, if you can use outside research or information to inform potential investors about the size of the opportunity, you should take advantage of that. It's very useful if there is an industry study that provides an estimate of the market size. A published industry source not only lends more credibility to you and your business, but also reduces the amount of time you have to spend finding estimates of market size. If you're in a completely new market, you'll have no option other than creating your own market size estimate. One of the common mistakes early-stage companies make is in overestimating the size of the market opportunity so that it becomes ignored in any presentation. You have a clever product that "every homeowner" will want (for example), but that doesn't play well with investors and they'll want more sophisticated thinking on your market opportunity.

Instead, figure out the size of the total addressable market (TAM). You can arrive at this number in a couple of different ways. A top-down approach focuses on readily available macro indicators. For example, if your business is chasing an opportunity in the real estate space, it's easy to find macro numbers on the size of the US real estate segment you are in and you can provide some color on how fast it is growing and what percent of that segment you feel is addressable for your business.

The other approach, bottom-up, uses individual metrics to arrive at a number if you were able to sell your solution to every single possible client. Continuing the real estate example, if your clients are commercial real estate tenants, you could use data that show there are 500,000 commercial tenants in the US (for the easy math) in your market and the average service price of your product is $5,000 a year, the total addressable market is $2.5 billion. From there you would (probably) refine that number to something more reasonable, like reducing the TAM by region or some other factor.

The CFO plays a major role in driving alignment among the leadership team in creating a realistic answer to the size of market opportunity. If this is not well thought-out, if it's wildly optimistic, if there is no data supporting your assumptions, you could lose investor confidence.

Chapter 6

Financial Plan

Finance

The pitch deck should have two types of financials: current financials and projections.

Current financial statements should include at least the current year-to-date (YTD) income statement with detail at least showing:

- **Revenue makeup.** Often you will want to break out revenue into "types," for example, breaking out subscription/recurring revenue from transactional and service revenue or breaking out hardware revenue from subscription revenue.
- **Gross margin.** Each industry will have gross margin benchmarks that are useful regardless of the stage of the company.
- **Costs.** Knowing the costs of the following is very useful:
  - Research and Development
  - Sales and Marketing
  - General and Administrative

  Investors like to benchmark these measures, although these will be very different depending on the type and stage of the business.
- **Operating margin/earnings before interest, taxes, and depreciation (EBITDA).** This is another metric that can vary widely depending on the industry, stage, and growth rate. Sophisticated investors will be able to easily decipher this metric and account for industry, stage, and growth rates, so include it even if the number looks bad to you.
- **Balance sheet and cash flow.** These can be included if there is an appendix or highlights like cash balance, cash burn,

and runway on the financial slide. In addition, there are a handful of other possible reasons to have more of the balance sheet in the deck:

- material capital expenditures,
- large working capital variability or requirements, especially around product inventory,
- commercial debt,
- significant leases or other off-balance sheet liabilities.

Financial projections are more important for later-stage companies than early-stage startups but you need to have some projections. This will help the investor understand your thoughts on the unit economics, margins, and needed investments. This is also one of the best ways a CFO can be a good partner with the CEO. Your role can range from sitting down and crunching through the model with the CEO, to creating the model yourself and iterating after getting input from the CEO on the key drivers and assumptions that make up the forecast.

For an early-stage company the financial projections can provide the investor a picture on how you see the company's finances evolving as the company matures. Specifically, you'll want to consider the gross margin, show what the operating margin can become as the company matures, and provide a rough idea of time to profitability with total cash burned. Even though most investors realize that the projections will quickly become dated as you have yet to figure out all of the key unit economics and product-market fit, it's your thinking about the projections that is critical to receiving financing. Another good idea is to use "mental math" to quickly sanity check the story that the financials are telling you. Does this gross margin improvement make sense? How much of the increase in revenue is falling down to profit and does that make sense? Does the employee base assumption make sense, given the client count increase?

The projections for later-stage companies will need to be much more defensible as there is now some history. For later-stage companies, your projections play a much greater role in valuation, which determines nearly everything in the financing moving forward. Your discussion around the projections should include something on key initiatives to improve growth rates and margins. For example, a major initiative could be that customer retention will improve by 5% each year and the discussion should include WHY this is possible and the economic impact of that improvement. Having realistic and defensible projections will require the CFO to deeply understand the business and be a great partner to the other teams like Sales, Service, and Marketing.

Chapter 7

Unit Economics and KPIs

At any stage of growth, it's important to identify the unit economics and key performance indicators (KPIs) and have a clear understanding of the story they tell about the business. Each business and industry could have a different set, but some of the common KPIs that cross industries include the following.

- **Customer acquisition cost (CAC).** For venture-backed companies looking to grow a business, this is one of the most important metrics, for both B2C and B2B companies. Your CAC will evolve and change over time (either higher or lower as you get past early adopters) so the earlier you start measuring it, the better off you'll be. Even though your CAC will evolve, investors will want to understand how fast the payback period will be against a CAC.
- **Customer lifetime value (LTV).** Lifetime value is an important companion metric to CAC and will include other common metrics like gross margin, average product price, customer growth dynamics, and customer retention rates. There's no one-size-fits-all LTV approach since it includes other measures, but it can be helpful to include it in financial projections and be able to substantiate how and why you came up with the metric.
- **Customer retention rates.** You will have a few rates that each tie to the key customer segments you're targeting. Things to consider are customer size, region, or product set, or other segments that tell a different customer retention story. Net Retention Rate (sometimes called Net

Revenue Rate), or NRR, is the customer retention rate, taking into account revenue net of upgrades, downgrades, termination, and cross-sells. Gross Retention Rate (GRR) only includes downgrades and terminations. Both metrics are useful to track, and many later stage investors (and possible acquirers) will spend a lot of time on these metrics.

- **Average price.** This can include average new vs. existing customer pricing, and you might consider segmenting by region, product, or other key factors specific to your business.
- **Assorted segment analysis.** As alluded to above, investors often want to understand metrics by segment and even if they don't explicitly ask for segmented data, the best investor presentations will give them that information.

Chapter 8

Investor Ecosystem Research

During the scaling process it is useful for the CFO to understand the dynamics of the venture capital ecosystem and develop relationships with firms that are a good fit for the company. Some things to research or think about before you go too far down the path with venture capitalists include the following (and again, Brad Feld and Jason Mendelson's book, *Venture Deals*, is a terrific resource for this topic).

Venture investors usually have a thesis they want to follow and often will not deviate from their thesis. The reason for following a strict path is often that the venture investors raise money from their limited partners (LPs) based on that thesis. The LPs typically look to diversify their investment across many asset classes, so if they have invested with a venture capitalist because that venture fund had a thesis of "early-stage biotech," the LP investors will want the fund to follow the thesis so their asset diversification is clear.

Venture capital often invests by stage, and the growth stage will dictate the minimum and maximum check size. For example, some venture funds will never write checks for less than $1 million, and others will not go over $500k.

By looking at a VC's portfolio, you can get a sense of the industry/business models the firm has invested in. For example, VCs may like to focus on SaaS companies and others on Marketplaces. It can be very helpful to build relationships with venture firms that focus on your industry or model.

It is also helpful to get references about an investor before you go too far with them. This is usually the CEO's job, but you as CFO can be a good partner and help them to ensure the investors will be helpful Board members and are able to appropriately invest in future rounds. It is good to understand how much "dry powder" a fund has allocated to your company to understand their likelihood of investing in future rounds.

Chapter 9

Pricing and Valuation

Finance

Early on for startups, valuation is not really done by some standard formula—there's not a published resource that you can use, or a seminar you can attend that will help you develop a valuation model. But there are some things to keep in mind as you go about developing your valuation model.

Seed round valuations are sometimes as simple as "Selling 10% of the company for $500k," which implies a $5 million valuation. Obviously, the percentage and the price will depend on the team, the idea, and how far along the business was bootstrapped. More often for startups, the seed rounds will actually delay the valuation question by structuring the investment so that the valuation will be set as a discount to the next financing round. That is, you won't really have a valuation if you discount to the next round.

Lead investors in Series A or Series B rounds often want a convertible note or SAFE structure ranging anywhere from 10% to 25%. There are many (many!) exceptions to this general range, but an early-stage VC will want to own enough of your company so there is still material ownership after some future dilution.

As the company matures, there are more standard valuation approaches. Often valuations can be based on multiples of revenue or profit (EBITDA) and things like growth rates, gross margins, and operating margins will all factor in a valuation as well. As revenue growth rates slow down, you will more than likely be valued on profitability metrics rather than revenue multiples.

When raising a venture round, you probably won't have much control over some of the key investor rights and protections, such as participating preferred, pro-rata rights, anti-dilution protections, and assorted required approvals. Again, we can point you to Brad Feld and Jason Mendelson's book, *Venture Deals*, for a more in-depth discussion of investing documents. In most instances the terms are determined in the early rounds and then in later rounds investors will want the same rights and protections. If you are able to structure your Series A and Series B to have simple preferred securities with some of the standard protections and rights, you'll be able to keep it simple for the later rounds. Keeping it simple is better for the common holders and it also helps negotiations around future rounds remain straightforward. You won't need to unpack the "value" of unusual rights and preferences and can instead use more typical valuation metrics.

There are a number of key transaction items to consider that investors will want to negotiate with you. These include:

- **Liquidation preference.** This refers to the payout order in case there is a liquidation event such as the sale of the company. Often it is simply the amount of money invested in the preferred equity of the company.
- **Participating preferred.** Investors with preferred stock would get their liquidation preference first, then participate as if their preferred shares were converted into common shares. If there's a term like "1x" it means they get their investment back and their shares become common. "2x" means that an investor would get twice their investment back, then their shares would become common shares. There can be higher multiples depending on the investor and what you negotiate. This is less common in the 2020's but a good thing to be aware of in case it is part of a term sheet.
- **Pro-rata investing rights.** Pro-rata rights provide an investor in the company the right to participate in a subsequent round of funding and still maintain their current level of percentage ownership in the company. For example, if an investor owns 10% of the company prior to a round, and the proposed new round is $10 million, they would have the right to put up to $1 million of the round, ensuring that they would still own 10% of the company. This is a common

term to deal with in raising capital and it is also something that ends up being part of the new round of negotiations. Investors will often waive a portion of their pro-rata right if it makes sense in the context of the round.

- **Board seats.** Early rounds typically come with investors being able to name a member of the Board of directors. As you scale, new investors become less likely to get a Board seat, and the Board starts to add independent members who aren't investors or employees. Independent Board members play an important role with a unique point of view different from a CEO or investor.
- **Approval of new debt and equity.** It is pretty common for preferred investors as a class to have approval rights on any material amounts of new debt or any new equity, including any increases in the employee option pool.
- **Information rights.** Investors will have the right to monthly/quarterly financials, the cap table, projections, and typically a right to see a copy of the annual financial audit. Stages after the early stages will often require an annual financial audit by an outside accounting firm. You generally want to limit formal information rights to larger venture investors if you have a lot of small investors on the cap table. This will save you a lot of bother over time as you pick up shareholders via employee exercises and acquisitions.

Chapter 10

Due Diligence and Corporate Documentation

Finance

Keeping corporate records, files, and compliance documents is an important area that many early-stage companies ignore and these tasks fall largely to the CFO. Think of it as how you store your ingredients in your kitchen. If you are going to make a complicated meal with lots of *mise en place*, if you are not organized you are going to have a rough time making the meal happen. Sure, you will *eventually* find everything and put the meal together but it is going to take longer than you think, you will likely miss a step and generally have more stress. Conversely, if you have a well-organized kitchen with clear labeling and thoughtful storage, setup is a snap and the meal will taste better with less stress! You will find yourself in a similar position when having to produce diligence documents during a transaction. It can be a breeze, where you are just uploading already prepared folders into a data room, or you can spend way too much time searching email folders, shared drives, CRM uploads, and worst of all asking your outside counsel for help (at $500 an hour) to find old documents. As I mentioned earlier about maintaining Board meeting minutes, it's important to start early and focus on maintaining your processes as you sign more clients, employees, and vendors.

Even if it's not required, it can be a useful exercise to do an internal audit of your current state of documentation. How do you know where to start? The easiest way is to ask your outside counsel what a typical due diligence request looks like and then assess your readiness against that list. Below is a very non-comprehensive list, which is a decent overview of the important items that will set you up nicely:

- Corporate documents
  - Incorporation documents
  - Bylaws
  - Shareholder agreements
  - Stock option plans and grants signed by employee/company
  - Stock ledger with appropriate signed documents
  - Board minutes since inception
  - Board books from the last 12 months
  - Complete transaction closing sets (financing rounds and M&As)
- Employee documents
  - Organizational chart
  - Signed offer letters and appropriate termination agreements
  - Signed commission agreements
  - Form 5500 for 401(k) plans
  - Employment agreements
  - Benefit plan contracts
- Intellectual property
  - Patents and trademark documentation
  - Domain names
  - Any licensing agreements
- Financial documents and analysis
  - Bank account information
  - Five years of financial statements
  - Accounts receivable ledger
  - Accounts payable ledger
  - Fixed asset list
  - Key metrics history
  - Sales pipeline history with relevant segmented win rates and sales cycle
  - Relevant revenue/sales information. For example, if you are a recurring revenue business, you will need to be able to produce month-to-month analysis showing starting

recurring revenue, revenue added, revenue lost, ending recurring revenue.
  - Historical client count, revenue, and retention rates
- Outside agreements
  - All client agreements with understanding of any non-standard clauses like Most Favored Nation status or marketing restrictions
  - All material vendor agreements
  - All real estate leases
  - All non-disclosure agreements
  - All compliance filings in US and other countries

Once you assess your status based on the list above, you can begin to use that information to help create several plans. For example, you can create a plan to amend any real missing items in your history. It may be that you don't need to have a copy of that lease that expired five years ago, but if you are missing large segments of current client agreements or Board minutes, you will need to have a plan to address it.

You can also create a plan on how to build a process going forward so you will not create any new problems, especially around storing corporate, employee, client, and compliance agreements. When you are going to be in a transaction, you will receive from the other party's lawyer a long list of disclosure requests. You can imagine that you are going to have to organize a bunch of folders in a data room for all appropriate parties to review. How quickly will you be able to do that? What are going to be the big problem areas? Having this thought of well before a transaction and partnering with other team leaders will help you avoid a lot of future stress.

Finance

Chapter 11

Using External Counsel

Even when you end up hiring in-house lawyers, there are a few areas where you will still use external counsel. These areas include real estate, intellectual property, litigation, and transactions. There are a lot of critical clauses in your assorted corporate agreements that you'll want an experienced advisor to help with—either in creating them or in helping you understand their impact on the company. There are a couple of useful things to keep in mind when using external counsel for a transaction.

First, before you even think of contacting outside counsel, have an outline of all of the tasks and responsibilities that you want them to analyze or review. Lawyers, for all of their experience and intelligence, can be a bit reactive when it comes to a mountain of work that is under a deadline. Especially during a transaction, a law firm will often involve many partners, associates, and paralegals to get the transaction done in a timely fashion. By creating a project plan at the very beginning you will at the very least have some sort of transparency around the tasks and timeline. If this is your first time on a topic, the process may be a bit iterative with your counsel, but they should be able to quickly work with you on the key pieces.

Second, send over your detailed outline and project plan to your outside counsel so that they can provide you an estimate on the total fees for the transaction or at least give you a heads up when the fees reach a certain number. You don't want the shock of opening up an email and seeing a six-figure bill when you were expecting something in the low five-figure range. An estimate is helpful not only for your legal accrual on your financials but helps provide transparency and avoids nasty surprises.

If you have an in-house counsel, it is a good idea to bring them along for the ride whenever possible. Because the CFO is often directly involved and is managing the external counsel, the internal counsel may miss out on some opportunities to learn about transactions, or on contributing their financial knowledge.

For some deals, venture capitalists will put in the term sheet that the company will have to pay their legal fees as well. If you have to agree to this, I would suggest instead that you work with investors to see if they're open to the idea of capping the amount you will pay. Of course, this cap will be the exact amount you will end up covering. But at least it's a known number that you can budget for.

Overall, keep your external counsel relationship in perspective. They are there to provide legal advice and leverage their deep experience and expertise to help you make good decisions. But the decisions are yours, you have the final say.

Chapter 12

Operational Accounting

During the startup journey, establishing a scalable accounting back office will pay dividends later on. It's a lot easier to create an accounting back office early and modify it as you grow than find yourself scrambling to put something together under a tight deadline. An alternative to doing all the work yourself is to outsource some standard tasks, and many firms outsource bookkeeping, for example, but keep other activities in-house. The benefit of this plan is that (1) you will not need to focus on basic vendor payments, expense logging, and revenue recognition, and (2) the third-party company will typically set up a default chart of accounts and expense tagging approach that makes the most sense for your type of business. But there are few areas you want to keep an eye on even in the early days.

Early on, you will likely use a lightweight accounting system, especially if you're using a third-party bookkeeper. During the onboarding of the bookkeeper, you will want to cover a few areas, most notably the chart of accounts. Review the chart of accounts to make sure you have the appropriate detail in them in order to do your required financial reporting. A useful process is to lay out what you would like your income statement and balance sheet to look like and ask yourself what questions you will need to answer. For example, how will revenue need to be broken out? Each journal entry needs proper support ideally attached to the entry in the system, or stored in a systematic way. Account reconciliations need to be performed regularly and reviewed, and documented. Those two simple things will make the audit so much easier so you're not scrambling to figure out "wait, why did we book this, what does this tie to?" And then from there you can get more advanced on what the audits/diligences need. You'll want to make sure you have accounted for things like subscription revenue, setup revenue, service revenue, transaction revenue, and any other revenue streams you might have. As the business scales, the analysis will require that sort of segmentation of the economics. On the cost side you may want to consider how to break out accounts like Marketing which might have advertising, brand, and digital components you'd like to track separately. Other areas to think about might include legal costs, and you could track transactions, intellectual property, contracts, or litigation. On the balance sheet, I'd suggest keeping it simple early on but make sure you have enough detail on your more important accounts.

Once you have a good sense of the accounts, you'll next need to consider the dimensions. For example, do you want to have a dimension for product and region? It is helpful to have the dimensions thought out as these will enable you to keep a scalable chart of accounts as you add new products to the business.

A second thing to organize within operational accounting are system integrations and even the earliest setups will need some integration with their financial system. The most common system integrations that startups will have to deal with are travel and expense (T&E) software and billing software. It is worth spending some time early on making the integrations work as it will save time with the accounting close as well as give you accurate and timely information. I'd also suggest creating custom tags in your expense software that match up with your accounting system and chart of accounts. Also, if you are going to be charging sales tax, it's super helpful to integrate it early on with other systems.

Ah … sales tax. Fun, fun, fun. Sales taxes are a complicated challenge and they become more challenging the longer you ignore them or put them off. Many companies, including ones I have been involved in, ignored the sales tax issue until the company became "bigger" because small companies really don't hit state radar until $100k or so in revenues as they trigger the amount in the Wayfair sales tax ruling. But if you do put it off, it can become a huge and expensive problem. States have become more aggressive as remote work has become more prevalent and recent court decisions mean that you have to pay sales tax in more states than ever before. Currently, if you are doing more than $100k (or 200 transactions) with a state, you will need to address sales taxes in that state. Some businesses may not have to pay sales tax, but again, states are becoming increasingly aggressive in their audits. One of the best things you can do is work with a technology partner to collect and submit sales taxes and integrate everything in your financial systems.

One task that you'll do repeatedly—monthly at a minimum—is the accounting close. Having an organized monthly accounting close is an effective way to ensure you are working well with your partner or internal accounting team. Every company has a different close process but most efficient ones share a few things. First, create an accounting close checklist, something that's repeatable and that has each task and the responsible party that can be checked off as complete when done. You can set up a simple spreadsheet or you can also use a project management tool or accounting close software to help provide more information. As your company scales, it is helpful to move to a software solution which will ensure your visibility and accountability during the process as you grow. I have found this to be one of the easiest quick wins to put in place as you scale.

Another item to include in your operational accounting system related to the close is a reasonable time goal each month. Of course, your goal will depend on resources, business requirements, and complexity, but it's always helpful to establish a goal. Say your goal is five business days. The organization knows that the Financial Planning and Analysis (FP&A) reports will not be done until after that five-day period so this provides a bit of cover to your team. It also gives the team (or your partner) knowledge that if they're going to miss that deadline and need more time, they'll have to communicate with you to avoid surprises.

You may very well get "surprises" from time to time but, especially around the close tasks, you can manage that by creating effective lines of communication and by understanding the dependencies of your company. The accounting team will be dependent on a number of things in order to have an effective close, ranging from accrual information for services used by your company where you haven't received an invoice, to timely expense reports or sales closes. You may have to send reminder emails to others in your company or, better yet, have them involved in the actual closing checklist.

Accounting Rules Implementation

Once you get the basics of your operational accounting systems, reporting, responsibilities, and approach set up, you'll need to tackle accounting rules. Generally, in the early days, you can get away with non-GAAP reporting, but like many other things, you are putting off future cleanup and missing an opportunity to have the organization embrace generally accepted accounting principles or GAAP financial reporting. So even in the early days it's worth implementing basic GAAP and the following few areas are important to get into good shape when you're starting up:

- **Capitalization of fixed assets.** You can make capitalization a complex process or simplify it. I suggest simplifying it and the easiest thing to do is to choose a dollar amount and anything greater than that amount should be capitalized according to GAAP. For example, if you choose $2,500 as your limit, then any fixed asset like a computer or furniture would need to be capitalized. After doing an initial pass on fixed assets within the company (deciding what's above or below your limit) you can then quickly research GAAP guidelines on that type of asset for how long you need to amortize it. It is also helpful to maintain a fixed asset ledger and many accounting systems will do that for you. In addition, for software development, there is the topic of capitalization of software. I was always a fan of the simple approach of expensing as much as you can, especially for early-stage companies. As the company scales, you may have to capitalize software development. The important things to keep in mind if you do this are: (1) alignment with your Board and investors, as there are implications on financial metrics; (2) to have a consistent and documented process so there is clarity and transparency (auditors will want this); and (3) close partnership with the Product Development team so the inputs to the process are understood and easily captured by the team, hopefully by using already existing processes.

- **Accrual of annual pre-paid and liabilities.** Most startups will have a lot of systems they use to run the business. We had almost 15 within two weeks of starting Bolster, which is fairly typical of a startup. Many of these systems either require payment up front for a year or give large incentives to pay up front. If you do pay up front, you want to make sure you are creating a pre-paid asset and then expensing 1/12 of it each month, rather than just expensing the entire payment. You also don't want to forget expenses you have incurred but not yet invoiced or paid, like legal bills, commissions, or bonus payments.
- **Revenue recognition.** It is super useful to think through revenue recognition as early as possible. Some more advanced financial systems will have deferred revenue recognition and modules tied to revenue received and billing but it's more likely as you start up that you'll create a spreadsheet as your revenue recognition schedule. To do that easily, create each row to function as a revenue schedule and include items like the client, invoice number, total amount, currency, etc. This way you will always be able to show a waterfall schedule for your booked revenue. It can also be helpful to have a related sheet for your planned invoices, so you can easily produce a number that represents the "booked but not yet billed" revenue. Depending on your revenue streams, it can also be helpful to break out revenue by type, for example, subscription revenue vs. transactional revenue.
- **Stock compensation.** Often in the early days, companies ignore stock compensation as part of their financials. That's OK because once a company goes through an outside financial audit, the audit firm will help them calculate the expense. But if you are using modern cap table software and performing 409a valuations appropriately, the software will be able to give you the monthly compensation expense with back up. So, it will save you time and money if you start to do this early on.
- **ASC 606.** ASC 606 is a framework developed by the Financial Accounting Standards Board (FASB) and the International

Accounting Standards Board (IASB) for businesses to recognize revenue more consistently. It also ends up impacting sales commission expense. It considers the length of time of the service, if the revenue is transactional or recurring, and a number of other factors. It can also impact your IT systems, HR policies, and more so it is a good idea to research the implications for your business as early as possible and discuss the impact with your financial auditors.

- **Sales commission.** Sales commissions can be tricky from an accounting perspective. In order to be ASC 606 compliant, you need to consider a number of factors including service period, plan period, and client terminations. These factors can end up with some balance sheet implications since you are deferring the recognition of the expense. It might be helpful to use software to help make accurate accruals and help you work with your auditor.
- **Sales taxes.** While implementing and following GAAP rules early will save you a lot of hassle in the future, putting in sales tax collections processes will save you a lot of bother AND money. As I mentioned earlier, states have become a lot more aggressive in going after even small companies for improper sales tax collection. Your best bet is to find a technology partner and automate sales tax collection. This will save your team time and a partner focused on sales taxes will be able to keep you up to date on rates, rules, and regulations.
- **Major expense categorization.** Most companies as they scale start to care more about how their expenses are categorized, particularly high-level categories like cost of goods sold (COGS), selling and marketing, research and development, and general and administrative. As the business grows and raises money through a series of rounds all the way to being a public company, investors and prospective investors will use these numbers (and changes to these numbers) as a way to benchmark you against similar companies. It's helpful to do a bit of research to understand how other companies that have similar models to you classify their expenses. Certainly, what you include in COGS will be critical because the gross margin is an

important financial metric used to understand the financial health of the business. In addition, when you are growing and are not making a profit, it is important that you have a sense of what the operating margin *could* be once you are mature.

Chapter 13

Treasury and Cash Management

For a startup, the most important part of cash management is just being able to accurately forecast it and not run out. But there are some basic things to keep in mind as you scale:

- **Parking money for low yield.** Although it is almost never a good idea to chase yield, and increasingly difficult in 2021, you can sometimes put excess cash in a money market or a very safe and liquid vehicle that provides essentially the same safety as cash but provides a bit of income. And it is very important to be aligned with the Board on the philosophy and operationally of how you are approaching cash management.
- **Cash reconciliation.** Performing a bank reconciliation to match the cash balances on the balance sheet to the bank account is a helpful control early on to keep an eye on different types of expenses, bank fees, and in transit payments, collections, and outstanding checks. Over time some of these numbers can be material and if you are not doing a reconciliation, you may receive some surprises.
- **International funding.** Companies that start to expand out of the US into other countries for sales and marketing will need to fund these operations by sending money to those entities to cover payroll and other expenses. It's important to consider timing and how much cash to leave in the foreign bank accounts, which will vary based on your business dynamics and foreign bank rules. Depending on how you are accounting for revenue recognition and

customer billing, you may also have to build in sophisticated transfer pricing agreements and corporate business unit dividends in order to optimize your cash. Some countries can be challenging when trying to get cash back to your US accounts.

- **Debt after financing.** Commercial banks that market to venture-backed companies are most likely to offer a debt facility, both a term loan or revolver, right after you have closed an equity fundraising. You might feel that this is a time when you don't need more cash, so you'll dismiss the revolver since it adds the expense and hassle of debt. But don't dismiss it too quickly because there are some scenarios where this is a good time to add a line to the business to start to build the relationship with the commercial banks and have some cash to cover any cash balance volatility. That being said, you shouldn't add debt just because you can, so you'll want to support the cost and complexity of a debt facility with clear operating reasons.
- **Cash controls.** In the early days there are usually not enough people to install proper controls around cash. But as you scale, you should consider putting in controls so no one person is able to send money outside the company or pay for personal expenses without some oversight. There are many, many examples of relatively small-scale theft in small companies. For example, you will want to ensure that the person who enables the corporate card payments doesn't approve their own credit card usage. In one story from my early days (before I was on the Finance team), the accounts payable accountant ran up charges on car services and shopping on their corporate card and then paid off the card without any oversight. One easy thing you can do is have the people who are able to create wires be unable to approve them, and vice versa. Or you can put in a control so that wires/checks over a certain amount require two signatures. Bank reconciliation is also helpful to have in place to ensure some oversight.
- **Cash burn.** It is also useful to have a clear understanding of cash burn and cash runway. I have used metrics like months of payroll or months of last month cash burn as metrics for cash runway. In addition, it is also helpful to include in

cash forecasts known large cash variability. Some examples include early in the calendar year payroll which will have more payroll taxes than end-of-year payrolls, new office space expansion, seasonality around billing and collections and large vendor annual payment cycles.

Chapter 14

Building an In-House Accounting Team

Outsourcing your bookkeeping can certainly work a lot of the time with the right partner and processes. But at some point, you will need to hire an internal team. At first, this will include a general accountant able to do it all, but eventually you will need to hire specific functional roles including Accounts Payable and Accounts Receivable.

Here are some common roles to hire as you scale a startup:

- **Controller.** At a startup, the controller will manage the entire accounting team and processes, be the point person for all audits and compliance and tax, and generally be responsible for ensuring that the financials are accurate and timely. When the company is smaller, you can have someone a bit more junior who is able to grow with the company. If you are scaling rapidly and complexity is being added quickly to the accounting function, it is a good idea to find someone who has done it before, giving you a critical partner in cleanly scaling accounting.
- **General accountant.** This is often the first accounting hire. A person who can do it all in the early days, and especially keep an accurate and timely set of books is a good first hire. Ideally they will be able to keep the chart of accounts organized as the business grows and generally do a little bit of everything with the bookkeeping. It is usually helpful if this person has an accounting degree or some experience with an internal audit firm. They will be making a lot of small decisions early on that will end up being the team process

for the next couple of years. So, the cleaner you can make it, the better!

- **Accounts payable (AP).** At some point, the volume of work paying vendors, employee expenses and other day-to-day clerical duties will become too much for the general accountant. The primary area to focus on for this role is proper timing of payments to vendors (not too early, not too late), accurate tagging of expenses to the proper accounts and departments, and effectively managing employee expense reports and expense attribution. This role may also help out with payroll and other reporting.
- **Accounts receivable (AR).** Similar to the AP role, there will be a point where it makes sense to hire someone internally to collect payments from customers. For startups, this person will typically be responsible for generating invoices, maintaining the AR records, dealing with the many, many ad-hoc and custom invoices the team asks for, and of course, collections. An effective AR role can easily pay for themselves through quicker collections and lowering bad debt. Both the AP and AR roles can be effectively carried out by junior people as long as they have excellent self-management skills.

As you scale, there are a few additional roles you'll want to add to the accounting team.

- **Revenue controller.** Some companies end up with complicated deals, invoicing, and revenue recognition. In these cases, it may make sense to consider a senior accountant whose primary responsibility is to review each transaction to ensure they are invoiced correctly and that the deferred revenue and revenue recognition schedules are accurate. This is typically a fairly complicated job so it is not something you'd hire a junior employee to do. You may never need this role as it is only really found in mature organizations with complicated deals.
- **Transaction specialist.** There could be a few different titles for this role, but essentially as you scale, there may be a role for accountants who help process and approve individual

transactions. This could include ad-hoc requests for changes to invoices, checking the data in the systems against agreements, and approvals for pricing and packaging requests. There are a lot of different elements in transactions and at some point it may make sense to have this role to keep the data timely and accurate.

Chapter 15

International Operations

When an organization is ready to expand globally, they may be trying to expand its sales footprint, create more product development capacity, or find manufacturing capabilities. If you decide to expand internationally, there are a number of important things to do right at the onset.

- **Local partners.** Find a local partner that can help you navigate the country compliance, filings, tax, pensions, accounting, etc. If you go the Professional Employment Organization (PEO) route, where you use a third party to co-employ your employees, and provide benefits and payroll, a lot of these issues will be taken care of for you in one place. This is usually cost-effective until you get to more than a handful of people. Once you expand operations beyond that handful of people, you'll want to make sure you have partners for accounting, audit, payroll, compliance, and legal. For audit, if you are using a large audit firm with a global presence, you can often use their local office as the audit firm. But it is typically treated as a separate agreement between you and the audit firm and it is almost certainly going to be more expensive than going with a small local firm. Although there are some efficiencies in partnering with a big audit firm, the choice of firm typically comes down to a tradeoff between the extra money you'll pay and the number of efficiencies you'll get.
- **Corporate organization.** Creating a corporate entity in a new region creates all sorts of complexities. This is an area where

you definitely need to have guidance from your primary corporate counsel on your overall corporate structure and you will want local guidance in setting up the entity in that country. You will also need to have an idea on how to recognize revenue and treat intellectual property (IP), because these two areas can differ from region to region. You will want to have solid documentation that can show how cash flow moves across entities and how the transfer pricing agreements treat intra-company loans, dividends, use of IP, etc.

- **Intercompany documentation.** Other than the official corporate documents, the most important item to have ironed out early on is the company's transfer pricing agreements. Roughly, this agreement will detail how the different corporate entities transact with each other in terms of products, IP, and people. You will definitely want to work with counsel in order to put this together but here are a few things to keep in mind.
  - If you have more than one transfer agreement across multiple organizations, you will want to have a consistent approach in maintaining these agreements.
  - You will want to clearly state legal ownership of any pre-existing IP and how to handle the creation of new IP.
  - Often employees may travel to or spend significant time at the office of the foreign entity and the transfer agreement should have clarity on how this is treated.
  - Ultimately the transfer pricing agreements are important documents for the different global tax authorities, so keep this in mind as you scale.
  - Transfer pricing agreements will be required in any due diligence process so make sure they stay current and have a clear renewal process so that they are always active.
- **Revenue attribution.** If you end up creating a new entity and have to produce financials, you will need to determine how you are going to treat revenue. This is obviously a complicated problem, and you'll have to work with an experienced advisor. Your options will be different by country, but at a high level, the two common approaches for technology startups are cost-plus-percent method or a directly invoiced

method. With the cost-plus-percent approach, you are simply taking all the costs you are recognizing for the entity and then having revenue be some percent higher. So, if you were cost-plus 6%, you would have a 6% profit and you would pay the corporate tax to the local tax authority on that 6%. The benefit of this approach is that it is very simple and allows you to continue to bill your customers out of the entity in the United States. But you cannot do this in all countries and it may not be optimizing your tax situation. Another common approach is to simply bill the clients sold and serviced in that country directly from that entity. This means you will need to build invoicing and collection functionality and follow the revenue recognition rules by country.

- **Cash management.** Companies will need to fund international operations by sending money to those entities to cover payroll and other expenses, assuming that the new entity is not billing and collecting itself. It's important to consider timing and how much cash to leave in the foreign bank accounts. And depending on how you are doing revenue recognition and customer billing, you may also have to build in sophisticated transfer pricing agreements and corporate business unit dividends in order to optimize your cash.

Chapter 16

Strategic Finance

In the early days of a startup, your role as CFO is mostly about establishing the systems I've outlined in Chapters 12–14, finding the right partner for bookkeeping, and making sure everyone is getting paid, and has some benefits. That's a tall order for anyone, but those are the basic things that you'll have to set up if you don't want to be a blocker in your company's growth. Once revenue starts, you'll have to add more services and functions, things like billing, collections, sales taxes, and other back office and foundational items. If you just do that—build the basics, hire for roles within that structure, and add functions as you need them, you'll create a capable finance organization. But you won't create a *great finance organization*, you won't be able to improve productivity or make a big impact in the company. To create a *great company*, you'll need to think strategically about how Finance can be a partner to the organization.

For example, here's something I wish we had done from the very beginning at Return Path: *develop consistent naming for products.* Sounds like a little thing. And as a startup, it is simple because you might only have a couple of products. But the number of products grows rapidly and if you don't set up your systems for automation and scalable data classification, and link those to your accounting system, you'll run into problems as you grow. Product and Sales team members love (for good reason) to create new packages and products and try and invoice them as one basic price. Often Finance will either just accept whatever Sales or Product tells them OR they will go to the other end of the spectrum and create a roadblock or friction on making the changes. Instead, the best thing Finance could do is work with Product and Sales early to create a product code nomenclature that can scale as new products are added and that can be useful and scalable in your accounting systems, too. The mindset shouldn't be "No" or "Do whatever you want," it should be "Let's figure out a solution so we both get what we need to grow the business." The business will have much better reporting 12 months from now if the product naming follows almost an algorithmic nomenclature rather than following the naming whims of a Head of Sales or CEO.

In one early stage company we had well over 100 product "names" with active revenue attached to each. While that by itself may not *technically* be bad, there was no logic or grouping, and it was nearly impossible for a new hire to make sense of any of it. To create useful reporting by product, we had to manually create a table organizing these product names in a way that made sense, and it was always a bit tenuous as new names were created. Compare this to a naming system where, depending on what part of the product code you looked at, you could quickly understand revenue on a number of dimensions like region, service level, or high-level product name. This is just one example of where Finance has a unique view (in this case, corporate reporting) that will be useful to another function (in this case, Product) while they are doing something core to their world.

In a startup there are many areas where the CFO will be able to add a lot of value as a partner, and I've highlighted a few important contributions the CFO can make to each functional area. Overall, the key is to think of Finance as an enabler of better decision making using data and analysis.

Sales

In the beginning, when sales first start happening, the key as a CFO is really just to be close to sales data and processes, and to help ensure that the "interest to order" process is clean and understood. This is also a good time to build a close relationship with the Head of Sales so that when there are challenges to overcome in the future, there is trust on both sides. You want to be able to know that when there are questions that pit the sales teams vs. what is best for the company, that the Head of Sales is considering long-term company issues like precedent setting and fair payments. And you want the Head of Sales to know that you are able to strategically understand the role that Sales and sales compensation structures have in effectively running a sales team.

One way to start building a relationship with Sales early is to help the sales team develop key performance indicators and put in motion some understanding around the cost to acquire a customer (CAC). Each channel will have extremely different costs to acquire customers and you'll want to understand all of the pieces to know strategically whether you should keep using that channel. At Return Path we generally used two channels, direct sales, or indirect sales, where a partner resold our products. In direct sales we sourced the lead and closed the sale with our internal sales team, while with a partner channel they sourced the lead from their client base and closed the deal, sometimes with our help. One of the things we found out after we dug into the data was that if we analyzed all of the costs that went into a sale, including the reseller revenue share, the assistance we had to provide, and the channel-specific marketing costs, the cost to acquire a customer through a channel partner was much higher than through a direct sale. Beyond the higher costs, we also found out that customers coming through the channel partner had lower retention rates. So, higher costs and lower retention. It turned out that our reseller channel was delivering much less value than we thought. Early sales teams, sometimes rightly so, focus primarily on the top line. So, an effective way to be a partner is to help the team understand all of the economics of different go-to-market strategies so that better decisions can be made.

As the company scales, another area where Finance can be an effective partner is by developing the concepts of cohorts and segmentation. A cohort is a group of people that are similar in one dimension and you analyze that collection of people as one group. The most commonly used cohort is acquisition date, and it's helpful to understand how different cohorts behave under different circumstances. For a startup starting to scale, customer retention by cohort is an important analysis. It is a lagging indicator but measuring the change in retention rate by cohort can help you make better decisions on the product, service processes, and team structure.

Segmentation refers to creating groups by looking at a group of data broken out in a number of different ways, for example, looking at data by region. As you scale, customer acquisition cost, lifetime value, and retention rates can all be very different by region. A company may need to invest well ahead of revenue in order to establish itself in a region so customer acquisition costs may be high, dragging down your overall company numbers. And in some of the regions, local language may be critical to client happiness, so if you are still ramping up that part of your service offering, it could also drag down overall retention rates.

The point is, Finance can partner with Sales and help them both think about concepts like cohort and segmentation, and also provide Sales with analysis and tools to understand the impact of the business on cohorts and segments.

Service

Often, by necessity, startups build Service teams in an ad hoc manner with all hands on deck helping in any way they can to keep their clients happy. As startups start to scale and hire a dedicated Service team and leaders, it can become a major cost center and a major source of value. The startup CFO can get involved early to partner with Sales to ensure the team is making good decisions on team structure and prioritization. There are generally two major areas in the early days where the service team needs to be monitoring. One is the service team cost as a percentage of revenue (in order to drive a healthy gross margin) and the other is the retention rate metrics by cohort and important segments. On the gross margin, the finance team can help accurately measure the metric and benchmark it against similar businesses. Unlike operating margin, which will be very different for a startup versus a mature company, the gross margin should be fairly similar for a startup versus a mature company, or at the very least it will be in the same ballpark. Giving the team a target, say, "you have about 15 points of revenue in order for us to have a healthy gross margin and be comparable to our peers," helps the team set reasonable targets for revenue under management per rep, and learn what the overall team cost structure should look like. On customer retention metrics, you want to help the team measure the gross number and the number by cohort and appropriate segment, especially by region and channel.

Marketing

Other than perhaps the Head of Sales and CEO, the startup CFO's most important partner on the executive team is the Head of Marketing. Why? A combination of the size of the budget and the sometimes lack of clarity of short-term impact of the investments. There are a few areas where the CFO can be a partner to the Head of Marketing starting from the early days.

- **Pricing and packaging.** Pricing products and services for startups is, more often than not, a bunch of guesswork and seeing what sticks. The CFO can work with Marketing to help benchmark pricing and help model out different assumptions and implications on the business in terms of margins and growth. Pricing models for startups typically are either a value-based approach where the product is priced at some percent of total value the product will deliver to the customer, or a cost-plus approach, where the pricing is dependent on how much it costs the company to deliver it. With either approach, clear analysis with testing is important. The CFO can also help the business understand what sort of packaging will work and how features should be bundled. Packaging can include all-in-one packaging all the way to completely modular packaging where each deal is customized. Regardless of the packaging, helping the team clearly state the value proposition will help the sales and marketing teams accomplish their goals.
- **Customer acquisition cost (CAC) and customer lifetime value (LTV).** Overall, it is helpful for the CFO to help the organization with all of the key unit economics of the business. For Marketing, two of the most important are CAC and LTV. There are a number of ways to calculate both,

so initially, just ensuring the calculation makes sense for the business is a great way Finance can help. Establishing targets for both of the metrics as well as the ratio and payback period is also helpful as it will likely require some benchmarking analysis.

- **Marketing initiatives return on investment (ROI).** Marketing teams are continually looking at the ROI on their efforts, both in aggregate across all their initiatives and on individual campaigns. In a startup, the best way Finance can partner on this topic is to help with revenue attribution, marginal cost assignment, and segmentation. There are of course many ways to do all of the above, but the key goal to keep in mind is what Finance can do to help ensure the data will drive good decisions. This means providing accurate and useful analysis and also providing timely information that can be delivered to the right people. It's also helpful to work with Marketing to make sure that the analysis you provide is using the proper data sources and is aligned with Sales and Product.

Legal

Don't be surprised as a startup CFO if all the legal issues end up on your desk. It's common and often there's really no one else other than the CEO who can handle legal issues. The primary things to be concerned about are: (1) effective management of outside counsel, including knowing when to use them; (2) building useful and reasonable template agreements, including a master services agreement with appropriate insertion orders and a mutual NDA; and (3) document management and storage process. If you don't have a legal background or related experience, it is worth spending a session or two with outside counsel educating you on HOW to manage a contract negotiation or approval, draft a series of good templates, walk you through each term and show what's OK to give on, and help you understand risk areas.

The area where you can save most money is by working closely with outside counsel during transactions, especially if the company has to cover the legal costs of the investors. Spending just a small amount of time looking at your outside counsel's specific hours may give you the chance to lower the bill. For example, if your counsel is stating that it took them 20 hours to file a Form D for the financing (which is a one-pager and takes minutes to file), you will want an explanation.

The startup CFO also ends up being the lead negotiator along with the CEO on most of the early legal decisions. This makes it easy in some ways as you and the CEO can quickly do benefit/risk math, but keep in mind that as you scale and create in-house counsel, you'll want to transfer that ability to them.

At some point during the company's scaling up, and depending on how many transactions and other legal touchpoints the company has on a monthly basis, you will want to hire an in-house lawyer. The trick I used to figure out when we needed to hire someone was simple. For a three-week period I came up with five to six categories of my function, one being legal, and assigned a category to each 15-minute time block. After the three weeks I saw I was spending over 35% of my time on legal matters. This was not a good situation for me or the company, so we hired a lawyer. For your first legal hire, typically you want to hire someone you can transfer the day-to-day negotiation of customer and vendor agreements to and someone who can manage your outside counsel for IP matters like trademarks and patent filings. So, you really don't have to hire a full-time general counsel early on. And you will usually keep the relationship for outside counsel around transactions.

Good CFO/Bad CFO

Rob Krolik, Venture Capitalist and former CFO

Jeff Epstein, co-CEO, CFO, Apex Technology

There's a stark contrast between an effective finance chief (CFO) and an ineffective one. Before you hire or promote a CFO, you should be aware of those differences.

- A good CFO knows how to communicate, manage teams, and knows the details behind the numbers.
- A good CFO has a sound understanding of all key finance functions including accounting, planning, fundraising, risk management, tax, facilities, insurance, and treasury.

- A good CFO paints a financial picture of the company's next 12–24 months to help the senior executives see the future and plan accordingly.
- A good CFO responsibly manages cash. They understand when the balance will be low and what to do about it (whether to slow down cash burn or raise capital).
- A good CFO obtains input from other senior-level executives, helps them understand the needs of the company versus the executive, and architects a financial plan that balances those needs.
- A good CFO reads the tea leaves of the sales team and the overall market, then helps course-correct to ensure the company has future viability.
- A bad CFO blames others, always has an excuse, and doesn't take personal responsibility. A bad CFO will say things like, "The executives spent too fast" or "The CEO is too optimistic" or "Our Board members always think of me as the bad guy" or "VCs won't give us funding" or "Our strategy doesn't work."
- A good CFO understands the root cause of what goes wrong by highlighting the issue and proactively providing information to the Board of directors, CEO, and other executives to help them see the path forward.
- A good CFO is a great manager. They hire all-stars in their respective fields (e.g., accounting, tax, facilities). They trust them to do a great job while maintaining open communication and having regular check-ins—trusting but verifying.
- A good CFO has a network of professionals to hire or solicit feedback from in order to make the organization better.
- A good CFO knows how to delegate since it would be impossible to be an expert in each area under their purview.
- A good CFO provides a vision of where the finance organization should be in 18–24 months to help the company scale and then supports the finance team to get there. Other executives do not consider a good CFO as only an accounting or tax expert, but as a critical strategic partner.
- A bad CFO gives Engineering, Product, or Marketing "advice" to the respective executive.
- A good CFO provides valuable data and insights when another executive is over or under budget and shares unbiased analytics that will help solve problems and respect boundaries.
- A good CFO speaks visually, with pictures and analogies, not just analytically. They communicate with short, pointed, nontechnical accounting or financial explanations in making a point.
- A bad CFO wraps themselves in jargon and focuses on what people can't do and is always ready to say "No."

(*continued*)

- A good CFO has high integrity and factually reports the numbers. They are a risk manager who helps manage the lows and highs by anticipating them and providing warnings to the company.
- A bad CFO looks for loopholes and manipulates the numbers to tell whatever story they want.
- A good CFO plans ahead and encourages staff to tell the truth quickly—whether it's good or bad news—without fear of reprisals.
- A bad CFO is continuously in a state of chaos and never wants to hear bad news from their staff.
- A good CFO acknowledges they don't have all the answers and engages their team in solving any issues.
- A bad CFO feels best about themselves when they have all the answers.
- A good CFO explains overages and shortfalls methodically, using waterfalls to show the ins and outs of how actuals vary from expectations.
- A bad CFO will not roll up their sleeves and dig into the nitty-gritty.
- A good CFO will pick up a piece of trash on the floor when they walk by without saying anything to anyone.
- A bad CFO acts like a cop and wields their power over others.
- A good CFO has the highest ethics and acts as the moral compass for the organization, calling out bad behavior without focusing on the individual, all the while holding their team to the highest standard.
- A good CFO focuses on shareholder value balanced with employee and societal needs. They discuss the company's financial picture with all employees to help create transparency, to help inform decision making, and to provide context to decisions that have already been made.

One final difference: a good CFO is a voracious reader.

Thank you to Ben Horowitz's article "Good Product Manager/Bad Product Manager" for inspiring this article.

Chapter 17

Other Areas to Partner With

Finance

While Sales, Service, Marketing, and Legal are the areas where a startup CFO can make a big impact immediately, the scope of your responsibilities are company-wide and you can make Finance a valuable partner to the entire company. Here are some tips on what you can do that will help your company go from average to great.

- **Budgeting.** For a startup, the budget is used to help determine resource decisions. Sure, it is always helpful to compare actual vs. budgeted revenue (and bookings), but for many startups, the budget is what drives investments in areas like new hires, marketing, and other material spending. And, of course, it is critical to use for cash forecasting. As you progress during your fiscal year, how you are doing against your budget will drive your allowable expenses, especially new hires. So, you will want to establish a formal budgeting process as soon as reasonable. For a startup, the process should include the following:
  - **Building the budget.** You should work with the other group heads and all those who will have control of spending and sales. Building a simple template for people to fill in sales and spending assumptions as appropriate will make this work much easier.
  - **Budget review.** Once you have a budget, and have it approved at a Board meeting, you should review it monthly with the team. Transparency with the team

and the Board on how you are doing vs. expectations, and why, is a critical part of scaling. This Actual vs. Budget process (AvB) can be used to open up planned investments and hiring or as a revision of expected cash burn.

- **Integrate the budget into processes and systems.** For example, working with the HR team to approve new hire requests vs. the budget, or marketing processes around investment in advertising will help Finance be a true partner. And the more you can work the budget and associated analysis into your systems, the quicker and easier your monthly reports can be created.
- **Budget revision.** Many startups will need to revise their budgets. Early on, your budgets may only be for two quarters, but you need to track carefully and communicate with the team and the Board on any budget revisions. As you scale, you will want to avoid any material revisions and try to stick to an annual budget process. But one that can be used to take advantage of any opportunities. This is another example of Finance as a partner. If you find something that is working for Sales or Marketing, or a good merger and acquisition opportunity, you don't want to put the brakes on it just because it was not in the budget.

- **KPI analysis and reporting.** Early on, the CFO is often the owner of any reporting and key performance indicators (KPIs). You can take this opportunity to build a business that uses data to drive decisions and are actionable. You can partner with the appropriate groups like Product, Sales, and Marketing to determine what they are. Additionally, you can help or own the systems that store or measure the metrics. Establishing an effective business intelligence tool early on can absolutely help you as you scale, to ensure that the data you are using for reporting and decision making is consistent and timely. You want people closest to the challenge or problem making the decisions and creating effective KPIs and systems, since all of these are an important part of making sure your company can do this as you scale.
- **Company communication.** The startup CFO will be the person to communicate financials and other corporate

numbers to the company. Often you will want to provide useful sound bites and the implications of metrics/financials that others can understand, not just rattle off numbers. You will also be the key person who keeps the executive team and other leaders informed on basic financials, burn rate, and investment mindset; you will educate the rest of the company on the pieces of financials that are important; and ensure no surprises for the Board/CEO, especially on burn rate and cash runway. In general, it is a good idea to work on your public speaking skills. Some basic things I learned early on was the importance of telling a story with the numbers and taking your time while speaking. Everyone will have an important one or two things to keep in mind that are specific to them. For me, it was to slow down and remember to pause at the right points and to take a breath!

There are a number of other functions a startup CFO may manage or partner with the CEO. Some of the more common ones include:

- **Corporate systems.** Early on, it is important to establish baseline systems for all corporate functions. This includes creating centralized management of corporate systems, including license management, agreement negotiation, and data sources of truth. A helpful exercise is to maintain system flow chart/map and usage metrics and eventually establish and manage the Business Systems team.
- **Corporate IT.** The key here is bringing your organization and management skills to the function. Some of the things you will want to do is establish basic purchasing guidelines, depreciation guidelines, inventory management, back-office email, shared drives, calendar, and other systems. And you will want to properly resource the IT help desk, build IT documentation, and guidelines including security.
- **Facilities and real estate.** For the startup this is really managing leases, the facilities team, office space, and conference room systems. Some other helpful things to create early on include the office space budget; remote work management,

local real estate lawyers (I would have a different law firm for each country) and local rules, and multinational support.

- **Human Resources.** For many companies the HR function will report directly to the CEO but for some startups the only HR needs are operational. In that scenario you will again be called on for your operational expertise and will partner to help build the HR systems, onboarding processes and basic payroll operations, and to establish health care and other benefits. Often the people and benefit costs are by far the biggest expense for a startup, so it is important to understand all of the pieces. There are also a lot of different regulations by jurisdiction, so make sure to track state business registrations and compliance filings.

Chapter 18

High Impact Areas for the Startup CFO as Partner

As a startup CFO, there are several opportunities to engage with high-impact functions that will add value immediately and set your company up with a foundation to scale quickly. A key area to focus on first is revenue operations. Regardless of whether or not revenue operations reports up into the CFO function, the CFO role will be critical in building and scaling the function. The revenue operations function within the finance team and this role is typically responsible for all of the less glamorous items like establishing processes, systems, and watching over the data. Yes, it's head's down most of the time, but the impact is significant, especially as you scale, develop more accounts, hire more sales reps, work with channel partners, and expand geographically.

The core mission of the revenue operations team is to help leverage the salesforce by removing friction in the sales process and providing the accurate and trusted data when the team needs it in order to make great decisions. In a startup you're going to have friction, most likely from misalignment between Sales and other functions, especially Product, Marketing, and Finance. It can be healthy to have some friction, or it could end up killing you, but the startup CFO has a unique view on how all of these teams can work together. Your key role is not to referee debates or choose sides, but to ensure that data and systems are available to everyone so at least people are looking at the same data sources.

For example, take the common scenario where the marketing team believes they are delivering enough leads to the company but the sales team keeps asking for more leads. This is the type of situation that is ripe for escalated friction, and it's a great opportunity for revenue operations to reduce that friction. When this happened in the past to a revenue operations team I was managing, we were able to show that although the marketing team was delivering a lot of leads, the sales teams were ignoring them because they didn't think they were worth pursuing, frustrating Marketing. Once we dug into the data we saw that now, and in the past, many of these leads were not in regions where our product would work. And there were also a lot of leads that should have had follow-up but were ignored. So, the sales ops team built some simple qualification automation around the region of each lead and then Sales and Marketing were able to look at the same data and align on the quality of the leads. Without access to the same data, Sales and Marketing would not be able to understand the bigger picture, would not align, and that friction would continue and maybe cause problems further down the road for the entire company.

A second high-impact area where the startup CFO can make a big impact at the get-go is with the customer relationship management (CRM) system. The CRM setup is critically important as it will be the source of truth for most of, if not all of, your customer data. It will be the platform that integrates with a lot of the other systems as you scale, especially your product environment and financial systems. These systems can get expensive and complicated very quickly, so as a startup it is a good idea to focus on simplicity and add complexity only when it is worth it. Early on you will want to map out your interest to order workflow (discussed below in detail) and identify each important data entry. It's important that these data entries are consistent with single sources of truth that eliminate duplication, encourage automation, and allow for scaling as you grow your business. The most important things to focus on early are customer/prospect names (that they are singular and consistent), opportunity/deal process and naming, and pipeline philosophy. Although it is important to be fast and scrappy while scaling a startup, putting a little extra thought and discipline into your CRM setup will be worth it.

Early on, working with the Head of Sales (and sometimes the Head of People) on commission plan design is a way for the

CFO to be a huge value-add partner. Although there are a lot of approaches to designing commission plans, effective ones, at least for software, share a few key things.

First, there has to be alignment between the CFO and Head of Sales because you want their perspective and buy-in on the dynamics of the commission plan. The commission plan is at its heart a tool to motivate and focus the sales team, so you need to be aligned with your Head of Sales. As CFO, you wouldn't dictate outbound strategy, so don't try to dictate compensation strategy. Your role as the CFO is to know if the company can afford it, to provide metrics if it is working, ensure it is reasonable to administer and fair to all parties and generally competitive in the market. There are a handful of pieces of the plan that will crop up year after year, like commission rates at different tiers, implications of beating targets, channel conflict, and you'll need an ally dealing with the many commission one-offs that arise. You don't want to put yourself into a situation where you have to make every commission ruling and you won't have to if you're aligned with the Head of Sales. They'll be able to filter and solve the ones that are obvious and consult with you on the ones that are more challenging.

Second, there needs to be fairness between the company and the sales reps which is generally not a problem in normal times, but under extreme ups and downs, it will be challenging. Commission plans that are not stress-tested run the risk of having situations that are not fair to the employee or the company. Many plans that do not consider extreme (or even somewhat not normal) outcomes can break down and quickly not be fair. A useful technique is to use Monte Carlo analysis to analyze thousands of possible scenarios under different assumptions for each salesperson's possible range or outcomes. You'll want to avoid commission structures that have too much, or not enough, possible variability. If you have super high variability, it will be unfair to one party, either the salesperson or the company; if the variability is too low, you've basically created a structure of deferred compensation, which is not motivational at all for a salesperson.

Another area of fairness to watch out for, and this is where a good partnership with your Head of Sales comes in, is to avoid double commissioning sales reps. Although it may make sense to double compensate internal and *external* representatives in order to eliminate channel conflict, you want to avoid plans that have full commission paid to two internal representatives.

Third, I'd create a structure with payment at dollar one and no commission caps. In general, plans that don't have caps and some payment on the first dollar of sales, have less gaming and manipulation from sales reps. Rely on your Monte Carlo analysis to ensure fairness with the lack of a commission cap. This advice sounds obvious but, believe me, there will be times in the life of your startup when even a dollar matters, so you have to create your commission plan and stick with it.

Fourth, a commission plan has to be easy to understand and calculate by the sales reps. Ideally, the sales reps will know while they are trying to close a sale how much they will make from that sale. At times, this goal may be at odds with the fairness goal, since sometimes to ensure fairness you'll want to introduce complexity, but commission calculation should not be a black box. It should not be onerous for the sales rep and it ought to fit into the "sales math" that your Head of Sales has developed. Typically, if you have a commission calculator available for each sales rep, they will make use of it nearly every day and as they gain experience, they'll be more savvy about where they are in the sales process and what commission they'll get.

Fifth, be thoughtful and clear about sales targets and tiers. Sales reps need clarity on what their different targets and sales tiers are and how they were calculated. It's important to make sure: (1) you are aligned with the Head of Sales, and (2) you take an opportunity to present targets and tiers to the sales team with Q&A. This will go a long way to reducing friction later as you scale.

Sixth, make sure you have effective communication, timely payments, exception management, and complete documentation of your sales commission plan. Effective sales plans need great back office operations that can keep up with sales, otherwise you'll be a blocker. A key area is to make sure that you provide timely communication to the sales team. You want to avoid the all-too-common situation of providing sales reps with their targets AFTER the period has already started. Timely calculation and especially payments are another critical area to ensure that you are actually receiving the benefit of a sales plan. Sales teams can quickly become discouraged with a company that pays them late. Exception management and a dispute resolution process are something many startups do casually for much too long and clarity and transparency of these decisions are also important to keep the sales team motivated. And it is critical that the company has the sales reps sign annual sales plans that document all the important pieces of the plan. This is not only a best practice for an organization but also required for compliance in some regions.

Early on you can generally pay commissions when the company receives payment. At some point when you scale up, you will likely want to move that toward payments on bookings. But you only want to do that once you have a clear understanding of refunds, bad debt, and payment terms, and of course, that you can manage the cash flow implications.

The overall best way to add value to sales can be as a partner to creating and managing the sales commission plan. Commissions are a motivating factor for sales reps and if you can create elements of the plan that are fair, can work in good and bad economies, and that can scale with you, you'll go a long way to helping to smooth the path for growth in your company.

Interest to Order: Order to Cash

I've mentioned interest to order several times and this is definitely a high-impact area that early-stage startups need to figure out as quickly as possible. It's critical for your company to understand the workflow that includes the entire process from a lead all the way to collecting the invoice. This workflow will involve a number of different systems, which include the email automation, website landing pages, your CRM, and your financial systems. Essentially anything that touches information from a lead ("interest) to collections ("cash") needs to be understood and managed by the finance team.

A helpful exercise to get started is to flowchart out the process. Include anything that involves a decision, every bit of information entry/transformation, and manual processes that finance or other teams undertake. This will help you determine: (1) which areas will have a problem scaling, and (2) where your critical sources of truth are for your corporate information. Early on, if you understand where your data sources of truth have to be and you put controls and automations in to ensure they remain accurate, it will enable corporate reporting to be accurate and scalable. Often in early stages of startups, much of this process just appears organically and there are multiple versions of the same data, such as client name, sales rep, and product. This is suboptimal and leads to a reporting process that requires an (often highly paid) person whose sole job is to manipulate the data to make up for the lack of a single source of truth. If you can flowchart everything and create processes and checkpoints around key tasks, you can totally eliminate this.

The flowchart should show the following elements:

- Reporting and Analysis
- Pipeline Management and Reporting
- Forecasting
- Training
- Ad-hoc
- International
- Sales Enablement
- Other Systems

Chapter 19

Board and Shareholder Management

It's easy as a startup CFO to focus on the hundreds of internal processes and systems that you'll have to create, and push any Board or shareholder tasks out. After all, your Board might only meet a couple of times a year, so why let that get in the way of doing the real work? Don't fall into that trap! The Board can be a huge strategic advantage for an early-stage startup and the CFO plays a critical role in Board dynamics. Early on you should work with your CEO to establish processes and checklists for Board meeting materials, records, communication, and sources of truth for shareholder and equity information. The earlier the equity records and processes are moved away from spreadsheets, the better.

For the most part, the startup Board Book is a job for the CEO, but this is an area where the startup CFO can be a good partner. You can help the CEO establish an effective Board Book template and then produce the monthly reports for the Board with enough time for them to review the materials *before* the meeting. You don't want to provide materials for the first time *at* the meeting or you'll waste everyone's time.

The Board Book process we have used for a long time roughly looks like the following:

- Agenda
- Official Business (approval of minutes, option grants, etc.)
- Primary Reading section. This would be anything meaty you want the Board to review or read and often includes an "On My Mind" section for the top things the CEO is thinking about.

- Standard Reporting. Includes financials, KPIs, etc.
- Appendix. Items you want to show the Board but may not want to discuss. Could include things like 409a valuations, product roadmap, etc.

As you scale, typically the meeting itself will evolve past just going through the Board Book and hopefully focus on strategic issues rather than a simple reporting out. And as you scale, the CFO may take ownership of coordinating the Board Book production and distribution. Ideally the Board Book could also go to a wider range of people than just the Board, including senior management. You can refer to Chapter 35 in *Startup CEO* for a lot more on the Board Book.

Corporate lawyers will often insist on fully executed resolutions and minutes during a due diligence process and they'll want those documents fast. So, start by storing these in a logical way to make your (and your team's) life a lot easier in the future. You also want to work with your outside counsel to agree on the minute taker, the source of truth for the records and resolutions, and cap table management. Generally, I prefer to have a person from the company as the minute taker; that way the company owns that responsibility and you have more control over the process and can make sure the minutes/records are complete and timely. Often, if you have your counsel take minutes, you don't see a copy of them until right before the *next* Board meeting. I like to take minutes during the meeting and sign them right after.

Another fun thing to do is include some sort of fun quote or comment in the minutes. People often don't read minutes (carefully) but once in a while they do and it is nice to give them a fun surprise. Once we had Shake Shack delivered for our Board lunch that included a couple of trays of shakes. Most of us had one, but our Board member Brad Feld ended up with two. I made sure to note in the minutes that Brad had to lie down on the couch soon after.

Once you scale a bit, you'll probably increase the Board size and change the cadence to quarterly. For quarterly shareholder mailings or calls, you'll need to understand requirements in corporate documents and be able to establish a source of truth for shareholder contacts. I'd also highly recommend a very structured meeting before, after, and during the Board meeting. We use 30-minute quarterly updates that follow a script, send out quarterly email updates, and have an email system of record established.

Another item to keep in mind as you scale the business and raise money is the concept of information rights. Information rights are rights shareholders received per the shareholders agreement that all shareholders are party to. Typically, there will be some rights the preferred investors receive that the common shareholders do not. This can actually be a helpful clause because as you grow, and employees turn their options into shares, you may want to have the ability to control the distribution of information, such as the cap table and key financials.

Chapter 20

Equity

Finance

Equity Management

Equity management is another area that is important to organize and where you should have discipline as early as possible during a startup. There are a few areas to pay careful attention to:

- **Stock compensation form documents.** It is useful to have form documents for option and stock grants, grant exercise documentation, and share transfers. Note that if you operate in different countries, you will likely need to have different stock compensation plans.
- **FAQ for employee questions.** There are always a handful of questions employees have and having clarity on them will help employees value the stock compensation. Some of the most common include:
  - *Tax implications of the grants,* which can be very different if the grant is an option grant (Non-qualified vs. Incentive Stock Options), restricted stock, or restricted stock units.
  - *How to roughly think about the value.* You should not give an exact value, but provide guidance on how to think about the valuation.
  - *How and when to exercise options.* Employees will often want or need to exercise their options when they leave the company, but also may have opportunities to early-exercise for a more favorable tax treatment.
  - *What happens when an employee leaves the company.* For many years, most venture-backed companies have followed a standard 90-day post-termination exercise period. No matter why the employee is leaving the company, they

have 90 days to exercise their vested stock options otherwise lose them forever. This created unfair situations where employees, many without a lot of discretionary cash, were forced to either spend a material amount of money on the stock of the company that has no market liquidity, with tax implications too, or lose the value that was part of their compensation for years. In the last few years, many companies that have significant employee stock option plans have started to use other vehicles, such as restricted stock, or a post-termination exercise plan (PTEP), that provides a number of years after termination for the ex-employee to exercise. Although not perfect, it is fairer to the employee.

- *How an employee receives additional stock compensation.* Typically, employees will receive more stock compensation if they receive a qualified promotion or continue working for the company. The grants for longevity may be a small amount every year or two, or mirror grants as they finish vesting. Almost every employee will want to know the answer to this, so it is a very good thing to include in an FAQ.
- Note that *employees in different countries* may have different tax implications for employees as well as post-termination treatment.

- **Board approval and communication.** You will want to keep the Board informed of your stock compensation and how much of the reserved employee pool is remaining.
  - Develop a clear philosophy and formulaic approach to most grants. Typically, an employee grant will be based on their seniority/market competitiveness of the role and cash compensation. This way when you include stock compensation grants in Board Books for approval, you can list the ones that were based on the formulaic approach previously discussed with the Board and call out ones that fell outside that approach, typically for executive hires.
  - You should work on a stock compensation budget that generally should last for one to two years (or the first 20–30 employees for early stage), depending on the stage of the company. This provides visibility for you

and the Board about how fast the company is granting vs. expectations as well as make clear to all when a plan refresh is likely required. Often your VCs will have a rule of thumb of how much dilution they would expect each year from option grants. So, it is a good idea to poll the group to ensure good alignment.

  - Produce a standard report that goes with each Board Book that shows a list of the stock grants with name, title, and grant amount, calling out any grants that fall outside the normal formulaic grant guidelines. Include in a report a second page that has the current status of the company's cap table with all of the fundraising rounds, all employee stock plan refreshes, and total amount of stock plan pool remaining. Doing this with every Board Book creates a level of transparency and accountability (for both the Board and the company) on the employee stock plan.

- **Equity management system.** For years, startups relied on Excel spreadsheets to manage their cap table and equity plans. Although free and flexible, the lack of corporate controls, disciplined processes, and central source of truth means even the best spreadsheet approach will eventually break down. Modern startups now have the ability to use a number of software packages, some with free versions, that will provide controls and sources of truth for your equity management. In addition, some will let you process the entire stock compensation process in their software, provide you stock compensation expense for your financials, and give you a liquidation waterfall analysis. By establishing an equity management system as early as possible for your startup, preferably day one, you will be able to scale cleanly and without the inevitable pain that comes with managing equity on a spreadsheet.

Secondary/Tender Transactions

Secondary transactions are where existing shareholders sell their shares to another party, often with the company being a facilitator of the transaction by being an intermediary. The most common case of this is during a larger fundraising event where some of the early investors or founders want to sell a small amount of their overall shares to give themselves some liquidity.

Whereas a standard fundraising transaction is where all of the cash is kept by the company (sometimes called a primary transaction), in a secondary, a material amount of cash is not staying with the company. This ends up with a lot more complexities to consider. These include:

- **Will this impact your 409a valuation?** Often these transactions end up with a price per share for the common that is higher than the recent 409a price. Impact on your 409a should be understood before execution of the deal and communicated to the appropriate parties.
- **Tax implications on the company and the employees.** Depending on the form of equity (stock, incentive stock option (ISO), or nonqualified option (NQO)), there are different tax requirements on the company and employee. For example, the company will likely need to have tax withholdings if a current employee is selling NQO options.
- **The regulatory requirements vary.** Depending on the scale and number of people involved, it may trigger the transaction to be considered a tender offer, which has legal disclosures and a blackout period.
- **Waivers from preferred investors** in the cases of right of first refusal and right of co-sale restrictions.
- **Internal communications.** In most cases, these transactions end up with some people being able to sell some shares and some who are not. You will want to be aware of employee morale and the importance of balancing transparency, privacy, and fairness.
- **Third-party systems.** With primary transactions, it is typically very straightforward. With secondaries, especially ones that trigger a tender offer, you will want to use third party marketplace software, which will greatly help with the required shareholder communication, flow of funds, and overall compliance.

- **Keep your partners informed early.** Your outside counsel will be the primary partner involved in a transaction, but you will want to make sure that your audit partner and their valuation team are along for the ride and informed along the way so there are no surprises come audit time.

409a Valuations

409a valuations have become an important requirement for startups. Essentially, the minute you have an option grant or if you start the company with an investment on day one, you will need to have a third-party 409a valuation. The 409a valuation establishes the fair market value of the common stock, taking into account a number of classic valuation variables such as recent fundraising, benchmarks, industry valuation multiples, etc. For most startups, the 409a will be valued using a recent financing as a base and then taking discounts for marketability. As a sanity check, there is usually a percent of the last preferred round that will seem reasonable to the valuation company (and the Board). This will change greatly depending on the stage and the type of preferred security, but roughly the range could go from 25% to well over 50%.

Once you have a valuation, you can grant options using that price as the strike price. You will want to have the valuation fairly close to the date when the Board approves the option grant. As you scale and build a quarterly heartbeat of Board meetings, you can simply do a 409a once a quarter to be effective close to the Board meeting.

Most of the time, the only valuation you'll need for operational purposes is a 409a. However, during an acquisition or when you are buying a company or assets, you will need to have a valuation done for accounting reasons. When you do, it is a good idea to build clear documentation and support as these valuations will be challenged much more by your financial audit firm during the annual audit.

Chapter 21

Mergers and Acquisitions (M&A)

Buy-Side

Once a startup starts to find their product-market fit and understand the clear drivers of revenue, acquisitions become a real possibility as a way to quickly grow the business. For a CFO, the main tactical efforts will include the financial merger model, the price, and how to pay for it, and managing due diligence. Primarily the CFO is the key partner (and sometimes voice of reason) for the CEO, who for startups, almost exclusively drives the buy-side M&A efforts. As the company scales, you might hire someone to focus on M&A and the CFO becomes a key partner.

The Merger Model

The merger model is to help give your company a clear view of the before and after picture of the merger. The easiest way to do this is to compartmentalize the analysis. First, create a base model of the two standalone companies pre-merger. This will include the actuals and forecasts. Then create another that merges the two, and some work that shows cost savings/synergies from the merger, and impacts on revenue, including the key drivers. Typically, that will include things like improved pricing, win rates, and client retention but could include a number of items. The key thing is to make sure the metrics that are driving the decision to make the merger are clear. At that point you can provide the CEO and Board some important financial metrics like payback period, total invested, and what certain metrics need to be for you to feel good about the decision. For example, when we looked at buying a new product (rather than building it), we thought that the two metrics that would improve were win rate and client retention. We were not going to change the price. So, by assuming a purchase price and cost synergies, we could illustrate how much those two metrics had to improve in order to justify the acquisition.

Figuring out the best way to pay for a purchase is an important means for the CFO to be an effective partner. You can use cash, stock, or even earn-outs. Modeling the different scenarios and presenting the options to the CEO and board are critical skills. Many startup mergers are paid for by using company stock. Those are certainly the best use of cash flow, especially if the transition costs are not huge. But the more you use stock, the more you end up with a bunch of small stockholders that you don't know and have no idea how much of a hassle they will be. One way around that is, as part of buying a company, insist that the sellers create a single purpose vehicle with a managing partner that you know. That way you only have to deal with one person. If not, you may end up with 10s and maybe 100s of extra very small shareholders that will increase the amount of shareholder management you will need to do in the future.

If the acquisition results in increased free cash flow (after all of the cost synergies), you may be able to use credit financing to help minimize the amount of stock you need to put in the deal. This tactic is usually used for later stage companies that have established teams that can be leveraged in a merger.

Buy-side acquisitions are also a time that the CFO can be a valuable partner for other parts of the company, helping build the transition plan and providing clear guidance and targets for things like new hiring, analyzing impact on new bookings and client retention, and other product, sales, and service plans.

Sell-Side

Selling a company (or division) is usually one of the busiest times for a startup CFO. Not only do you have to be part of all of the many, many early meetings with possible buyers and bankers during the early courtship, but you are on every email, call, and planning meeting as the Board and management team decides whether or not to sell. You and your finance team are often working late trying to respond to the requests from possible buy-side teams on historical numbers and financials. How you and your team react to an opportunity to sell the business is the culmination of your years of setting up scalable processes that lead to quick responses to diligence requests and modeling of scenarios. A prospective buyer will almost always want to see your financials and key corporate information in a different form and segmentation than you are used to. So how you can react to this surprise and still continue to operate the day-to-day business will be a true test of your team and your systems.

Key items before due diligence:

- **Investment bankers.** If you are at a very early stage, it is unlikely you will be using an investment banker to help you with the process. As you get bigger, a banker can be helpful for you as a resource for some of the modeling, a coordinator for a lot of the diligence process, and most importantly, assisting you in navigating the industry landscape and the prospective buyers. They can also provide guidance to the CEO and Board during negotiation of valuation terms and deal documents. As you choose a banker, the most important thing you can vet is their relationships and knowledge of your industry and the corporate development groups of prospective buyers. They can quickly give you a sense of valuation and strategically how you should approach your particular sale situation.
- **Financial models.** Almost every M&A process will require a new set of financial models. For buy-side, it involves merger models and the impact of adding a new company or asset to your business. For sell-side, it is usually normalizing your financial history for how your company looks today. For example, if you sold a division in the past couple of years, you will need to build pro-forma financials (mainly the income statement and key bookings numbers) as if your company did not have that sold division. It is also building a multi-year forecast. Typically, to run a startup you don't need to have a forecast past your current budget. To sell, you will need to re-create some of the way you forecast your

key sales and cost drivers and build a two- to three-year forecast.

- **Due diligence.** You will (unfortunately) typically have to respond to due diligence requests even before you get a term sheet or letter of intent (LOI). These are usually high level regarding corporate documents, organization charts, financials, forecasts, and key business metrics like bookings and client base. Most of this is hopefully on hand and can be handled entirely by the finance team. Once an LOI is signed, the due diligence expands to include many others in the organization. As CFO, you will need to quarterback the diligence process for your company. To prepare for this, the best thing you can do, well before the sale process, is to review a typical due diligence request and assess your readiness. Once it starts, the major areas can include the following critical requests:
  - *Legal:* All material agreements, client agreements, NDAs, IP items like patents and trademarks, data licensing, partnership agreements, key vendor agreements, real estate leases as just a start. Much of this can usually be negotiated from a sense of materiality or importance. Typically, the buyer/buyer's lawyers will press on certain items they consider important.
  - *Human Resources:* Employment agreements, stock compensation agreements, employee termination agreements, signed sales commission agreements, and payroll.
  - *Corporate areas:* All historical corporate documents such as shareholder rights agreement, stock purchase agreements, and Charter. They will typically push for every version of these and the required signatures.
  - *Finance areas:* All recent tax returns, audit information, state filings, DBA filings, and definitely Good Standing certificates in any state you are registered in.
  - *Data security:* Buyers may hire third-party firms to test your company's data security process, test your systems, and audit your information storage policies and procedures.
  - *Accounting:* Even if you get annual audits from a global account firm, buyers may want to hire a firm to do a

Finance

"quality of earnings" analysis which will include not only analysis of things like your accounts receivable, accounts payable, fixed assets, really any and all accounting, but also the trends of your sales bookings and the likelihood of them continuing in the future.
- *Product, client and service:* Buyers will also spend a lot of time analyzing your client base, including client concentration, segments, and retention, but also want to know about your roadmap and development process.

The CFO will need to coordinate a lot of this work with the different groups in the company that will respond to the requests. The use of secure data rooms (your lawyer or banker will help you set that up if needed) will be your friend in helping you have one place to manage the multiple possible buyers and keeping it easy to follow which buyer saw what data. The CFO will also need to negotiate with the buyer which diligence is really necessary vs. what's on the checklist. Buyers and their lawyers will look for any opportunity to get more and more information so you will need to work with your side to keep it reasonable.

Other sell-side items that are typically more for later stage companies:

- *Flow of funds analysis:* The flow of funds is exactly what it sounds like, detailed instructions of where every dollar goes with wire instructions. This will include paying for bankers and lawyers so you will want to be aligned with your outside counsel on their fees. This will need to be approved by all parties involved so it is a good idea to set the structure up as soon as is reasonable.
- *Working capital adjustment:* Because you will be operating your business during the sale process, your working capital could be higher or lower than usual. Especially if you have a seasonal business or stopped paying invoices during the process. So typically, you will agree with the buyer on a reasonable working capital number that is usually based on some average balance over the last twelve months. Then the final purchase price will be adjusted based on whatever the working capital number is at the close versus this agreed

to number. Additionally, there is often a 90-day working capital escrow that can cover any expenses that come up that weren't, and should have been, included in the closing balance.

- *Fairness opinion:* Consider doing this to protect against shareholder lawsuits.
- *Shareholder communication:* Required mailings, waiting period, compliance filings, compensation for insiders.
- *Seller representations and indemnification:* Work with your bankers and lawyers to understand which types of representations are seen in the market and how you can use rep and warranty insurance to simplify this part of the deal.
- *Shareholder representation and M&A platforms:* Once a deal is closed, there may still be a lot of communication, including tax form preparation, with seller shareholders to go and guidance needed on any liability and working capital escrow issues. By hiring a shareholder rep and using an M&A platform, you outsource a bunch of that work. This becomes valuable if you have a lot of small unknown shareholders. If you are a small group with a couple of investors, this will not be necessary. But if you have 400 shareholders of all sizes and sophistication, it is very helpful to hire a shareholder rep.

Chapter 22

Bonus Section: What We Used for Our Internal Systems When We Started Bolster

The company was formed on April 1, 2020, in the midst of the COVID-19 pandemic. As everyone was going to be working remotely for the foreseeable future, it was very important to establish the tools and operating systems as soon as possible. We had the advantage of receiving some support from our early investor High Alpha's venture studio on a few of the tools and processes. We started to put these systems into place on day one and completed almost all of this in the first couple of weeks. This of course is very different than twenty years ago when we started Return Path, but I also feel it has come a long way the last five years. This section will become dated as systems are improved and new choices are established but I thought the overall process will have some interest.

Selected Systems

Communication: Slack

We created a free account. Planned to upgrade when we hit feature requirements. In the early days, we kept channels to a minimum. General was reserved for rare company-wide events that don't need context. Water Cooler for open conversations, random fun facts, and pictures. Then let it be organic with curation as it grows.

We established integrations with Trello, Outlook, and Zoom. The Zoom Integration enabled quick video chats to happen with a simple/Zoom command. The Outlook plugin enabled a nice schedule within Slack and put a calendar flag in your status to let other Slack users know that you are currently in a meeting. With the Zoom integration, if a Zoom meeting went on a long time, it kept the flag active. When the meeting was over, the flag cleared.

Project Management: Trello

We created a free account. Trello has a very good free platform for a small number of users and boards.

Focused one board to manage the activities of the eight of us. The first column was a "How to use this Board" card, the "Weekly Goals" card, and the parking lot for cards to be assigned/discussed. Then each column had cards for each functional area with appropriate people as members on the card. We used labels for "Need to get done this week," "Blocked," and "Ready to Review" to promote asynchronous review and work.

Email and Calendar: Exchange

Cost: $4/month per user.

This was a bit of an unusual approach as most companies will just start with the entire GSuite. We decided to start with Exchange as some members of the team had a strong preference for the email/calendar functionality of Exchange and Outlook. It generally works well if you are OK with Outlook and Outlook has strong iOS apps.

File Storage/Management: Google Drive Enterprise

Cost: $8/user per month plus $0.04 per GB per month, estimated $900 a year.

Certainly going with different systems for email and document management is not ideal. We wanted to give it a try as we had a strong preference for Google Drive. The OneDrive/Sharepoint experience we had in the past was not a positive one and effective file-sharing/storage is important to get right on day one. We established about eight shared drives that mirrored the functions of the business. We also tried to ensure that we started right away with a logical document storage process, so we established a number of official folders for things like corporate documents, financials, and invoices. We also were able to install the Google Drive desktop apps so that the access to the files was as easy as if they were local files.

Video Communication: Zoom

Cost: $2,000 per year for up to 10 licenses.

We felt Zoom had clearly the best technology and workflows for a startup. We could have gone with a $1,000 plan but we wanted some of the features included in the Pro license. We planned on integrating Zoom inside our platform so may have to adjust the license in the future. This was a critical piece of software to help us effectively work together and conduct nearly 100 external meetings in the first month of operations so we wanted to ensure we had the best technology. We did not go with any of the other features of Zoom like Zoom Room or the softphones.

Web Analytics: Databox

This was purchased at the suggestion of our investor High Alpha as it works well with Hubspot.

System Password Management: 1Password

Cost: $50 for the year and first 6 months free.

To ensure centralized storage of system and administration passwords, we created a 1Password account to store passwords for all of our corporate systems and production technology on AWS.

Product Storyboarding: StoriesOnBoard

Cost: $300 for the year for one edit license (and many viewers).

Given how distributed we are and our future product development, we thought we would find a tool that helps storyboard/lean canvas very helpful. Plus, it integrated with Trello.

Financial System: QuickBooks

Cost: $65 a month.

We were planning on having a third-party handle the back office for bookkeeping so just went with the basic approach they were familiar with. The key adjustment was revising the Chart of Accounts to match the business and ensuring the key integrations with Bill.com and Expensify were completed. A quick note that although we did not set up any sales tax software at this point as we were months from any revenue, and perhaps longer from the type of revenue that would require sales tax collection, we did ensure that when were ready, we could plug into our systems quickly and painlessly. Fortunately, just about every sales tax solution will plug into QuickBooks and we made sure our internal accounts were ready.

Accounts Payable: Bill.com

Cost: $1,000 per year.

This may be a little early for this software given the small numbers of vendors we will have and we are months away from revenue. But given the cost, we thought we might as well integrate early on and help develop complete documentation of all of our AP management and give us the option of using their invoicing in the future.

Expense Management: Expensify

Cost: $1,000 per year.

Another nice to have but, given the cost and the ability to start from day one with consistent expense tagging and documentation, we felt it was worth it. Easy to set up. We have a policy where the COO approves everyone and the CEO approves the COO. Expensify also has the ability to directly refund to an employee's bank account once the expense is approved so the process will scale well.

Equity Management: Carta

Cost: Startup package, price varies.

We had a lot of experience with Carta and put a lot of value on having the equity ledger on Carta from day one. Initially not using them for either 409a or Board resolutions. So just need the basic package for about six months and plan to eventually move to them for 409a valuations.

CRM: HubSpot

Cost: Their startup package, about $1,000 for the year.

Typically, companies will go with Salesforce.com, but one of our investors had a lot of experience using HubSpot for startups and our early workflows for our CRM were basic enough where the tool itself wasn't as important as building a solid foundation.

HRIS: Gusto

Cost: $1,600/year.

We wanted to use a lightweight and basic HRIS at this stage that could scale for our early planning and members of the team had a positive experience with Gusto. In addition, Gusto was able to issue 1099s for any contractors we used and can integrate our payroll and benefits administration in all of the states we were planning on being in for the foreseeable future.

Calendar Management: Calendly

Cost: Free or $10/month/user for more features

This has solid integrations and ease of use. We needed to set up a lot of calendar events in the early days as we had to do 100s of market interviews and found Calendly helpful as it integrated with Outlook and Zoom.

Payroll: Gusto

Part of the HRIS.

401k Provider: Guideline

Cost: About $1k. $39/month + $8/PartEmployee per month

We liked Guideline's approach of no AUM fee and really just a place to buy Index funds. This approach minimizes the overall expense ratio of the plan and fits well for what we wanted in a 401k. If we had not already funded the business, we likely would have waited before spending money on this benefit. It also integrated seamlessly with Gusto.

Insurance Provider: Vouch

They had a really easy workflow to get the basic General, EPL, E&O, and D&O. The price was less than what we had thought it would cost to get the basic insurances needed for our operations. Very effective for a startup.

Key Processes Established

Documents to track
Document storage
Domains and aliases
Purchasing and asset management
Corporate cards
Social accounts to build
409a/83b valuation
Trademark search
Brand positioning
Vendor tracking
Asset management tracking
Chart of accounts
Incorporation/EIN/W9

The total is under $10,000 for the first year.

Chapter 23

CEO-to-CEO Advice About the Finance Role

Matt Blumberg

What comes before a full-fledged CFO? Lots of startups have nothing more than an outsourced bookkeeper or one junior staff accountant. Sometimes a founder or a founder's spouse even steps in on this front. As startups scale, they are likely to hire a more senior accountant, maybe an AR/AP/Collections staff member, or even a Controller or VP Finance.

Signs It's Time to Hire Your First CFO

You know it's time to hire a CFO when:

- You wake up in the middle of the night concerned about cash—not just that you're running out of it, but that you aren't clear on how much you have and how fast you're spending it.
- You are spending too much of your own time managing fundraising, debt, investors, and cap table questions, and issues.
- Your Board asks you about some small-to-mid-level analysis or metric like CAC, customer profitability, margins, or ROI, and you don't have a great answer and aren't sure how to get one.

When a Fractional CFO Might Be Enough

A fractional CFO may be the way to go if your business model is simple … some combination of a limited number of complex accounting issues, a limited number of customers/invoices/transactions, and an insignificant difference between the income statement and the cash flow statement. If what you need is someone to oversee a gradually growing team, a slow-paced implementation of higher-order systems, basic financial analysis or modeling, or the occasional fundraising event, a fractional CFO may get the job done, for several years.

What Does Great Look Like in a CFO?

Ideal startup CFOs do four things particularly well:

1. They spend time learning and steeping in the substance of the business—understanding the product and spending time in-market with customers and partners. They do not believe their function is "corporate" or a service function. They insist that the people in their department do the same.
2. They are deliberate about regularly reviewing homemade systems, processes, and spreadsheets and looking for opportunities to streamline things, reinvent them, or move them into systems. Once most things are automated and in systems, they are constantly evaluating whether or not the systems are serving the business well enough and are looking to integrate systems across the company. They are not afraid to tear down and reinvent systems and processes that they themselves set up in the past.
3. They have the right balance of pessimism and optimism and are strong at communicating both. While they are proactive and timely about delivering bad news to you and the Board, their orientation isn't around "no" and bad news. Their orientation is around investment and return and always thinking about things going on around them in the company through the lens of realistic opportunity.
4. They can fly at multiple altitudes at the same time, noticing the smallest detail that's off while thinking about business models and strategy. While most executives need to be strategic and tactical at the same time, the CFO needs

to be like that more than most—mostly because the details and tactics are frequently life-or-death for your startup.

Signs Your CFO Isn't Scaling

CFOs who aren't scaling well past the startup stage are the ones who typically:

- Play the role of the Late Night Hero over and over again. There's always a pending crunch time that requires their personal attention and a ton of manual work—the monthly close, the audit, the budget, commission planning, compensation cycles. CFOs who are mired in doing all of these things personally and manually haven't built the systems, teams, or processes required to scale the business.
- Allow accounting teams to swell in size. "Throwing bodies at the problem" and allowing people to pile up on a team instead of taking a process innovation perspective to Accounting is easy because it's the path of least resistance, but your CFO would never allow other teams to do that, so why should they permit it on their own teams? Accounting teams in particular tend to be the most traditional, paper-based, teams and don't need to be.
- Get forecasts wrong, or don't even try to do them. Especially while your startup is in burn mode and constantly calculating its runway and months until the next required financing, regular and accurate/conservative forecasts are critical. Even without a ton of revenue visibility on forward-looking sales, good CFOs should have enough of a grip on expenses, cash flow, and order-to-cash dynamics to produce good, rolling 12-month cash forecasts.

How I Engage with the CFO

A few ways I've typically spent the most time or got the most value out of CFOs over the years are:

- Doing mental math together. My CFO and I are always attuned to key metrics and from time to time project them forward in our minds. We are constantly checking to see that our financial and operating results mesh with our mental math. When looking at our cash balance, we look back at the last financial statement's cash number and mentally work our way to the current statement: operating profits or losses, big swings in AR or AP, CapEx, and other "below the line" items. Do they add up? Can we explain it in plain English to other leaders or directors? The same thing applies to operating metrics—the size of our database, our headcount, our sales commission rate.
- Spotting the wrong number on the page. I'm sure this has driven CFOs crazy over my career, but I have some kind of weird knack for looking at a wall of numbers and finding the one that's wrong. It's some combination of instincts about the business, math skills, and looking at numbers with fresh eyes as opposed to being the one to produce the numbers in the first place. But it's part of the partnership I have with my CFO that improves the quality of our work and quantitative reasoning.
- Working hard to tell stories with numbers. The best CFOs are the ones who are also good communicators—but that only partly means they are good at public speaking. Being able to tell a story with numbers and visuals is an incredibly important skill that not all CFOs possess. Whether the communication piece is an email to leaders, a slide at an all-hands meeting, or a Board call, partnering with a CFO on identifying the top three points to be made and coming up with the relevant set of data to back the numbers up—and then making sure the visual display of that information is also easy to read and intellectually honest can be the difference between helping others make good decisions or bad ones.

Part Three

People and Human Resources

Cathy Hawley

Chief People Officer

Human Resources has evolved from transactional, compliance-focused administrative work to more strategic, proactive work that touches all aspects of the business. The work is much more rewarding and impactful now. This section focuses on the "new HR" which I'll call "People," and also focuses on a particular framing for People leaders who want to help a CEO and leadership team create a people-centric, values-driven, learning organization and culture that optimizes the contribution and experience of employees and their impact to the company. Trust me, as someone who has experienced traditional HR and also worked with CEOs who are *only* business- and not people-focused, strategically driving a people-focused company is much more rewarding and impactful. If your founder/CEO has interest in building a values-driven company, you'll find this Part relevant to your success. I've been lucky to work with Matt at both Return Path and at Bolster, and his mentorship and guidance helped me be more impactful in my People roles, and also expand my impact. At Bolster, I'm working with two sides of our marketplace: members (executives who want to work in on-demand roles) and clients (CEOs who need to bolster their leadership teams with on-demand executive talent).

I started my career in a very traditional HR role as the HR Manager at a 100-person truck stop. I did everything you'd expect in that role including hiring, processing payroll, writing an employee manual, dealing with workers' compensation claims … if it was remotely connected to compliance, I did it. The truck stop was hierarchical, had traditional gender roles, and no one was consulted on any decision since the CEO made all the decisions himself. Although I knew I couldn't thrive there in the long term, I did learn everything about transactional HR that they don't teach you in college, like how to actually run payroll, negotiate benefits, manage difficult employee situations, and manage risk. My role was to focus on the company first, by making sure that we mitigated risk and reduced costs. Many HR people started their careers in similar environments, and, without the benefit of a more open-minded CEO or company, they become the HR Directors of comic strips.

Luckily, my career shifted after that and I later had the opportunity to work with technology companies whose cultures were focused on people—on trusting that you've hired adults who don't need "managing," and focused on providing development opportunities for people. At these companies my role was to focus on both the business and on people, and help us make good business decisions that were best for people.

Both business-first and people-centric/inclusive companies exist today. If you're an early-stage startup, and have a CEO who cares deeply about people, help them create a people-centric/inclusive environment. Building a sustainable company that people love to work for will drive engagement and business results. When someone feels included, and can bring their whole self to work, when they don't have to worry about being judged for having different opinions or ways of working, they do better work. They can focus on the task at hand rather than worrying about what people think, or whether they will be heard if they have an opinion.

At Return Path, we didn't explicitly start with building an inclusive culture. Luckily, the founders and early team members did have a bias toward inclusion and they built a lot of practices and processes that led toward an inclusive environment. At Bolster, we are intentionally building inclusion into our strategy and culture (and in our case, it's related to our actual product offering, too). You'll see people-centric and inclusive practices throughout this Part that are different from those you've seen at many companies and you'll see how they lead to a more inclusive environment. It's not easy to be inclusive without the support of your CEO since you'll continually have to educate and influence leaders and employees to embrace a different way of working. Having your CEO embrace and lead that education and influence with you is powerful. In my experience, there are more CEOs right now who want to build an inclusive culture than there are HR leaders who have the skills and experience to do this, so the quicker you can embrace the mindset and build the skills required, the more impact you can make at your company and in the world!

Payroll, benefits, and people operations are, of course, urgent in a startup. I'll touch on those later in Chapter 39. These are table stakes and don't highly differentiate you, so I'll start with what I think is your most important role: building an inclusive culture.

As Chief People Officer, you will have far more tasks to complete than time available to you, and I have highlighted the critical things needed to scale: articulating your values, building diversity, equity, and inclusion into your foundation, building your team, setting up structures and practices that lead to the culture you desire, leadership development, recruiting, people operations, onboarding, talent management, organizational design, and team development. Each chapter provides ideas, tools, and issues for the startup and scaleup phases so you can both learn about your current situation and also see what's on the horizon. I realize that the journey from startup to scaleup is not linear and there are times when you need to reduce headcount and I have a chapter on best practices for doing that (Chapter 35).

Chapter 24

Values and Culture

Driving alignment on values and culture will guide you in nearly everything you do. Your values help you find and hire the right people, reward and recognize people and behaviors, influence your organizational structure and operating system, and help you make decisions. Your values also shape the things you don't want to do, or shouldn't do: they prevent you from hiring people who don't fit your company, they guide you on the markets you'd like to enter, or ones you don't want to enter, and they frame how you select vendors and customers. Your values also influence the culture that you build.

Culture is the sum of the everyday behaviors of employees—how work gets done, how people interact with each other, and how they exemplify the values. For example, if leadership team members make every decision themselves without including others, you'll soon find that the culture of the company is one in which everyone waits for a leader to make a decision before they take action.

You may run into a CEO who just wants to build a business and isn't interested in the culture. They may believe that business takes priority over culture, that culture is just "soft stuff," or that it doesn't add value. There's a quote attributed to Peter Drucker that "culture eats strategy for breakfast"; as Chief People Officer, it's important for you to help people understand the importance of culture to *support* the business strategy. I don't advocate for culture *over* strategy. It's just that many executives focus *only* on strategy, and you can help them add culture to their perspectives. If you're in this situation, you can help paint the picture of what the company will look like if you aren't focused on culture and values from the very start. You are going to have a culture, and the options are to build it intentionally or let it happen by chance!

Every person you hire at an early stage has an outsized impact on the culture. If their values don't match those of the CEO or executive team, they won't be successful in the organization, or they will build a toxic subculture that is hard to dismantle. Until they leave the company, there will be tension and unproductive conflict which wastes time, resources, and impacts productivity and engagement. For example, suppose you have a CEO who values transparency and direct conversations and you hire a leader who is more political and is not transparent with their colleagues. That individual is unlikely to be successful, and the reverse is true as well.

Being able to hone in on the values and culture your team wants is an iterative process, and the starting point will differ depending on how strong a vision the CEO/founders/leadership team have about culture. The number of people involved in the initial process can make a difference, too. If you have three people, you can become aligned more quickly than if you have a dozen people. To get things going, have conversations with a small group of people, just the CEO and founders or, if the team is small, the full leadership team. Start with a conversation about what matters to the stakeholders: What do they care most about and what's the best culture they can envision? Or, you can start by asking what legacy the CEO/founders would like to see the company have. Here are a few topics you can ask about:

- How do you think about hierarchy and decision making? Do decisions need to be made at the top, or do you want people to feel empowered to make decisions themselves?
- What is the level of transparency that you want? Do you hold this as a value? Would you rather be fully transparent about everything, or operate on a "need-to-know" basis?
- Will you embrace a remote culture, or do you want people to work in the office? (This topic is especially important right now, as we manage through the global pandemic. It's an opportunity for companies to rethink their priorities and explore whether a remote culture works for them.)

- How important is diversity to you? If important, what does it mean to you?
- How do you feel about corporate social responsibility? As a startup, how much effort do you want to focus on this?
- How innovative do you want or need to be?
- Do you value speed over quality?
- What's your financial model? Will you reinvest any profits in growth? What responsibility do you feel for sharing profits with employees?
- Do you value a work hard/play hard model or do you want to focus on more of a work-life balance?
- Do you have a preference for building everything in-house, or are you willing to partner or outsource?

Your role as Chief People Officer is to help continue that conversation, to facilitate it so that alternative voices are heard, until you have enough content to articulate that culture. Once you're comfortable (as a team) on the culture, shift your focus to values. Look online to find lists of company values—either through articles or looking at different companies' websites and pull together a starting list of values for the team to explore. The list should be broad and encompass some of the diverging viewpoints you heard. You'll want to have another conversation with the CEO/founders/leadership team about values. Your aim in this conversation is to narrow down the values list to the most important three to seven values.

Getting alignment on culture and values is the best use of your time as a startup Chief People Officer and the results will carry the company far, since the culture and values impact every facet of how you work together and how you approach and solve problems. Your company's values need to be ingrained within the people in your company and it's something that you, as Chief People Officer, will measure, evaluate, champion, and help course-correct as your company scales.

As you grow, you'll want to build deep skills in the company on how to talk about your values and tell stories that demonstrate your values in action. This helps to make values part of your company's DNA and when you grow, the values will grow with you. While people may have an intellectual understanding of values, they'll also need to be able to see them in action concretely, and you'll have to help them recognize and reward values-based behaviors. At Return Path, we built values into every people practice: we interviewed people on values alignment, we had story-telling sessions during onboarding, all our rewards were based on a value, we highlighted people who embodied the values, and we evaluated values-based behavior in promotions and performance.

You'll also periodically evaluate whether the stated values still accurately reflect the actual and desired behaviors. Generally, values don't change over time but they might if you've experienced a change in leadership or if you realize that you missed or misstated some values at the beginning. It's not uncommon for the leadership team to refine the values over time, to help them be more clearly understood by employees. This is especially important if you're growing very fast, if you have remote workers, or if you're geographically dispersed.

At Return Path, about 10 years after our founding (in 2008), we experienced a large growth spurt, including international growth. We weren't sure if the mission and values we started with reflected who we had become. We ran a company-wide values exercise to better understand what was most important to employees. Everyone at the company joined one of a series of small-group brainstorming sessions where we discussed current values and mission and recommended changes. Matt took the results and recommendations from all those sessions and crafted our revised values framework. It turned out that the revised values weren't too different but were reframed in a way that was easier for people to understand.

You can find many examples of values on company websites; here are the final values from Return Path; you can see that Matt had some fun with the acronym!

Our Path to Going **A**bove and **B**eyond the **C**all of **D**uty for **E**veryone. We are …

Owners

Unconventional

Results Oriented

People First

Agile
Transparent
Helpful
Appreciative
Business Focused
Collaborative
Data Driven
Equity in Opportunity

A few other companies that have impactful values include:

- Sendgrid: "We value a culture centered around 4 values: Hungry, Happy, Honest, and Humble."
- Twilio: In the buckets of "How We Act," "How We Make Decisions," and "How We Win."
- Moz: Transparent, Authentic, Generous, Fun, Empathetic, and Exceptional.
- Bolster!: We wanted to make sure that our values were very clear and easily remembered, and also raised the bar. Some values that we articulated at Return Path are just table stakes for us here, so we were diligent about articulating values that are really clear and explicit. We also distinguished between values (rarely changing) and operating principles (dependent on stage). Our values are:
  - Always Have Heart: We treat everyone with respect and value the unique individuality of our employees, members, and clients. We are helpful and thankful, inclusive, and assume positive intent.
  - Be Transparent: We help others understand our thought process; show our data, and facilitate honest and open feedback.
  - Continuous Growth: We are intellectually curious, and this helps us to grow. We strive to be and to develop high-performing individuals, teams, company, members, and clients.
  - Do the Right Thing: We always err on the side of high integrity and putting company, members, and clients above self, even when that is difficult.

The work that you do as Chief People Officer on your company values and your culture will be some of the most impactful work you do. You'll need to revisit the conversation over time to make sure the values and culture are still relevant as you grow. Values generally don't change over time, but your culture may. Either way, you'll have to be intentional about revisiting conversations on culture and values.

What I Look for in a Chief People Officer

Scott Dorsey

Managing Partner, High Alpha

My dream HR and People leader embodies a unique blend of characteristics, skills, and experiences. Just like the CFO, this startup leader needs to be honest, ethical, and have strong character. Trust is the foundation for this role.

Building on this foundation, I am then looking for the right mix of tactical and strategic HR experience. On the tactical or transactional side, the leader needs to be proficient in getting key systems and processes running smoothly—payroll, compensation, benefits plan, performance review process, performance improvement plans (when needed), employee handbook, employee onboarding, etc.

Stretching further, I am looking for this leader to build HR into a strategic function, not just a transactional function. First step is helping the CEO develop core values and a culture framework. And most importantly, working to reinforce these values through actions, behavior modeling, and excellent company communication. Next up is recruiting with urgency, discipline, and high standards of excellence. Is anything more important than attracting the best talent possible and helping them achieve their full potential? Building diverse teams and an inclusive culture has never been more important.

Most of all, I am looking for the startup Head of HR to be an amazing business partner to the CEO, to the leadership team, and to the company at large. This person needs to take care of small problems, bring positive energy, surface new opportunities, keep a pulse on the organization, and build trusting relationships. All these tasks are a critical part of the charter. In many ways, the HR and People leader is the glue that brings every functional group together. And cross-functional collaboration is imperative to company success.

As an investor and Board member, I sleep better at night knowing that our portfolio company has a star HR and People leader who is extraordinarily trustworthy and reliable and will make the CEO, team, and company better every day.

Chapter 25

Diversity, Equity, and Inclusion (DE&I)

I noted earlier that building an inclusive culture will reap many rewards in the long term. In every aspect of the People function you can build practices to interrupt unconscious bias and create a more inclusive workplace. Intentionally building DE&I into the foundation of your organization is much more impactful than trying to retro-fit it later. I use the term "DE&I" rather than just diversity or inclusion, because it's not enough to bring in diverse talent to make sure that your organization is representative of your location. If you don't compensate people equitably, then you won't retain or engage your newly diverse workforce. If the people you hire don't feel included, they can't have the psychological safety that is required to be the best team member they can be. It's a lot of work to uncover and interrupt our cultural biases, but it's critical for a number of reasons:

People

- It's the right thing to do.
- There's a business case for diversity, demonstrating that more diverse Boards and leadership teams have higher valuations and growth, and diverse teams are more innovative.
- Building a just and equitable company where people have access to opportunities is becoming more expected by the people you want to hire. It is the new standard for being a great place to work.
- DE&I won't just happen on its own. Even with the best intentions, there is systemic and individual bias. To overcome these, you must be intentional about adapting traditional

> methods of sourcing, recruiting, compensation, promotions, and performance management. You need to counter the biases in our systems and help people understand how to proactively engage in inclusive behaviors.

If you don't start out with DE&I as the core cultural value at your company, what can you do, as Chief People Officer, to get there quickly? One tactic we found helpful at Return Path was to partner with an organization that was an expert in shifting to a DE&I-focus. We partnered with the National Center for Women in Information Technology (NCWIT), which is where we learned a lot about DE&I (and not just about women in technology!). Our partners at NCWIT, and especially Jill Reckie, helped us to think about DE&I strategically and comprehensively. Leveraging our learning and partnership with NCWIT, we built an internal DE&I team that was incredibly impactful. A senior People team member led the overall initiative which included volunteers from across the organization split into work groups, focused on different parts of the employee lifecycle. In many cases, the People team functional lead participated with the work group so they could implement the programmatic recommendations. Even with strong systems set up, we always looked for ways that unconscious bias was filtering in. Given that we all have our own personal experience and journey and it's hard to deeply understand another person's journey through our own personal lens, it was really important to include diverse voices and perspectives in these work groups.

See www.Startuprev.com for a chart showing our work groups and examples of the work completed in each.

Chapter 26

Building Your Team

As a startup you'll need to be scrappy and hire people who also can wear multiple hats. Your first hire should be someone whose skills complement yours, and they also need to understand that startups don't have processes and procedures in place that can be followed or modified. There is more uncertainty and things change rapidly, even with the best planning. And your first hire needs to understand that the People role is every bit as entrepreneurial as any other role. While you'll create a strategic People plan that aligns with the company's strategic plan and covers recruiting, organizational design and development, and operations, your first hire must be comfortable shifting between roles, embracing change and uncertainty, and have values alignment with the company. Ideally, you can provide a pathway for growth and opportunities to expand their skills, allowing this individual to grow with the organization and be your right-hand person as the company moves from startup to scaleup.

As the organization grows, you'll need to continually evaluate whether you are sufficiently staffed for the next stage of growth, and able to achieve your strategic plan. I recommend that your team's growth plan be tied to the growth in the number of people in your company and not revenues. In an established company with a repeatable revenue stream, it's easier to plan for people growth, but in a startup the need for people usually outstrips revenues. If you wait to hit revenue milestones, you'll find that you are severely understaffed and when you do hire people, they will spend their time catching up on things left undone rather than being focused on growth. You'll also need to hire people who can be scrappy and not be perfectionists. I hired a really smart woman once who cared a lot about her job, but didn't move quickly and needed everything very orderly before she could switch tasks. She spent days putting all our employee files into folders that required two-hole punches in the top of each page, and a separate tab for each type of paperwork, which then also meant a lot of time per file setup. That type of heavy organization doesn't work in a startup, even though her responsibilities included ensuring that we had the right paperwork for each employee.

After your first hire, you should think about scaling your team, and building for scope and breadth by hiring specialty roles. Your second hire is just as critical as the first hire and I recommend that you consider hiring someone from inside the company, someone who understands the business and demonstrates interest in people—maybe someone who connects with others at the company, welcomes new people, and offers to help them understand the company. This person may never have thought about a People Operations role and most likely is not trained for it. That's OK! You will be better served at this stage of growth providing your own training on the technical skills.

Early on at Return Path, we promoted our office manager, Amanda McDermott, to run payroll and benefits. Her predecessor stayed at the company in another role, so was able to provide training and support, and Amanda also completed some online courses to learn more. She had the right customer-focused mindset and process orientation, and she also knew the employees and the business. Learning payroll was easy for her and she continued to grow in the People Operations role as the company grew. If you're not ready yet to promote from within, keep that in mind for when you become larger. I've done it both ways: hiring a traditional HR person and shifting them to a people-focused role, and hiring a people person and training them on technical skills. I was more successful with the latter, as I didn't have to help people "unlearn" their old habits of being more compliance focused. If you've been creative about your People practices, I would recommend this approach.

As a startup with a small team, you'll be managing many roles and responsibilities. As you grow, these roles will need to be filled by others, including People Business Partner, Talent Acquisition/Recruiting, People Operations, and later Onboarding and Leadership Development. You'll generally maintain your role of building an intentional culture, building values into your DNA, coaching the executive team, and building and growing your team. The following roles are critical for success (Table 26.1).

Table 26.1 Critical roles.

Function	Role	Description
Organizational Development	People Business Partner (PBP)	The lynchpin to the organization and a critical role to fill if you want to scale the company. A strong PBP understands the business, knows the products or services you're offering, understands the sales cycle, the challenges, and the opportunities. They know the leaders and team members well. The PBP helps leaders, employees, and teams organize, execute, and develop. They coach people at all levels of the organization and run critical People programs such as performance reviews, talent reviews, and leadership development.
Organizational Development	Leadership Development	Initially filled by the PBP, will eventually own overall leadership development and change management. Works closely with PBP on execution of both.
People Operations	Talent Acquisition	Working closely with hiring managers to understand talent needs, they write job descriptions, do competitive market analyses, and look to the future for gaps that need to be filled. They aren't just putting "butts on seats": they are helping to build teams. Talent Acquisition also manages your employment brand, which can help your company attract the right people.
People Operations	Onboarding	Responsible for the full onboarding lifecycle, from pre-hire through to 90 days. Often responsible for the overall plan, and helping Operations, Talent Acquisition, and Managers to execute it.
People Operations	Operations	Manages all the systems, repeatable practices, processes, and compliance. Depending on your size and complexity, this role can be combined with Talent Acquisition and Onboarding initially, although they are fairly different skill sets, and if you don't hire specialists soon enough, your team will be overwhelmed and unable to keep pace with your growth.

There's one caution here for companies that are experiencing a lot of growth: it's easy for any team in your company, including People Operations, to become so maniacally focused on their part of the business that they become insular and siloed. That sets the stage for subcultures to emerge. You'll need to ensure that the People team has a really strong operating system to ensure they are focusing on the most important and impactful work, that the team collaborates effectively with each other and the rest of the company, and that all team members keep each other informed.

As your team grows and develops, you'll want leaders for each of the different functional areas. If you're growing rapidly it's easy to justify bringing in a senior manager to work with a functional area, and if you're relatively small, it's easy to justify not hiring a manager. But what if you're stuck in the middle? Too small to hire someone but too big to be without managers? One solution I've found to that problem is to designate a leader for each functional area and keep everyone reporting to you—the leader and everyone on the functional team. While it can be a little confusing at times, I recommend having every function explicitly owned by someone, even if one person covers more than one function. Be explicit about ownership, so you don't forget important parts of your strategic roadmap.

If you're scaling the People team, the logical first managers will be in People Operations and Organizational Development and you will have systems and people in place for the entire organization. Once you've designated leaders for those two functional areas, make sure that your operating system is updated to include a leadership meeting so leaders can stay aligned on strategy, and guide their teams effectively. Of course, you'll want to continue to have full team meetings as well, so that all functional team members still have relationships and visibility into everything happening on the team and in the business. You'll also want to ensure that you invest time every quarter on team development and deepening skills and emotional intelligence on your team. Your team members need to embody the values of the organization and be role models in their behavior. Keep in mind too that it's really common for a People team to care so much about others in the organization that they neglect themselves. As the Chief People Officer, you need to be aware of this tendency and know your team well enough that you can spot when people are too emotionally invested in the goings on of a company. While you can't completely eliminate this tendency, you can identify it quickly and take steps to minimize the impact. If you have a strong operating system that includes regular leadership and team development, one-to-ones with members of your team, and a level of trust where people can admit that they are struggling, you can help your team stay focused and balanced.

As you continue to grow, the sub-functions may be ready for their own leaders or managers but there's a tradeoff in scaling by creating subgroups with managers and that is usually less speed and nimbleness. I try to keep the hierarchy as light as possible and prefer to create team leader roles instead of manager roles. I also use the same employee/manager/organizational structure guidelines for the People Operations team that we use in other parts of the organization. That way you have structural alignment company-wide and that makes promotions, compensation, and career pathing similar across the whole company.

A powerful way to extend your reach and share responsibility for cultural stewardship is to get a network of volunteers from other teams at the company to drive different people-related programs. Make it a requirement for each employee to "give back" to the community. At Return Path, we had volunteers who ran social events, community service programs, our well-being program, and our diversity, equity, and inclusion programs. The People Team managed the programs globally, gave direction, guidance, and budget for local committees.

If you are creating, or already have, a values-driven company, the People team is critical to the success for the business. I'd even argue (although my colleagues might disagree!) that the People team are the true drivers of a company and impact both top-line growth and productivity. Values and culture impact hiring, turnover, engagement, morale, and productivity. There's a measurable effect of culture and values on innovation.

Chapter 27

Organizational Design and Operating Systems

People

Be intentional about your organizational design from the beginning, and evaluate it periodically to ensure that your design principles still align with your company values and stage of growth. Use the data you collect from exit interviews, turnover, employee surveys, and employee conversations to determine if the design and principles are still relevant.

Your initial design work will be with the CEO, and will also engage the leadership team. Your work on organizational design is not theoretical, or something that stands alone. Just the opposite! It goes hand-in-hand with culture and values conversations. Your organizational design will never be perfect and will always have tradeoffs to consider. There's a saying that you should "never let perfect be the enemy of good" and that's the case with organizational design. As long as your design aligns with your culture and values, it ought to serve its purpose, which is to help people be as productive and engaged as possible and to ensure that there is the right flow of communication between people and teams.

There are a few things to build in from the beginning, including practices around manager span of control, manager role, cross-functional teams, and how you think about hierarchy. People have preconceived notions about all aspects of organizations, and having conversations with the leadership team to help them think about the implications of organizational design principles is critical. It's really easy to grow too hierarchical where managers have small spans of control or if you hire managers who have a "command and control" style. You can help leaders get creative about career development when a team isn't big enough for a leader, and employees want to take on leadership roles. Employees in this situation can expand their role by managing workflow, team operating systems, or projects, without becoming the team manager.

In the beginning, you may not have executive leaders for all the different functional areas, so some executives will lead functions that are new to them. You may have a COO who is also leading Sales, or a CFO leading Technology. Help ensure that leaders who are responsible for several functional areas hire functional experts or get strong mentorship in the areas where they aren't as knowledgeable.

It's best to evaluate organizational structure every six to twelve months for effectiveness, and to make small adjustments, or do a full redesign and restructure if absolutely necessary. Often as you grow, you realize that you've prioritized role function over product/division, and you may need to swing the pendulum the other direction. While every organization has different needs, many organizations are moving toward flatter, more networked, team-based structures. These structures need even more clear leadership and operating systems to be most effective.

At DoubleClick, my division managed global services for all product lines. Each product line would separately modify their investment in services based on the success of their products. When the key executive and I started in the division, small regular layoffs were standard. The Product line would reduce their investment, and then the Services organization would reduce their staff. This resulted in a lack of engagement and productivity, as people were constantly wondering if they were going to be next to go. In addition, changes were made without any notice. One day, you'd be working alongside a colleague; the next day their desk would be cleared out. I worked with the new executive to understand the impact of these reactive changes, and we agreed on a new process. We created strong career pathing so people could move between roles and product service areas, and we also created a planning process for each product area to follow, so we were able to predict need and manage our staffing appropriately. At one stage, we decided on a large-scale change in roles, and proposed a consultation process that included transparency and engagement of employees. We were modifying over 100 roles, and this would result in a reduction of 20% overall. Within a week, we communicated the changes, asked people their preferences, assessed skills, and made all the role-change decisions. Some people voluntarily left the organization and others took on new roles. There was some risk in this approach, and many of our internal clients were worried about losing their best service people. The result of the transparency and employee input was that we retained everyone critical to the organization, and people were more highly engaged than before the reorganization.

At Return Path, we made changes to the organizational structure periodically. At one point, we realized that one of our new product areas wasn't getting the right level of focus from the different functions. It was a new product and wasn't driving a lot of revenue, and when each functional team prioritized their work, this new product was last on the priority list. We moved the product into its own cross-functional team that was managed separately from all the other functional teams to ensure we created the right focus within the company. Another time, we knew one area of the business needed to be divested, so we moved all our teams into cross-functional business units, which made the sale much cleaner from both a balance-sheet and people perspective.

When you change your organizational structure, there will be a lot of questions from employees and, perhaps, some anxiety on what those changes mean for them. To effectively lead an organizational redesign, you'll need to make sure you follow your company values and principles as closely as possible, to speed up the transition, because when you make decisions that impact people's role, manager or compensation, every moment employees spend thinking about the change is a moment when they are not being productive. Be as inclusive as possible in the decision-making process and people will embrace change more readily. Common thinking is that the primary role of a manager/leader is to make decisions *about* people. That common thinking is limiting and defines the "command and control" model of leadership. If at all possible, it is more effective to make decisions *with* people, to be inclusive and collaborative. An inclusive approach takes longer and can be more difficult and it is also much more effective, more motivating, and leads to better results in the long run. People are more engaged, the more they are able to co-create their environment and success measures. The leader's role can be to ensure that everyone is aligned and moving in the same direction.

Alongside the organizational design, you'll want to work with the CEO and leadership team to create a company-wide operating system. By "operating system," I don't mean a bureaucratic structure that serves to control people or burden them with reporting requirements just so senior leadership can be informed. I mean a transparent process that shows how often, where, and when your team will meet, how they will determine and prioritize their work, how they will communicate and connect with other teams at the company, and how they will hold each other accountable for results. A strong operating system that changes as the business changes helps ensure that all teams are aligned and working toward the same goals, and holding each other accountable for results.

One of the most impactful changes we managed at Return Path was a company-wide agile transformation project. For years we used agile methodologies in our engineering department, and in December 2013, the week before my sabbatical, Matt challenged us to shift the entire company to agile practices by June 2014. When I returned in mid-January, a team of four of us from the People and Program Management teams ran this project (along with our normal roles!). This was one of my favorite and most impactful cross-functional teams so I'll call out my team: Mike Mills, Caroline Pearl, and Jane Ritter, with support from Dan Corbin. By June 2014, we had completed pilots with 12 teams, developed a framework for five team types to leverage agile practices, and trained 50 facilitators in how to effect change in their team's operating practices. In addition to the individual team changes, we modified the entire company operating system to leverage agile practices. We increased productivity by 13% in one year; we measured this by looking at a number of different metrics, the most relevant to other organizations being revenue per person.

Along with a company-wide operating system, help to build your leadership team operating system. When you are small, this may be as simple as a tactical leadership team meeting twice a week, a strategic meeting once a week, and a company-wide meeting once a week for everyone to share progress on goals and hold each other accountable. Patrick Lencioni's book, *Death by Meeting*, is a good resource to help you think through the different types of team meetings and how to run them.

You can also help establish these norms across the company. I recommend helping teams leverage agile practices: build a strategic roadmap, engage stakeholders in your work, align on prioritization, remove roadblocks in real time, collaborate on big projects, and keep a backlog so you aren't just responsive to every request that comes to you. It's really easy to fall into the trap of working on the most urgent rather than the most important. There are always fires to put out and having a strategic roadmap and a good operating system is critical for success.

Your operating system is not fixed in place for all time; you'll need to evaluate it periodically, and test to see whether it still serves its purpose, and redesign as appropriate. As you grow and change, your operating systems will need to be updated for your new organization structure. At the startup stage you may be able to pull everyone into one room for an informal company update. But when you grow beyond the one-room company stage, you'll want to partner with your marketing/corporate communications team to develop more robust communication practices to ensure alignment.

At Return Path, we had an operating system that changed every year, except for the annual "un-Roadshow." The un-Roadshow was a company-wide kickoff to help everyone understand the final output of the annual strategic planning. Each team contributed to the plan, and no teams had seen the full and final plan. The un-Roadshow always included in-person time with the full leadership team for people to ask questions and have unscripted dialog to ensure they fully understood how their roles fit into the strategic plans for the year. The format was changed every year to keep the programming as unique and engaging as possible. The rest of the year included quarterly planning and retrospective meetings with all team leaders, distribution to all employees of the Board Book, and all hands meetings to discuss progress against goals. We followed up with roundtable discussions with leadership team members. It's a lot of work to maintain a strong operating system, and requires partnership with the CEO. It may not be your responsibility, but you'll definitely play a role in its creation and maintenance. See www.Startuprev.com for the final company-wide operating system at Return Path.

Chapter 28

Team Development

Almost all companies require collaboration within the team to get work done. Even if cross-functional collaboration isn't required for each person's primary responsibilities, teams still need to collaborate on effective practices and processes to achieve their shared goals. Many companies only focus on team development at the leadership team level, and then on team "building" for other teams. All teams benefit from having support and intention around how they operate and interact. Skills built around team development for the core team also help each individual be a better team member when they work cross-functionally.

At startup level the most important team is the leadership team, as they impact the entire company. It's absolutely critical that the leadership team build a high level of trust and the ability to have productive conflict. There's always going to be conflict; it's unavoidable. But it doesn't have to be debilitating or result in damaged relationships. If there are high levels of trust, conflict can be extremely helpful to the team. There are many tools and assessments you can use to help with team diagnostics, and the most impactful, and simple, tool I've used is Patrick Lencioni's book, *The Five Dysfunctions of a Team.* (Lencioni's book, *The Advantage,* combines his thinking in both the prior books I mentioned (*Death by Meeting* and *The Five Dysfunctions of a Team*), as well as introducing an annual planning framework that we adopted and modified at Return Path with great success). Do yourself and your company a favor and take the 5-minute assessment every quarter, evaluate the results as a team, and agree to the development you need to make the team successful. The type of development the team requires depends on the results of the survey: if you score low on trust, work to deepen relationships. If you score low on conflict, understand the underlying reasons, and then offer coaching or development on conflict. Having a highly functioning leadership team sets the stage for having a highly functioning organization. Without that, it's very difficult to keep teams aligned and working together effectively.

In addition to building an effective leadership team, you may also facilitate strategy development and planning discussions or offsites. Some of the most impactful work you do will be to help your leadership team have deep conversations on business strategy. Strategy development discussions are not just straightforward conversations and it's crucial to build an agenda that helps drive toward your goals. It's often helpful to partner with someone outside the team to help build agendas and facilitate offsites, so that you can participate fully. At Return Path, we worked with executive coach Marc Maltz for almost the entire lifespan of the company. Marc helped us to build and adapt the leadership team over time, coached our CEO, individual team members, and the team as a whole.

Once you have a high functioning leadership team, it's much easier to build high functioning teams who report to them. Teams at the startup stage are often very cross-functional and the more you can help all individuals build the skills and practices to work effectively as team members, the faster they can make an impact on the teams that they join.

Picture a scaleup of 50 people. All employees report to a leadership team member. Some teams are small and each individual wears multiple hats. Even larger teams don't always have a standard role. Everyone knows each other. If these cross-functional teams don't have the skills to communicate with each other effectively, and they let their egos get in the way, even small conflict becomes painful and takes an emotional and productivity toll. If the teams all have the tools to set up effective operating systems and are able to communicate effectively, small conflicts can easily be handled between teams or people, making the team quickly productive.

For teams that work together regularly, treat them as an intact team, and follow the same steps from the "Five Dysfunctions" process that you did for the leadership team. Because all team members have a common language and skills, often the focus is on team dynamics and calling those to the surface. These skills also help teams be more inclusive. Working with people who come from different backgrounds and cultures requires more effective communication and inclusive practices such as ensuring that everyone has a chance to speak up at meetings, and that quieter people aren't ignored.

When team membership changes with a new hire or a change in role, take the opportunity to talk about your team operating system and dynamics and build new relationships and dynamics where appropriate. As Matt says, "Every time you add or lose a team member, you have a new team." At Return Path, we had every new employee take a Strengthsfinder assessment, and they used that to introduce themselves to the team. That helped to build trust quickly.

Chapter 29

Leadership Development

As a startup, you generally aren't large enough to build or need a formal leadership development program. Work with the leadership team to ensure you are all aligned in expectations of leadership and management, and that each person on the leadership team is managing his or her own department effectively. Your role in a startup is still important and you need to drive conversations and learning and development around leadership skills so the leadership team is modeling strong leadership behavior to others in the organization.

You might have to coach the CEO and each leadership team member when you see or hear about behavior that doesn't align to values, and good leadership team members will appreciate that feedback in an honest and direct manner. We all have blind spots and identifying, giving feedback on, and helping leaders manage these blind spots early on will give you a good foundation on which to build. This can be difficult for you when it's the CEO or a co-founder who has a blind spot that impacts other people or when they act in ways that don't align with company values. These can be tough conversations, but if you have strong relationships and explicit permission to give feedback, your impact on the leadership team and the company can be significant.

As you grow beyond the ability of the leadership team to manage all direct reports, you'll need to hire managers and you'll need to develop and deliver management training for first-time managers. This is a prime opportunity for you to put in place the culture, values, operating system, and DE&I principles that will be the foundation with which you can scale quickly. Your management training needs to include necessary tasks such as approving expenses and PTO, managing compensation, and building an effective operating system, but it should be more than a list of do's and don'ts. You should also build leadership skills, such as emotional intelligence, listening, coaching, and having difficult conversations. You don't have to carry this weight alone. Ask leadership team members to develop and deliver training in their areas of expertise. This helps them deepen their skill and it gives them an opportunity to connect to the development of their team members and really embed the leadership behaviors that you want to reinforce.

As you scale, ensure you have enough People Business Partners (PBPs) to coach all managers across the organization. The PBP will help support managers' growth and development and continue to help you build your leadership bench. It takes a long time for a person to become a strong leader but you can shorten the path with intentional training, practice, and coaching. As you grow, many of the scaling practices in this section such as leadership development programs and leveraging volunteers across the organization can reduce your PBP:employee ratio.

In addition to training for managers, develop and deliver training on core leadership skills that you expect from every employee. Things like communication skills, interrupting unconscious biases, emotional intelligence, and receiving and giving feedback. When you help individuals develop these skills, the organization is better able to handle conflict and change. Collaboration, conflict, and decisions can be managed by the people closest to the problem, rather than needing to be escalated to formal leaders.

As you grow and start to hire second-level managers, you'll need to develop and deliver leadership training in the skills required to lead multiple teams. Emotional intelligence is even more critical here; the leader is even more in a "fishbowl" than first-level managers, as they are leading bigger teams or departments, and people have higher expectations of their standards of behavior.

When you get to this point, consider engaging an outside consultant or leadership development firm to support development and delivery. Ensure that the consultant is highly aligned with your values. Build and facilitate content together whenever possible; the consultant will have a big impact on your organization, since they will be impacting the people who role-model behaviors for others. At Return Path, we co-created one of our senior leader programs with the Refinery Group. Angela Baldonero, our Head of People at the time, worked very closely with them to ensure that they fully understood our culture, values, and leadership expectations. In addition to building high quality programs, the investment paid off in other ways: one of their consultants, Mark Frein, later became our Head of People, and another of their consultants, Russ Hamilton, later partnered with us to build communication and leadership skills throughout the company. The last program that we built with Russ, "Leading Teams", was one of the most impactful leadership programs I've ever experienced. The participants were from across the company, and the cohorts still regularly get together to connect and support each other, even though most no longer work together. See www.Startuprev.com for a table showing the components of our leadership development at Return Path.

Chapter 30

Talent and Performance Management

People

You have an opportunity early on to establish your philosophies around talent and performance. Help your leadership team understand the value of having open and honest communication around performance. Coach each leadership team member on having regular performance conversations with their team members so they are building skills and practices and modeling that behavior for future managers. Remember that in a startup it's likely that this is the first executive role for many on your leadership team and they haven't been groomed for years to hold that position. They're learning about the "art" of being an executive so you should help them understand how critical communication and listening are to being successful.

Run a semi-annual process to formalize performance conversations. These can be done on a form or shared spreadsheet, and can either be simple self:manager reviews, or can be more team-based. For the self:manager review, build in questions on career pathing to ensure that managers and employees are having conversations about professional development. It's not too early to groom your next leaders in the company, even though you won't yet have a formal career path laid out. You can also build in questions for managers to get feedback from their employees.

Consistently evaluate what is and isn't working with your current feedback practices, and then modify the practices and build systems around them. Be creative and meet the needs of your organization. If you have a collaborative, transparent culture, you can implement team-based performance systems. Google published an article on its Re: Work site that talks about

Return Path's innovative work around performance which I'd recommend reading (https://rework.withgoogle.com/case-studies/return-path-team-effectiveness/). It highlighted our live 360° practices which were focused on team and individual performance and development. If your company is more hierarchical, you can implement more traditional systems based on annual goals and self:manager reviews. At Return Path, we iterated on our review process every year to make sure that the practices were still working effectively for individuals, teams, and the organization.

As you grow, build support tools for teams and employees, including manager training and guidelines and training for giving and receiving feedback, and developing high and low performers. Build manager guides and training on biased language and phrases to ensure inclusion. Performance reviews are a great example where unconscious bias can creep in, for example, women are often given critical feedback in reviews that they are too "aggressive" and men with the same behavior are given positive feedback for being "assertive" or a "leader." Having guidelines like these to help interrupt our own biases is a good example of a process change that supports DE&I.

Ensure you are capturing data in a way that allows you to evaluate your people and their talents across the organization, so that you can ensure you have the right people to support company growth and succession planning. Run semi-annual department and company-wide talent reviews to ensure high performers are recognized and developing and low performers are developing or managed out. As you grow, you'll want to ensure that you have succession plans in place for all key executive roles, and a limited number of key senior, nonexecutive roles, where there is already a logical internal successor.

Chapter 31

Career Pathing

To support both performance and professional development, first map out what it takes to be successful at different levels within your company, aligning these expectations to your values. You can do this by interviewing high performers and people who are aligned culturally. At Return Path, we had five "RP Expectations" articulated across four levels from entry level to executive. You can use these expectations in hiring, for performance conversations and to support career development. See www.Startuprev.com for an example of the expectations at manager level, which includes development steps within each level. This may be too detailed for a raw startup; we crafted this when we had almost 100 employees. At Bolster, we've re-created leadership expectations for all executives, in addition to functional expectations at different stages of company growth, and built these frameworks into our software platform.

After you've mapped out expectations at a company level, collaborate with leaders in the largest departments and map out career frameworks for their teams. Ensure these are developmentally focused and clearly lay out what it takes to be successful in a role, and ready for promotion to the next level. Career pathing frameworks seem really hard for people to develop, possibly because roles change and if the framework is wrong, employee expectations aren't met. I'd encourage you to start simple, and continuously remind people that these are developmentally focused, will change over time, and are guidance only. The framework will show the difference in responsibilities and skills between the junior and senior roles on the team, as well as the compensation starting points for each role and geography. In sales, also include the quota for the new role. See www.Startuprev.com for excerpts from Return Path's service roles.

Once you have frameworks in place for the largest department, it's really important to share these more broadly, as career paths are often more like a jungle gym than a ladder—people often benefit from moving across functions to broaden their skill sets, rather than always following a defined path.

To go along with your career pathing frameworks, document your promotion processes and practices. Generally, moving up within a role is different than being hired into an open position. For the former, have one to two cycles a year for these promotions. Be clear on the criteria for promotion. If you only need a certain percentage of the department to be in a senior role, let people know that they need to meet the criteria for the role *and* the business must have a need for it. When someone applies for and is offered a position in another department, the promotion should happen when the role opens rather than waiting for the next cycle. Having a strong practice around posting all openings internally along with a robust promotion process to move between levels in a job family, and requiring managers to have regular career pathing conversations with their teams, support employee retention, and diversity, equity, and inclusion. Without these processes, it is easy for leaders to sponsor people who are similar to themselves, and therefore give those individuals a better opportunity for promotion. Building systems and processes to interrupt our unconscious biases is a proactive step to building an inclusive culture.

Chapter 32

Role-Specific Learning and Development

People

To support employees' success in their role and development into more senior roles, you will have to support development of job-specific skills. At early stages, teams aren't large enough to justify full training programs, and will benefit from having some of their repeatable practices documented, so before you hire a new employee onto a team, encourage the team to document their team's processes and practices, and this can be a starting point for building an informal training program.

For teams who are hiring more regularly or have skill development needs across a larger team, build training programs that start with the most critical role-specific skills. Identify and provide guidance to key employees who can mentor and train other employees. Often customer-facing roles need to be trained in your methodologies and in customer service skills. The training at this stage doesn't need to be developed or delivered by a learning professional, but can be enhanced with support from you or a PBP. When this work is more centralized, you get a perspective onto which skill development is needed more broadly across the organization.

As you grow, you may hire learning and development professionals onto your team, rather than relying on subject-matter experts on the functional team. This is another place that you can benefit from hiring an internal person who knows the business and is passionate about development, and then train them on learning and development practices.

When you get to this stage, step back and conduct a needs analysis across the organization, and partner with key departments to develop and deliver training programs. In some cases, you'll still use subject matter experts to deliver training, and in others, you'll build robust training programs that leverage online, in-person, and on-the-job modalities.

It's helpful to have a learning management system to track attendance and other key metrics so you can evaluate learning objectives and outcomes and adjust your programs as needed.

Chapter 33

Employee Engagement

As a startup, with a handful of employees, it's easy to understand the pulse of the company, but as you scale, it's difficult to know whether values are guiding everyday behavior, and if there is alignment between the aspirational culture and the experience people have on the job. Initially, make sure you talk to employees regularly and surface any specific patterns of behavior to the leadership team. That can be difficult, especially if the feedback is about the leadership team behaviors. If this is the case, I would collect data in as objective a way as possible, and pull it together to share with the leadership team, rather than try to rely on just what you are hearing informally in the organization.

As soon as you are able to afford it, I recommend a robust employee survey tool such as Culture Amp or Emplify to help you understand your employees' experience. Both of these tools have banks of questions (so you don't have to create your own questions from scratch) and industry benchmarks. This will give you data to gauge the culture and values. This data can be used with the leadership team to create a formal organizational development plan. An organizational development plan is similar to a strategic plan, with the focus on the company—the people in it, their interactions, their working relationships, and their perspectives on how things are being managed—rather than on the business. For example, if you learn through an employee survey that managers aren't living the values, you can implement a management training program to improve manager skills, more clearly set out management expectations, and develop a manager scorecard to give managers feedback on how they are doing against those expectations. Or if you get data that one specific manager isn't living the values, you can provide coaching and their leader can start to monitor performance more closely.

Once you have a baseline of employee engagement, you can increase the cadence of your surveys by doing periodic shorter surveys or by adding department-specific development planning sessions. These are opportunities for teams and departments to evaluate how their team is doing and agree on a development plan to help them become a more effective team. This is a perfect opportunity to have a People Business Partner collaborate with the leader to run these sessions, and then coach the leader and the team in implementing their plan.

Chapter 34

Rewards and Recognition

Think about the ways that you want to reinforce behavior and performance that contribute to your culture or your business. Build a simple recognition program that is aligned with your company values. Start with a manual system using your internal website or a form, unless you have a platform that offers this functionality. Just make sure to publicly recognize people for their contributions and allow peers a formal platform to say "thank you" and recognize each other's contributions.

At Return Path, we started our rewards system when the company was less than six months old with a wiki page showing six awards that aligned to our values, and every employee could give another employee recognition that came with a $25 reward. We manually processed the recognition part each week, and once a quarter sent a gift card to each employee with all their rewards. This worked for quite a long time, and only took a couple of hours each month and a few hours at year-end to manage. Later, we purchased a SaaS-based rewards system, and modified the program a little to encourage more cross-functional recognition.

As you expand, consider carefully what resonates with your values and culture. You may want to add team recognition and a more robust process for communicating recognition. Ensure that you have some guidelines in place to have equitable systems for different role or personality types. Not everyone likes public recognition, and people in client-facing roles are more likely to get recognition for small achievements than people in back-end roles like engineering. Use the same solution for all groups, so you can encourage cross-functional recognition, but be careful about too many rewards for the "most recognized" since that doesn't always reflect the highest performance or impact to the culture or company. Many companies have a sales awards program like a "President's Club." I've stayed away from those in favor of a system that recognizes people throughout the company, but they are popular awards with salespeople. At Return Path, we created an annual event called Performance Aspire that was a combination of development, fun, and networking for top performers across the organization, and attendees were selected after a peer nomination process from all departments. We had an amazing positive response from attendees, many of whom hadn't attended anything like that previously.

Chapter 35

Reductions in Force

People

At any stage, when the business is not performing well, or if you've shifted your business strategy and completely changed what roles are needed, you may need to manage a reduction in force. The traditional way to manage this is to have the leadership team sit in a room with spreadsheets and talk about headcount. Maybe they engage leaders in the conversation to help build a change plan. They then go and execute the change plan, walking people out the door immediately after firing them. We've all either seen or heard about someone getting walked out the door with their box by a security guard. The problems with the traditional way of doing a reorganization is that:

- It doesn't allow for any input from employees.
- It treats people as if they can't be trusted.
- It doesn't give space for the people left behind to process departures.
- It doesn't allow for any type of transition.

Traditional executives often justify this approach by saying that they have to "pull the bandaid off" and that it's too risky to let terminated employees back to their desks. They'll say that it causes too much disruption to consult people ahead of time. The reality is that it's incredibly damaging to the trust and goodwill that you've built up with your employees if you treat a reduction or downsizing as a "headcount" issue. It takes a long time to regain the trust, confidence, and credibility of employees and make them believe that you are a people-centric and transparent organization if they come into an all-hands meeting one day and find boxes in the lobby and rent-a-cops on hand to escort people out the door, after being fired in a group setting. It's easy to treat people well when things are going well but it takes much more leadership courage to treat people well when things are going badly.

Some suggestions for managing this type of difficult change:

- Decide as a leadership team the overall strategic direction and budget going forward.
- Engage leaders in the conversation and ask them to help co-create a plan.

If there's any negotiation in who leaves and who stays, create a short process to consult with employees (noting that this is legally required in some countries):

- Let them know the rationale and strategy.
- Ask for input on what the employee wants (do they want the new role/voluntary redundancy/apply for a different role?).
- Make decisions, and communicate to impacted employees first in a one-to-one verbal communication (yes, I mean, not in an email or in a large-scale video call).
- Give all impacted employees a generous package that includes:
  - Transition time: to transition their work and say goodbye to colleagues. Usually a week is sufficient, although some roles are much longer.
  - A severance package.
  - Career placement and continued severance if they can't find a job.

If you do this well, you'll reap the rewards in years to come. Return Path had a large layoff one year, and the next year, as the business grew again, we re-hired six people who had been laid off. All were excited to return, and that wouldn't have happened if we'd treated them poorly when they exited.

Chapter 36

Recruiting

People

Your first priority is to work with the CEO/founders/leadership team to create the culture and values that define your company. Once those are established and communicated, you'll need to focus on recruiting. After you've hired your favorite former colleague, your CEO's brother-in-law, and your cousin's neighbor, you'll need to figure out where you'll find the rest of your people! Since there's so much work to do to build the People part of the business, it's likely that your first hire will do some recruiting, and your second hire will be a recruiter. You might even get a fractional or contract recruiter, until you know for sure you have enough work for a full-time role. I strongly suggest that you stay deeply engaged in recruiting efforts at this stage, especially for leadership roles since leaders have a far greater impact on the organization than new hires at other levels.

Thinking about the overall recruitment process, strategy, and approach your organization is going to take will be helpful to do up front. For example, will you have a strong customer service approach with your candidates and hiring managers or will it be a softer touch model? Do you plan to grow in one location or will you be open to employees working remotely or expanding to new geographies where talent pools might be stronger? There are pros and cons of different approaches and there is no one best way, but thinking through the time, resources, and outcomes for your recruitment strategy before it ramps up is critical.

You'll also want to think strategically about what type of talent your organization will want to hire. Do you need industry experience or specific skills and experience? If so, will hiring from your competitors be part of your overall strategy? Do you need everyone in one location? How many people will you need to bring in at one time? Understanding your talent needs, and being aligned with the leadership team and managers is an important driver of growth. All of these can also change over time—you might start out with hiring more experienced industry people and then move to a junior-hiring model when you get critical mass in a department. During the early stages, everyone you hire must be able to wear many hats as they'll all be doing multiple jobs. It's only when you scale that your employees will be able to separate into singular, focused roles. These early hires must also be flexible as priorities change frequently and they'll be pulled from one project or initiative to another.

It's essential that there is a values match between each new hire and the company. Once you have a values match, find candidates with diverse backgrounds and experiences. According to the *Wall Street Journal*, "the 20 most diverse companies in the research not only have better operating results on average than the lowest-scoring firms, but their shares generally outperform those of the least-diverse firms, the research shows" and, it's never too early to start. Your early hires are critical to your success as they are the most likely to be hired into or promoted within the organization to positions of leadership. If you want to give your company the best chance of success, make sure that new hires (especially executives) align on values with the CEO and company. If the CEO values transparency, for example, and another executive has a "need to know" attitude, the organization will reject that executive.

Like an Organ Transplant

Matt Blumberg, CEO, Bolster

I've often said that hiring a new senior person into an organization is a bit like doing an organ transplant. You can do all the scientific work up front to see if there's a match, but you never know until the organ is in the new body, and often some months have gone by, whether the body will take or reject the organ.

New senior people in particular have a vital role in organizations. Often they are brought in to fix something that's broken, or to start up a new position that growth has created. Sometimes they are replacing a problematic person (or a beloved one). Usually, the hope is that they will also bring a fresh perspective and good outside view to bear on people whose heads are too much "in the business." In all cases, their role as leaders makes them have higher visibility and higher profile than most, and therefore more impactful if they succeed. It also makes them more problematic if they don't.

What happens that causes the body to reject the organ? It could be a few things, but in my experience it's usually one of three. Sometimes the execution isn't there—in other words, the person knows what needs to be done but isn't effective in getting it done, for any number of reasons. Usually, you feel like you were sold a bill of goods. Other times, specifically in cases where the person is coming into a new job that didn't exist before, it turns out the job was poorly specified and doesn't need to exist, or that the person coming in is the wrong person for it. Usually, the person feels like they were sold a bill of goods.

But I think in most cases, the cultural add just isn't there. And that's not really anyone's fault, although it "should be" something you can interview for to a large extent. These are the most painful ones to deal with. Decent to stellar execution (good enough to not end employment over it), but poor cultural fits.

How quickly does this take? I've seen it take a quarter. I've also seen it take a year. But in both cases, the warning signs were there much sooner.

A footnote on this is that as Return Path has grown, I've come to a new thought about this—it doesn't just apply to senior people. It applies to almost any new hire. It may be an outcome of having a really strong and consistent culture, or it may just be the natural extension of this axiom.

We once had an open role for a senior leader on my team, who would be my manager. We knew that we needed someone who was process-oriented as that wasn't a strength of anyone on the team. We found and hired a person with that skill set even though we had some misgivings about this person's fit on the team; they seemed a little stiff, formal, and process-oriented for our environment. The misgivings were spot-on: the hire was a train wreck. While they did have the process-oriented skills we desperately needed at the time, they didn't share our values around people and actually had the opposite value! They valued process over people while we valued processes to support people. That mis-hire was highly impactful in the wrong direction. It set us back financially and emotionally, we squandered valuable time, and we almost lost at least one valuable leader before we made a decision to un-hire the mis-hire! We wasted a lot of time and resources in onboarding, evaluation of performance, and getting feedback from employees, and we created a lot of stress in the team. We learned a valuable lesson that it's absolutely necessary to pay attention to the person's values and not just their skill sets.

To determine values fit, we built interview guides against our Return Path Expectations to help interviewers with this assessment of whether the applicant exhibits the behaviors required at our company for the specific level of role. We also conducted reference checking for each senior hire. You'll want to dig deeply into the candidate's work style, communication strengths, and weaknesses, leadership, along with the technical skills needed for the role. Most importantly, make sure you understand whether a candidate shares your company values and whether they will be a culture add. (I'm not a fan of culture "fit" as it can unintentionally exclude people from different backgrounds, which reduces the diversity of your culture.) Without alignment on values and culture, you're doing a disservice to the potential new hire and you're hurting your company's chances of success.

Even at the startup stage you'll want to build a simple recruiting process that all candidates will go through, including structured or behavioral interviews. Even if you get resistance from frenzied hiring managers and executives, it's important to have a process that is consistent and lightweight, one that is inclusive (see www.Startuprev.com for a DE&I chart with more specifics), and one that is efficient for the People team, hiring managers, and the candidate.

You will save a lot of time, energy, and frustration in the long run if you spend quality time planning for each hire. If you have good recruiting practices and processes, and use a tool like Gapjumpers, it could take as little as 15–30 minutes to whittle down hundreds of candidates to the top three who are likely to be successful in your company. Often every individual in a startup wants to interview every new hire. While that's admirable, it's unrealistic and you should highly discourage that. Instead, create a recruiting process such as the sample of the Interview Chart at www.Startuprev.com.

You'll also want to build some basic approvals and tracking processes such as the sample of Recruiting Approvals and Tracking Processes at www.Startuprev.com.

You'll also need to think about how you will source candidates. Often startups have a harder time attracting candidates because they aren't yet stable, known organizations, and there isn't an easy way for candidates to research what it's like to work for them. Many of your first hires may be sourced through internal referrals. While this is a good source of candidates, and you may be tempted to go with quick wins here, you have a real opportunity when you are small to build in diversity from the beginning, and sourcing from your internal referrals may limit your pool of diverse candidates. Once you've hired homogeneously for a long time, it's much more difficult to diversify your culture and your hiring. If you are hiring quickly, it can be helpful to supplement your team with an outside source, who knows your culture and value proposition well, and has the tools, resources, and experience in bringing you a strong and diverse candidate supply. Establishing a social media presence can create more name recognition and highlight your company as an employer of choice. So can having a presence and posting jobs with local meetups, startup events, universities, Indeed, LinkedIn, Twitter, Google Search or other more niche sites are useful. Look for automation in sourcing as a way to bring speed and efficiency to this process.

One way that we found a great talent pool was through the creation of a return-to-work program which targeted candidates who had taken time off work for caregiving, and then we supported them in their re-entry. If you get creative, there are a lot of people with potential to be successful in your company who don't have a "traditional" resume.

As the organization grows, as you go from a handful of employees to 100 or more, you'll need to add additional staff in the recruiting area and build more scalable processes. The stage when you need to take your team to the next level will be highly dependent on the employee growth plan. You'll have an understanding of your capacity for hiring, and you'll want to keep track of the strategic plan around hiring and growth, so you can build your team ahead of demand. At this stage, you will be using a lot of the same processes as before, but you'll ensure that the practices are all scalable.

For all your other processes, you'll start to move away from individual conversations with each hiring team. Instead, focus on building training and automated processes including hiring manager training that is mandatory for all interviewers. This can be a pre-recorded module or it can be an in-person class that you offer periodically. You can also write "how to" guides for managers to help them understand the recruiting process and their role in it. Take everything you've been telling hiring managers in one-to-one conversations and put it into a guide.

I prefer to be hands-on in recruiting, and to be highly involved and manage all recruiting in-house. The only exception is if we have overflow sourcing and there are times when you'll need to use outside recruiting agencies. It's worth developing relationships with a couple of local agencies and putting agreements in place before you need them. If you have a burst of hiring activity, you'll need to supplement your team with an agency that you trust, and who can bring you diverse candidates. Share your recruiting practices and processes with the agency and ensure they understand your culture and business.

As you grow, you'll need a coordinated effort to manage your employment brand. Although it's important at any stage to build a close partnership with Marketing in building the value proposition for talent, that work really pays off as you grow. Work with your internal marketing/corporate communications team to develop a full plan around employment branding, including partnering with sites that candidates use for research (i.e. Glassdoor, the Muse, Indeed, Built In). It's important to have a strong career site on your corporate website that features pictures and quotes from your employees, a clear Employee Value Proposition that you can advertise, and materials and swag for career fairs. Your career site will be a critical tool in your employment brand and should also include things like your values statement, benefits overview, what's great about working at your company, or any kind of content to showcase what an incredible culture and company you're creating or have created. Career sites allow candidates and former employees to leave comments about the company, and my approach is to use that feedback (especially if it's negative) for ways to improve your recruiting processes. We also responded to each comment publicly to make it clear that we heard and were working on the feedback.

At this stage of growth you may need highly specialized and hard-to-find positions, and I recommend posting your jobs to niche job boards. This is a great way to increase diversity in your talent pools. It's also possible to diversify your talent pool by participating in recruiting events, by sponsoring events that play to your company's needs (engineering fairs, pitch competitions, etc.) or by getting involved and sponsoring local community activities. You may meet with hundreds of people at local events and you might come back empty-handed, but you will have introduced your company and needs to many people who in turn are now able to articulate your company, vision, values, and needs to people in their network. Sometimes a wide net can generate qualified candidates reaching out to you.

There are a lot of details that go into creating a robust recruiting process and some metrics on the data you're collecting will help you both evaluate and improve your processes. Some sample metrics that are worth tracking include source data, cost per hire, time to fill a position, quality of people hired, first year attrition (both actual numbers and percentage), and candidate experience during the interview process.

Everything in this chapter has focused on external recruiting; earlier we talked about career pathing and promotions. Internal recruiting is hugely valuable; it's an important part of your growth to provide opportunities for your employees to map out their career. There's nothing quite so powerful as helping people develop skills, add responsibilities, and work with mentors to contribute to the company in greater capacity. At Return Path, we had a bias toward an internal promotion unless there were specialized technical skills that were lacking internally and we couldn't train. We then invested heavily in leadership development to help people grow and develop their careers. When Return Path was sold, the entire leadership team except the three initial founders had been promoted from within the company, thanks to a strong developmental focus and investment in people over the years.

Chapter 37

Onboarding

One of the more people-focused roles within People Operations is onboarding. Effective employee onboarding is critical, even in startups. Most people think of onboarding as getting a benefits overview and a computer. That's important, but that's really just "orientation." The first experiences an employee has with your company will help determine longer-term success. With a small team you may not have time to create a fully curated and personalized experience for every new employee; instead, put some structures in place to ensure hiring managers and teams are creating that great experience for new hires.

As you start to increase hiring, you can build a more robust onboarding program so that each manager doesn't have to re-create a program for their new employee. The onboarding manager can be responsible for setting up the programs that will be run for all employees, and then build manager and new employee guidelines that give them each a framework for job- and company-specific activities and outcomes in the first 90 days. In addition to building the guidelines, the onboarding team can manage or collaborate with other teams to manage things such as:

- Training in soft skills that are critical to success at your organization as discussed in leadership development above.
- First day setup. You don't want people to show up and no one knows who they are or to have to put a desk together. Get buy-in from office management so they understand the impact that the first day setup can have on the employee

experience, and are creating a great experience for each new employee.

- A way for new employees to be introduced to the company, whether that's a welcome email to the entire company or a short bio shared on TV screens, helping new employees introduce themselves in a non-threatening way, and also letting colleagues learn about them before they meet in person.
- Programmatic story-telling sessions led by the CEO and senior leaders to deepen employees' understanding of values and bring them to life.
- 90-day reverse review process, where the CEO or member of the leadership team talks to the employee one-to-one about their first 90 days. This is a good way for the leaders to develop connections with employees, get ideas from them while they are still learning, and also get feedback about the recruiting and onboarding process. It also highlights for leaders that personal engagement with employees is really important.
- A "buddy" program to connect a new and an experienced employee to answer questions, and help the new employee integrate effectively.
- A unique onboarding process including job-specific training for departments that have a significant number of people in the same role.

Onboarding vs. Waterboarding

Matt Blumberg, CEO, Bolster

A senior hire once said to me that they enjoyed our onboarding process during the first 90 days at Return Path and that at other companies the first few months were more like waterboarding.

We placed a lot of emphasis on onboarding—the way we asked employees to spend their first 90 days on the job. I've often said that the hiring process doesn't end on the employee's first day. I think about the employee's first day as the midpoint of the hiring process. The things that come after the first day—orientation (where's the bathroom?), context-setting (here's our mission, here's how your job furthers it), goal setting (what's your 90-day plan?), and a formal check-in 90 days later—are all make-or-break in terms of integrating a new employee into the organization, making sure they're a good hire, and making them as productive as possible.

Nothing has a greater impact on a hire's long-term viability than a thorough onboarding and if you don't onboard them properly, they may never work out. This is where all companies, big and small, fail most consistently. Remember your first day of work?

- Did you (or anyone at the company) know where you were supposed to sit?
- Did you (or anyone at the company) know if your computer was set up?
- Did you (or anyone at the company) have a project ready for you to start on?
- Did you (or anyone at the company) know when you'd be able to meet your manager? Probably not.

Take onboarding *much* more seriously, and you'll be astounded by the results. We had a Manager of Onboarding whose only job was to manage the first 90 days of every employee's experience. If you don't have a manager for onboarding, here are some things you can, and *must*, do to assure a successful onboarding process:

1. Start onboarding before Day 1. Ask new hires to create a "Wall Bio"—a one-page collage of words and images that introduces them to the team—before their first day.
2. Set up your new hire's desk in advance, including their computer, monitor, telephone, nameplate, and company swag.
3. Prepare an orientation deck, one-to-one, or session for Day 1 including mission, values, structure, and strategic plan.
4. Clearly set 90-day objectives and goals including the new hire's job description, first steps they should take, resources they should know, people they should meet, training courses to take, and their major objectives for their first 90 days.
5. Run a review process at the end of 90 days.

With *that*, the hiring process is done. Now, repeat.

See www.Startuprev.com for an example of first day/first week minimum requirements for new hires.

Chapter 38

Compensation

You'll want to develop a basic compensation philosophy and make sure there is alignment with the leadership team. While benefits and other non-monetary perks such as company culture factor into your overall Employee Value Proposition, for this chapter I'll focus on financial compensation. Compensation is both an art and a science, and people often have strong perspectives about how to compensate employees, and have strong emotions attached to this. If you aren't aligned with your leadership team and CEO, you'll spend a lot of time on this topic for new hires and during compensation reviews and promotions.

I'd recommend aligning on the following questions around your compensation philosophy and how you'll make decisions:

- How will you compete in the market on compensation? Will you pay below, at, or above market rates for base compensation?
- Will you have a commission plan for sales?
- Will you have a bonus plan for employees?
- Will you give stock options to all employees?
- How often will you review compensation and consider promotional increases? Will you pull out junior roles and do more frequent reviews for those positions whose market rate can change more quickly?
- Who is responsible for compensation decisions? Is it the Chief People Officer or the functional executive?

- How will you handle geographic differences in market compensation? Will you pay different amounts in your different locations?
- How transparent will you be about compensation? Will you tell everyone your philosophy, salary bands for each role, or every individual employee's salary?
- How will you handle executive compensation?

Collaborate with your recruiting team to understand the market, and determine rates for specific roles based on your philosophy. Once you have the starting pay rates for specific roles, I'd suggest you set a no-negotiation policy, especially for junior roles. A no-negotiation policy will help keep compensation equitable across people and roles and will also speed up the recruiting process since it eliminates the back and forth between you and the potential new hire.

On Decision Making

I prefer to have final compensation decisions made by a small corporate committee, consisting of the CEO, the CFO, and the CPO. This allows for more consistency and equity across the organization. In the last couple of review cycles I managed, the People team made all compensation recommendations based on our reviews of internal and market data. Managers weighed in on role changes and promotions, and we had more levels within each role with specific salary starting points, which allowed us to focus primarily on promotions and market adjustments, rather than performance. This reduced a lot of the friction in the organization, and allowed managers to focus their efforts on performance and development.

Chapter 39

People Operations

While focusing on the culture and company you are trying to build, finding the right people to help you, and building systems to keep track of everything, you also need to focus on the transactional aspects of the function. These two broad areas—transactional work and culture—reinforce each other and as a startup it's best to create both of them in tandem. All aspects of transactional work will be influenced by company values and when you're growing rapidly it's tempting to recruit strong transactional people to the company right away. There are lots of people with the traditional skill sets, but I've found that being intentional about building the People Operations team sets the company up for rapid growth. Building the People Operations team with an Employee Value Proposition (EVP) mindset will help you develop effective processes, set the stage for long-term growth, and attract the people you'd want in People Operations. The EVP is a broad term for the entire employee experience and in a nutshell answers the question, "What can the employee expect to gain by working at your company?" That answer is the foundation of your employment brand, which is not set in stone on Day 1, and rather changes as you grow. The EVP includes values and cultures, which can play a big role in why someone wants to work at your company.

Again, even while it may seem urgent to just get payroll and benefits started, you'd do well to consider your EVP even in the very beginning. Your EVP will change over time as you mature, as you have more resources, or even as you respond to the market through new offerings, strategic partnerships, or other growth models. Many startups initially start with lower base compensation and higher equity. That's great if you have a small number of employees, but what happens once you have hundreds of employees? What equity amount will you need to compensate for low salary? Other companies provide great benefits (over higher compensation or equity) to attract new employees. Some companies have low benefits to reduce per-employee investment. Any of these approaches can work, as long as it's aligned with your values and the EVP is clearly stated.

Setting up the operational side of HR can be thankless work. People only notice when things go wrong—and they will let you know! Many leaders treat People Operations as a necessary evil, as a cost-center that either needs to be done in-house or outsourced, and rarely as a valuable asset to the company. It's your job to help people see operations differently and if you help people change their perspective about this, if you can get them to think of People Operations as a key contributor to your Employee Experience journey, this work can be more rewarding and impactful. My former colleague, Amanda, whom I mentioned earlier, is a great example of taking a transactional role and making it people-centric. Amanda joined the People Team as Payroll Coordinator, a highly transactional role, but she changed what had been a transactional, reactive role into a proactive, people-focused role. Amanda created and modified processes to be more people friendly and effectively communicated changes that impacted employees. The impact on the company was huge and I urge you to hire people who care about the people they are serving, even if the role is largely administrative or data-intensive. My best People Operations team members didn't have HR backgrounds. They were interested in the work and had a background in customer service. They were passionate about helping people, and they did that from a transactional role.

As you grow, you'll want to gain efficiencies with your People Operations, so they can serve a larger population without growing the team; otherwise you can easily become bloated, move too slowly, and not offer enough value to the company. If you are using a professional employment organization (PEO), you should regularly re-evaluate it to make sure it still meets your needs and budget. Often, companies bring payroll and benefits in-house when they reach 100 employees. If you are hiring in multiple countries, you may want to use a PEO in each country until you get enough employees there to justify an entity. Make sure that your broker relationships are still effective and that they know about your growth plans. The best brokers will be an extension of your team so if they aren't providing what you need, give them clear guidance on your expectations and be ready to find someone else if their business model doesn't match your needs.

Be creative about benefits—keep on top of trends, and talk to your employees about what is important to them. Some things we had at Return Path over the years were:

- open vacation
- sabbatical program
- extended parental leave
- Well-being day
- payroll and insurance benefits.

Open Vacation

We realized that people's work and lives were becoming more intertwined, with work sometimes needing to be done on vacation time, and personal things needed to be done during work, and so we created an open vacation practice. With this, it meant a reduced administrative burden on tracking paid time off (PTO) balances. Employees could use PTO flexibly and could choose increments other than ½ or full days. We also included some guidelines for employees, like average and minimum annual PTO, approvals for longer than two weeks' PTO, and performance. You want to ensure that there's a balance between people and teams and that there are not abuses of its use. In many countries other than the US, you can use the "open/flexible" philosophy, but statutory leaves are usually sufficient.

Open Vacation: Was It Worth It?

Matt Blumberg, CEO, Bolster

At Return Path, we had an "open vacation" policy for years, meaning that we didn't regulate the amount of time off people took, and we didn't accrue for it or pay out "unused" vacation if someone left the company. I still get asked about this all the time, so I thought I'd give a couple of follow-up questions I usually get asked about it, and my response.

The first question I always get is, "Wow—does that really work? What issues have you had with it?"

No issues with it at all, other than it's a little weird to apply internationally, where we have 50 people across seven countries, since most of those countries have significantly more generous vacation policies/customs than the US. But we generally make it work.

The second question I get is whether people abuse it or not:

In all the years we've done it, we only ever had one person attempt to abuse the policy, one time. People do still have to ask their managers if it's OK to take time off, and they do still have to get their jobs done.

Finally, people ask me for general advice on implementing this kind of policy:

Continue to track days off and generate reports for managers every quarter so they at least know whether their people are taking not enough or too much—generally people will take not enough, and you will need to encourage them to take more. Also, our managers were "really" worried about launching this, so we had to do some hand-holding along the way.

My thoughts: The results of this policy for us have generally been great. People take about the same amount of true vacation they used to take, maybe a little more. They definitely take more half-days and quarter-days where they probably still get a full day worth of work done, without worrying about counting the hours. Best of all, there's a strong signal sent and received with this kind of policy that we trust our team members to do what they need to do in order to live their lives AND get their jobs done.

People

Sabbatical Program

Creating a sabbatical program early on is a good retention tool. Our initial sabbatical program was six weeks after seven years of service; we changed this to four weeks for every four years of service, which I'd recommend. The benefit is really appealing to employees, and it's also good for the company. When someone takes an extended leave from their job, they have to train others, and it often exposes single points of failure. It also gives others opportunities to stretch into new roles for a short period. When I took my sabbatical (an amazing trip to New Zealand and Australia—a once-in-a-lifetime opportunity), I trained others on the team on every aspect of my job and essentially didn't have a job to return to. That gave me the opportunity to lead a large-scale agile transformation project across the organization. Many people had a new lease on life and their job when they returned from sabbatical. And, we had a lot of fun celebrating people's return. Our President had his entire office filled with balloons from top to bottom, and many people left their "welcome back" artifacts in their cubes for months after their return.

Sabbatical: Was It Worth It?

Matt Blumberg, CEO, Bolster

I've written a few times over the years about our sabbatical policy at Return Path, and people ask me about our policy all the time. While it's not necessary to share every detail of the policy, a few specifics can be helpful.

The two biggest priorities in having a sabbatical are: (1) to ensure that people actually take advantage of this benefit, and (2) to ensure that people communicate and prepare for an extended absence. Here's a short list of "pre-sabbatical" tasks for employees and their managers.

As the employee

- Prepare your team. Make sure your goals and metrics for your time out are super clear and clarify who others should contact in your absence. Set expectations of management for coverage and look to see how your team members can take over some of the responsibilities. Give them stretch goals while you're out!
- Prepare your individual contributor work. Hand off all loose ends with extra details and make introductions via email if your manager/team member is going to have to work with external parties.
- Prepare your manager. Brief your manager thoroughly on everything going on with your team, its work, your individual contributor work, including one-to-one check-ins. Agree on a plan for coverage of team activities, one-to-ones, and big initiatives.
- Prepare yourself. Figure out how to keep your work and personal communications separate—your email (autoresponder, routing, disabling from your smartphone), your

voicemail if you use Google Voice or Simulscribe, etc. Block out two full days immediately when you return to catch up on email and catch up with your manager and team. Plan any personal travel early so you get good rates!

As the manager

- Prepare your team. Figure out what kind of coverage you need (either internal or external) while you're covering and make sure the rest of your team knows your time will be compromised while you're covering. Rearrange your calendar/travel and add new team meetings or one-to-ones as it makes sense. You don't have to do exactly what your employee did, but some portions of it will make sense to pick up, and if your employee works in another office with members of their team, you might want to plan some travel there to cover in person. Remember to undo everything when the employee's sabbatical is over.
- While you're in charge. Learn as much as you can by doing bits and pieces of their job. This is a great opportunity for the employee to get some value from a fresh perspective. Surprise your employee with how much you were able to keep things running in their absence!
- Prepare for your employee's return. Keep a running tab of everything that goes on at the company, critical industry news (if appropriate), and with your employee's function or team and prepare a well-organized briefing document so your employee can hit the ground running when they return. Block out an hour or two each of the employee's first two days back to review your briefing document.

My main takeaway from this advice? I am overdue for my second sabbatical, and it's time to start thinking about that!

Extended Parental Leave

While many countries other than the US already offer generous parental leave, we found that our employees really valued the extended paid time we offered for new parents. Initially, only employees with six months service were eligible. And then we had a new employee whose wife had a baby in his first six months at Return Path. His manager argued strongly for the employee to be offered the leave. The manager had recently taken parental leave and realized that it was important for him personally, and also helped him be more focused when he returned to work. We modified the program so that everyone was eligible.

Well-being Day

Every employee could select one from a variety of well-being packages each year. These packages were negotiated specifically for our employees, and included things like sessions with a nutritionist, a financial advisor, or a meditation coach.

Payroll and Insurance Benefits

There are some more transactional things you'll have to think about immediately in building out People Operations:

- As a startup you can either set up a simple payroll system such as Gusto, or contract with a professional employment organization (PEO).
- A PEO is a co-employer that manages all your payroll and benefits, which can save you time and money while you are small. They bring the benefits of economies of scale, as they are co-employers with many different companies. Their benefits can be more robust and affordable than what you can get as a single company, and they can ensure you are in compliance with all appropriate regulations and reporting requirements. The downsides to a PEO are that you lose some control of the employee experience and it can be expensive.
- If you don't use a PEO, evaluate your options and select and implement a simple payroll system. You'll want to ensure that the system can effectively onramp employees without manual intervention, and also collect some data that you can report on, such as employee turnover.
- If you don't use a PEO, evaluate and contract with a good benefits broker. A good broker will help you strategically and also tactically. They will help educate you about benefits options, understand your EVP, and advise on options that will support that. A benefits broker will also be a liaison to employees so you don't spend all your time answering benefits questions.

Processes

- Set up simple processes for collecting and making payroll changes. When you are small, you may get verbal requests for changes, and you should send a request by email so that you can track approvals for the changes. This is important for audit purposes and also ensures that your data is accurate and has the appropriate approvals. You'll need to work with Finance to determine payroll roles and responsibilities. The People Team is generally responsible for keeping employee data and responsible for ensuring any changes are accurately reflected. Finance is generally responsible for setting up tax entities, filing taxes, and processing payroll. A simple tracking system such as a shared Google sheet can be used to communicate and double-check payroll and benefits changes.
- Leave of absence practices such as family/medical leave, parental leave and disability leaves are becoming more complex, and you'll need to understand required leaves of absence for each country and state in which you operate. The Society for Human Resource Management (SHRM) is a good resource for this type of information. Once you understand the laws, you'll need to understand how the leave is paid and communicated to different agencies. Sometimes your broker will help with these calculations, and if one of your vendors offers leave processing, it's a benefit you should take advantage of.
- Create a simple process for collecting paperwork for employee files. Keep as much electronically as possible, working with IT to ensure that you have the appropriate level of security on those files. There are still requirements in some countries for paper files, so check requirements in each of your geographies.

Employee Documents

For new hire documentation, there are some new employee forms that are required by law, such as tax and employment eligibility forms. For the others, you have options for the presentation. If you look for standard forms online, you'll find a lot of them are written from a legal perspective. If you are trying to create a culture of trust in your organization, these aren't the best forms to start the relationship with. You need to understand what's required to protect the company (such as making sure it's clear that you are an "at will" employer in the US), and add those components to your paperwork. Write your offer and corresponding onramp documents in a conversational tone, welcoming your employee and giving them the information they need to be successful and excited about this move.

- If you decide that you need legal documents such as non-disclosure, inventions, non-competes and non-solicits, you will want to leave in the more formal legal language. Think carefully about which of these protections you really need, and which are enforceable in your locations. As you grow, it can be helpful to have clarity around what you expect of employees, and a basic form with all these common restrictions is fairly common. Non-competes are becoming less enforceable in some locations, unless the role is a very senior role and the individual is being compensated for not working.
- Your employee handbook can also be written in a conversational tone. Most companies find that some form of handbook is really useful, and many of the policies included are required by law, but that doesn't mean that you have to write it in legalese, like my boss at the truck stop wanted. Most sample employee handbooks will be full of legal information and written as "rules" people have to follow. I find that most of these sections are irrelevant, and can be removed, and others need to be rewritten from legal to conversational language. So, while you need an employee handbook, you don't have to make it difficult for people to understand. At Return Path we called our employee handbook the "People Pact" and it's available to view on our website.

Office Management

Whether it reports to you or not, office management plays a big role in the employee experience. If you're starting up I would fold

Office Management into People Operations and if you're at a later stage of growth and Office Management is not reporting to you, I would strongly advocate for that to happen. The impact of Office Management goes far beyond the person sitting at the front desk greeting visitors. Over the past few years, more and more companies have been moving toward remote-first environments, and your People Operations and Office Management team can help facilitate remote employees' success. The COVID-19 pandemic is already making many companies rethink their Office Management strategies.

- Your CEO, founders, and leadership team should have a clear idea on whether or not you will be office-based as a result of their work on values and culture. If you are working in one location but expand to others I would recommend that you immediately become "remote-first." That means that you will set up your processes and practices assuming that you won't be connecting in person. There are a lot of tools available to help employees stay connected and help managers keep track of progress, like Asana, Slack, and Zoom.
- If you're office-based, you'll want to collaborate with others on the office layout. If you're a company that collaborates, then you'll probably want flexible space for meeting and desks that can easily be moved so that you can work together. If you are hierarchical, you may want offices for all managers and if you aren't hierarchical, you may want to save all office space for collaborative meeting rooms. Given the realities of the COVID-19 pandemic during the writing of this book, you may also need to space all desks eight feet apart and provide partitions.
- If you are remote-first, you'll most likely have employees come into any location only occasionally, and you might want to set up a lot of areas for video-conferencing in offices, especially for ad hoc meetings. You'll also probably want to develop a plan for desk-sharing,
- In today's environment it's expected that companies will provide food and drinks to employees and you'll have to decide on those. Since you'll also own any Well-being offerings, I'd suggest offering healthy snacks and food.

Chapter 40

Systems

As you grow, you'll want to evaluate all your current systems and practices to ensure they are scalable. If you are growing quickly, inadequate systems will become a roadblock on your team's success and will fuel frustration in the organization. Your Human Resource information system (HRIS)/payroll system should be more robust so that you can seamlessly integrate with other systems such as your applicant tracking system (ATS). At this stage you should start to invest in greater functionality in your systems, such as performance and compensation management. Those things can quickly get unwieldy and cumbersome to run without a system that scales with you. You'll also want to start reporting on KPIs (key performance indicators) and employee metrics so you can monitor your employee data and catch any trends. You'll look at data over time like employee turnover, employee engagement, performance, revenue per person, and compensation equity.

As you start to grow, you'll need to ensure that all your systems are integrated. Often HR systems have single functions, but to get the best of that functionality, you'll need a standalone system for each. (This is changing and there are now systems that bring best-of-breed in multiple functions together.) By integrating several processes you'll reduce duplicate work for the People Operations team. When you are smaller, you can manually handle this work; as you get bigger, you'll want to streamline so you don't make mistakes and can leverage your team for more high-impact work. See www.Startuprev.com for guidelines about systems to consider.

Conclusion

If you can help build an organization from the startup phase, you have a unique opportunity to impact a company's DNA. Consider how all the different parts of the employee lifecycle fit together and support and reinforce each other. If you come into an organization that is in a later stage, you can still influence all these systems; it's just a little more difficult to unwind things that aren't supportive of the company values and mission. If you join a company that has significant dysfunction, which requires a lot of reactive work, insist on over-hiring for your team so you can stay focused on proactively changing the structures and systems, instead of being side-tracked dealing with difficult employee relations issues that stem from poor systems, training, and structures.

I said that I'd help you think about how to influence a CEO who isn't aligned with a people-centric approach, and there are things throughout Part 3 that can help you build a business case for a people-centric approach to each body of work. The best advice I can give you is to ensure you build a trusted advisor relationship with the CEO and leadership team, and take the time to have meaningful conversations about legacy, and the impact of the different approaches to leading your employees. It's a skill that you need to hone, because we still have a cultural perspective that stems from production lines that people need to be "managed" rather than "led." Many employees are still afraid of HR and think of HR as the "police" who they have to be "politically correct" around. I wish I'd had this perspective early in my career; who knows, maybe I would have been able to influence Jerry, the truck stop president, to create a better environment for all employees if I'd had the right knowledge and skills? He was a good man who cared about his people, and just had really traditional ideas about how to manage them.

And, my final piece of advice, about which I often need reminding: take care of yourself and your team. It's so easy to be like the proverbial cobbler, whose kids didn't wear shoes. Good People professionals care about others, which is what makes them good. That often means not making time for themselves. Make sure you have an operating system where you invest as much in your own team/team members as you do in other teams. Have quarterly offsites, encourage professional development, have a good operating system so you are planful and organized about the work.

You will have a significant impact on the organization, since your team is interacting with employees at all levels. If you focus on the values, structures, and culture, and you build an environment where people can bring their whole selves to work, you can have a transformative impact on the culture and the people within it.

Chapter 41

CEO-to-CEO Advice About the People/HR Role

Matt Blumberg

What comes before a full-fledged Chief People Officer? In most startups, the HR function starts out as tactical—you have to get people hired and paid—and frequently outsourced to a PEO. As the company grows, it probably in-sources payroll and benefits, hires a recruiter, and maybe has an HR Manager who handles the function.

Signs It's Time to Hire Your First Chief People Officer

You know it's time to hire a Chief People Officer when:

- You wake up in the middle of the night convinced that you're the only person in the company who cares about your core values.
- You are spending too much of your own time training managers and leaders, or working on interpersonal dynamics on your leadership team.
- Your Board asks you what your talent strategy is with respect to improving diversity, retention, and engagement metrics, while simultaneously decreasing average employee salary, and you don't have a great answer and aren't sure how to get one.

When a Fractional Chief People Officer Might Be Enough

A fractional Chief People Officer may be the way to go:

- If you have a very competent HR manager or director who has strategic inclinations but not enough experience operating as a strategic executive and who just needs a little more supervision in order to "level up."
- If you need someone to play more of a consigliere or team coach role to your executive team but don't want to engage a coach—and your day-to-day HR leader is getting the job done but is too junior to facilitate workshops for the senior team.
- If you have a very junior HR function or are in-sourcing it for the first time and need help setting up the whole function from scratch at an advanced size relative to other functions.

What Does Great Look Like in a Chief People Officer?

Ideal startup Chief People Officers do three things particularly well:

1. They believe their function is strategic. Throughout this Part of the book, Cathy writes about the ways in which HR/People can be a strategic function and not just a tactical corporate function. It's true of most functions, but for whatever reason, likely past experience, HR leaders frequently don't view themselves or their functions as strategic. If that's their frame of reference, then they will likely be tactical managers. If they believe they can move the needle on the business by improving engagement and productivity and efficiency, if they believe they can make the executive team more effective by helping you with team facilitation and coaching … they can do anything.
2. They call you (you, the CEO) out on things directly and firmly when they see you doing or saying anything that is a bit off, whether around language, inclusion, values, authenticity, or anything else. They along with you, are the

principal stewards of the company's values and culture. Even the best CEOs benefit from having a watchdog from time to time.

3. They think about investment in People in terms of ROI. It's one thing to run a killer recruiting function and fill seats efficiently, with high quality, as asked. It's an entirely different thing to start the recruiting process by asking if the role is needed, at that level and compensation band, or whether there are other people, fractional people, contractors, or shifts in lower value activities that could be put to work instead. Only Heads of People with deep understandings of the business can transform the function from a gatekeeper/"no" role into a business accelerator.

Signs Your Chief People Officer Isn't Scaling

Chief People Officers who aren't scaling well past the startup stage are the ones who typically:

- Are overly focused on the transactional aspects of the job. Don't get me wrong, there are many transactional elements to HR—payroll, benefits, systems, process, etc.—and they all have to go well or employees freak out. But the Chief People Officer who spends all their time on these issues isn't delegating well, isn't building a machine, isn't building scalable people and processes to flawlessly and efficiently handle the details.
- Won't speak up in executive team meetings. Chief People Officers have every right and entitlement to hold opinions about the company's strategy, products, operations, and financials. The good ones do—and they're not shy about speaking up publicly about them. The weaker ones, or the ones who are in a bit over their heads, don't because they either haven't taken the time to learn and formulate those opinions, or because they don't have enough confidence among their peers, to voice them.
- Have trouble managing/leading their own team. Since a good Chief People Officer is one who spends time coaching all the other leaders in your business on how to be effective

leaders, it's particularly worrisome when they themselves are not—especially with what is usually a relatively small function. This is a classic case of the cobbler's children walking around barefoot, and it's a sign of trouble for your HR leader.

How I Engage with the Chief People Officer

A few ways I've typically spent the most time or gotten the most value out of Chief People Officers over the years are:

- ALWAYS as a direct report. No matter who my HR leader is, even if the person is more junior than other executives, I will always have that person report directly to me and be part of the most senior operating group in the company. That sends the signal that the People function (and quite frankly, diversity, culture, and a whole host of other things) are just as important to me as Sales or Product.
- I insist on hearing about ALL People issues. First, I am a very "retail" CEO, and I like engaging with people in the business, at all levels, in all departments, in all locations. So I like to keep tabs on what's going on with people—who is doing particularly well and about to be promoted, who is struggling, who is a flight risk, who is going through some personal issue (good or bad) that we should know about. Even more than just me wanting to be in the know, like the example in the Weekly Sales Forecast meeting, when the Head of People knows that I want to know about all these details, they insist that all the different People Business Partners roll those issues up to them, which means they're in the know as well. The number of issues we have nipped in the bud, and the number of opportunities we've been able to jump on to help employees over the years because of this retail focus has been immense.
- As an informal coach for me and with my external coach. As I wrote in another sidebar in this Part, a great Chief People Officer can call a CEO out when a CEO needs to be called out. And that also means that great Chief People Officers

engage with CEOs deeply about how they are doing, help them process difficult situations, and help them see things they might not otherwise see. Being a CEO is a lonely job sometimes, and it's good to have a People partner to be able to collaborate with on some of the most personal and sensitive issues.

- Designing and executing leadership/management training. The best way to create a multiplier effect of employee engagement and productivity in your organization is to teach all leaders and managers how to be excellent at those crafts—and how to do them in ways that are consistent with your company's values. I always took a lot of time, in large blocks of hours or days, to either co-create leadership training materials and workshops with my Head of People, or to lead sessions at those workshops and engage with the company's managers and leaders in a very personal way. That always felt to me like a very high ROI use of time.

Part Four

Marketing

Nick Badgett and Holly Enneking

Chief Marketing Officer

What Is Marketing, Really?

Here's the thing, everybody thinks they're a marketer. Everybody. Whether they ever read a book on marketing or not, whether they ever took a course on marketing or not, everybody *thinks* they know something about marketing. And really, who can blame them? We're all marketed to throughout the course of our lives, from childhood to death and we all have opinions on what we like, what moves us to purchase something, and what turns us off. We all have brands that we follow, admire, or loathe. Maybe it's the visual identity that you gravitate toward and that tee shirt you love. Maybe it's a message that really nails it. It could be an advertisement or jingle, an email you received, something you saw on social media or TV, or the way you feel when you're in a store or on a website. Maybe it's the product itself. Regardless, we all have exposure to good, and maybe not so good, marketing.

The beauty of marketing lies in the sheer variety of approaches, methods, and mediums at your disposal, and it's also what makes it challenging. Marketing is unique and different to each one of us, and we all have opinions. And as a marketer, you're sure to hear them. How you respond to these opinions, requests, and ideas (lots of ideas) can often steer you on or off course. As a startup CMO your main problem is not going to be that there's nothing to do, or that you don't have any ideas but just the opposite: you'll have more ideas than you can possibly vet and more things on your plate than you can possibly complete.

Why Are French Fries Like Marketing?

Matt Blumberg, CEO, Bolster

My friend Seth has a theory about life called the French Fry Theory. The theory is simple—"you always have room for one more fry." It's pretty spot-on, if you think about it. Fries are so tasty, and so relatively small (most of the time), that it's easy to just keep eating, and eating, and eating them.

I've always thought that the French Fry Theory can be applied to many things, usually other food items. However, I came up with a new application today: Marketing.

So why are French fries like Marketing? You can always do one more thing. One more press release. One more piece of collateral. One more page on the corporate website. One more newsletter. Trade show. Webinar. Research study. Ad. Search engine placement. Vendor. System. Speech. Take your pick.

The world we operate in is so dynamic that marketing (when done well) is nearly impossible to ever feel like you're completely on top of. There's always more to be done, and the trick to doing it well is knowing when to say "no" as much as when to charge into something.

My hat's off to twenty-first-century online-industry marketers. To bring this analogy back to its starting point... their plates are full!

What you need as a startup CMO is a framework to organize and prioritize your efforts so you work on the tasks that will make a big impact. Marketing has evolved to include production, sales philosophy, relationship and customer focus, social, digital, and marketing technology. We expect marketing to continue to evolve and, as CMO, your responsibility is to identify and evaluate new tools, take on new responsibilities, and manage big expectations. Your ability to leverage data and report on marketing return on investment will prove to your company the value of marketing efforts, and the rest of the C-suite will take notice, too.

So, what is Marketing, really? It is the process or technique of promoting, selling, and distributing a product or service as Merriam-Webster defines it. But we'd like to suggest another way to look at it, a perspective that we've learned throughout our careers in startups. We believe that marketing has three primary responsibilities:

1. Build and maintain the company brand.
2. Generate demand for sales.
3. Support the company culture.

When it comes down to it, most everything that we do in marketing can be attributed to supporting one of these primary responsibilities, sometimes more than one, sometimes all three.

Chapter 42

Where to Start

Where to start can be daunting, but it can be simplified if you look at it through the lens of the three primary responsibilities outlined earlier. Focus on building the brand first because if you don't have a brand or a perspective on your brand or if you don't have the narrative or you don't have values or if you can't articulate your culture, then you don't have any reasonable way to go to the market. Imagine the following conversation:

Potential customer: What's the name of your company?

You: We haven't figured that out yet.

Potential customer: What is it that you're selling?

You: We're working on it.

Potential customer: What's the price of your product?

You: We don't know yet.

Potential customer: Why should I bother with you, your company, or your product?

You: We're good people. Trust us. We'll get there.

As excited as you might be to get out in the market and start selling your product, if you don't have clear thinking—from everyone in the company—about your brand and all that it entails, you'll never have the consistency and discipline to gain any traction in the market. You may be able to sell a few products, to generate some revenues, but without the brand you'll stall and you won't be able to scale, much less create a sustainable business.

Building and Maintaining a Brand

It's probably worth discussing what brand means because if you ask ten people what "brand" means, you'll likely get ten different answers. That's a problem for any company but especially for startups. Yes, the brand is your logo, your wordmark, your company name, your messaging, and other obvious graphics and phrases. It's also much more. It's the associations and the way a person feels when interacting with your brand, product, or service. At ExactTarget, we used to say that your brand is the sum of the conversations about it. We like to look at brand this way because it turns the brand from something that is static, that is "viewed" into something that is dynamic, something that engages people.

When we first started at Return Path, they were not a startup anymore but well on their way to scaling up with 400 employees and $80M + in revenue. One of the first things we did was a brand refresh, which is fairly common when new marketing leadership is brought in. In fact, that might be why a new marketing leader is brought in—to break up the old way of thinking and generate some fresh ideas. For us, it was time, and we had alignment and support from our CEO, our Board, and leadership team. We knew the brand was tired and needed a refresh, plus the business had evolved over the years, we had created new business units, and we wanted our brand to reflect that evolution. We worked with an agency over the course of a few months to support the initiative, revamping everything from the logo (which hadn't changed for more than 15 years!), to the color palette, to design elements, to the brand voice. It was rewarding and successful and a lot of work. However, we had the benefit of an already established and reputable brand to work with, we had the building blocks in place, and we just needed a visual refresh. But what about creating a brand?

You might think that creating a brand starts with a company name and logo, something a lot of people spend a great deal of time and energy perfecting. But before you spend any effort doing that, we'd suggest you work on creating a brand perspective. Without that strategy, you'll never be able to facilitate the conversations around your brand from partners, customers, and employees. Start by asking yourself and your colleagues general questions that help you understand how to shape what you want to be as a company, as a brand. Who are we? Who is our audience? What is our industry, and where do we fit in? How would we describe the personality and attributes of our desired brand? What value will we provide our audience? What do we do? How and why do we do it?

On the surface, these questions are simple, but answering them takes a lot of thought and you'll get better results if you ask a wide range of people to weigh in. You'll also get better results if you take time to let all the ideas sink in. This is not a 15-minute conversation or a short survey that you send out broadly, and it's not something that can get done in one meeting. But, if you have answers from a diverse group of people on these questions, you'll find that several recurring themes or concepts emerge. Those themes and concepts form the basis for your brand strategy. Notice that I didn't suggest that the CMO or marketing lead "creates" the brand strategy and then seeks to get buy-in from others in the company. That's the wrong way to approach it, that's a top-down, I'm-smarter-than-you approach that will never be effective.

We used this exact methodical approach when we were developing the brand for our new startup, Bolster. All of the company founders participated in the exercise, guided by our partners at High Alpha Venture Studio. We derived five concepts resulting from examining our company name and synonyms: (1) Support and Strengthen; (2) Boost and Energize; (3) Connection; (4) Grow and Thrive; and (5) Fast and Disruptive. From there, we explored visual representations of the concepts and ultimately, we were able to whittle down the concepts to two that will be the foundation for our brand strategy: (1) Support and Strengthen and (2) Connection. We had a few rounds of exploring these visual representations, which we used to select our logo and wordmark (see Figure 42.1).

Marketing

Figure 42.1 The Bolster logo.

We also used the same approach five years earlier when we needed to develop a name, logo, and other elements for a new non-profit we were spinning out from Return Path. The "Returnship" or "Return to Work" program helped people (mostly Moms but some Dads) re-enter the workforce after taking a break for caregiving with a paid internship reinforced by training and networking, and when the program became a standalone entity, it needed a clear identity. Our brand team started to create the new identity by creating an overall framework.

The new name and brand:

- should maintain a connection with Return Path;
- should not be gendered;
- should be inclusive;
- should feel aspirational and achievable.

After multiple brainstorming sessions and developing proposals around a variety of names, the founders selected the eventual winner: Path Forward.

When establishing your brand, collaborating broadly across the organization is key. You can achieve that by having each founding team member participate, or by collecting ideas and input from key stakeholders in the company. This collaboration can be extremely valuable in generating creative ideas and creating alignment around the eventual brand. It won't just be owned by the CEO or CMO, dictating to the rest of the company what the brand will be, but becomes a shared creation driven by the input and engagement of a broader base of the company. This level of collaboration is invaluable, and reflective of similar approaches in other parts of the business, including product development, creating company values, business development, and more. There won't always be unanimous agreement, but everyone is able to provide input, agree or disagree, commit, and move on.

Chapter 43

Generating Demand for Sales

There are many different personas that fit today's CMO. There's the thought leader, the brand marketer, the product marketer, the digital marketer, and others. Some would argue that the demand generation CMO is the most sought-after persona in today's business environment (for both B2B and B2C). After all, marketing—if done right—is no longer a cost center, but a revenue center. Demand generation is the set of sales and marketing tools, approaches, systems, and programs that generate interest for your product or service. They are the things that nurture prospects into customers and retain those customers over time. Unlike the traditional marketer who is measured by the number of leads generated, the demand generation marketer is measured by their contribution to revenue.

That sounds logical, doesn't it? That a demand generation CMO is measured by their contribution to revenue? But there's a lot that goes into figuring out how demand leads to revenue. John Wanamker, considered by some to be a pioneer of marketing, is credited with coining the phrase "Half the money I spend on advertising is wasted; the trouble is I don't know which half." Good demand generation is anchored by good data, and good data will allow you to understand your marketing and sales cycles. Good data also helps you identify and refine your ideal buyer and understand what's working and what's not working. You need good data to make informed decisions about marketing investment and without a rigorous, comprehensive way to collect, analyze, and act on data, you're left with nothing more than hope and luck.

Market Research

So, where do you start with demand generation? We'd suggest starting with early market research so you can get some solid data. You can't determine your demand generation strategy without better understanding your company's target market and understanding how customers will react to your product or service. The information you collect during market research will also help your business with product design, improve the user experience and craft your messaging. Don't be intimidated by the word "research." We know it sounds academic and conjures up thoughts of statistics, but there are multiple ways to perform market research, there are agencies that specialize in the craft, and there are numerous books focused on this topic alone. The key is to start doing it, and once you do that you'll build a competency in market research and it will drive everything that you'll do in marketing. So, don't wait on this task!

At Bolster, we decided to tackle market research by following the principles in Ash Maurya's seminal startup product book, *Running Lean.* We started by conducting one-to-one interviews with our defined audiences and, in some cases, had two team members present to make sure nothing slipped through the cracks. We first established personas for our audiences. Think about a persona as a collection of attributes (beliefs, values) that are similar within a group of people but different from other groups. In our case we came up with three different personas. Then, we conducted over 100 interviews with those personas, and we are still conducting interviews as we are writing this. The first round of interviews we called the "problem interviews" and our aim was to identify our early adopters and learn about their biggest problems and how they currently solve them. The next round of interviews we called the "solution interviews" because we wanted feedback on our proposed solutions to those problems. And what we were really trying to determine was the smallest solution that would work, the minimum viable product (MVP).

For both problem and solution rounds, we found it extremely important to remain objective. We began each interview by stating our hypothesis on the problem or proposed solution and either asked respondents how they currently solve that problem today or asked them to respond to our proposed solution. Then we waited. We didn't prod them for a response, we didn't rephrase the questions. We just sat back and listened. By listening, we mean *really* listening, and it is the key to making this a valuable exercise. We used a script to make sure that we're staying on task, but we also allowed ourselves to vary off script in order to truly listen to what our respondents had to say. We documented our results, reviewed all of the responses, and discussed our findings during our weekly strategy meeting. We continue to conduct these interviews, refining what we want feedback on as we go, and we anticipate that these will not stop for some time.

Competitive Intelligence

We probably should also share that we conducted a fair amount of competitive research prior to starting our interviews. We documented every player even remotely close to what we're doing, and captured every piece of applicable information we could to compare to. It really helped us hone in on our primary competitors, complementary players, and gave us a lens into several elements of the business that hadn't originally occurred to us before. Competitive intelligence is not the sole domain of marketing, and there are others in the company—people who are in Product or who are customer-facing—who have insight on competitors, the market, and trends.

Technology

There is no question that technology has changed marketing. Some people say that marketing hasn't really changed but that technology has changed marketing, and there's a lot of truth to that. A 2015 industry report listed 1,876 vendors in the marketing technology (martech) space but by 2020 the number had increased to over 8,000 martech solutions! Obviously, you won't be able to evaluate even 1 percent of the potential vendors out there, so where do you start? It's daunting, but I'd suggest starting simple and growing your tech stack as your business needs change. There are a set of requirements for all businesses that you'll need to address, such as budget and B2B versus B2C, but there are a handful of areas that you'll need a solution for early on:

- Customer Relationship Management (CRM)
- Content Management System (CMS)
- Advertising and Search Engine Optimization (SEO)
- Email
- Project management/collaboration
- Social
- Analytics and reporting

A good CRM system is your foundation and it's what you should start with, regardless of your industry or offering. This is especially true if you have limited resources. Some of the more robust CRM tools today have marketing automation, social, SEO, landing pages, analytics, and much more built in, so by finding a good CRM solution, you'll be covering a lot of the areas I mentioned above. You'll want to start with a CRM tool and then supplement that based on the inherent functionality that you can leverage, and what isn't going to work for you. Try to use what you have available first before purchasing additional software, then talk to other marketers. Building your tech stack isn't easy, but building it right from the start so that all of the solutions work together versus independently is the key. Remember that as you scale—as you build more solutions, enter more markets, employ more people—your marketing efforts will skyrocket, so keeping things as simple as possible will help prevent marketing from becoming the roadblock in the company's growth.

Metric Definition

As with most things in life, you're not going to be effective without goals and goals need to be measurable, right? You need to be able to clearly articulate your successes and failures and you need to be able to first define how you plan to measure them. You need goals. Figure out early the metrics that you will monitor and share, even if it's only a guess in the beginning. It doesn't matter, but you need a starting point because any metrics you develop can and will change over time. And similar to the situation in martech, there are plenty of metrics to choose from. Some common marketing metrics include leads, marketing qualified leads (MQLs), sales accepted leads (SALs), marketing return on investment (ROI), customer acquisition cost (CAC), or conversion rates. I can't tell you what the best option will be for your business, it's likely a combination of several of these, but establish your metrics early and measure them often, and over time.

Being able to report on trends will be one of the most valuable aspects of your marketing operations, and you'll need ample time for good trend analysis. At Return Path, we developed a sophisticated marketing operations function that was the cornerstone of our marketing department. We had real-time dashboards, weekly scorecards, bi-weekly pipeline meetings, and many other forms of monthly, quarterly, and annual reporting. It helped us plan for the year, showcase our results, adjust our tactics, and celebrate our successes.

While gathering all this data can be cumbersome and time-consuming, it's worth it because all the data on trends, all the data from pipeline, and all other data we collect not only provide benchmarks but also feed into our strategic plan. Yes, it takes time to develop, but once you start and have some runway, these metrics help determine both short- and long-term goals.

Chapter 44

Supporting the Company Culture

After your brand has been built and a demand generation strategy has been established, you can turn some of your attention inward. There's a significant role for a CMO and the marketing organization as a whole to play in nurturing the company culture and positioning the company as not only a great one to work *with* but a great one to work *for.*

From recruitment marketing efforts, to facilitating internal communications, to supporting employee engagement, there are countless opportunities for Marketing to collaborate with the HR/People/Talent team and lend marketing support to either spearhead or provide resources for employee-targeted initiatives. Very early in your startup you should work to develop a collaborative, proactive relationship with the People team. Don't wait for them to reach out to you for marketing ideas that can augment your culture, but set up a regular cadence with them to discuss and plan how Marketing can add to the culture.

Recruitment Marketing

As your startup grows and evolves, attracting the right talent will be a major focus of the Chief People Officer and the People team. Lucky for them, there is a lot of support Marketing can provide when it comes to positioning the company as an attractive work environment.

The first step is to develop a clear value proposition for potential employees. The People team can provide all the information you need about benefits, values, and the employee experience and

Marketing can apply their expertise to craft a compelling story to drive home what makes your company an exceptional place to work. That story and the defined value proposition will be the cornerstone of your recruitment marketing efforts.

Next, it's important to partner with the People team to establish a presence in places where candidates evaluate potential employers, such as Glassdoor. The same way Marketing maintains an active presence on social media sites, as the company grows, the People team will be actively monitoring these types of sites in order to provide insight into the company culture and respond to any feedback. Working with the People team to create appropriate messaging, design custom images, and build the required content for the site will help to ensure that the company is putting its best foot forward and delivering a consistent message to prospective employees.

It's no secret that your website is your greatest asset when it comes to demand generation, and it also has a major role to play when it comes to educating prospective employees about your company and culture. Creating a Career Page that is easily accessible from your website allows you to share all the things that make your company great—like your core values and work benefits—and also provides a channel to promote open positions and links to online applications.

Finally, Marketing can provide significant support to the People team in producing materials and swag to use in recruitment events. Just as brand consistency is important when it comes to building awareness and driving demand for new business, it is equally important when it comes to attracting and converting candidates. Work with the People team to think creatively about how to stand out in potentially crowded markets and provide insight that will inform and engage the talent you're looking to hire. From unique swag, to digital sessions, to email campaigns, there are all kinds of ways you can collaborate to develop innovative approaches.

At Lev, a marketing-focused consultancy where Holly works, the marketing and talent teams are closely aligned in recruitment efforts, especially in developing the Career section of our website. As with many companies, the Careers page is one of the most highly trafficked pages, and it's important to provide useful insights to inform and attract talent while also making it easy for those candidates to apply. The Marketing team relied on the Talent team to identify the key pieces of information they knew would attract talent and address common questions, while the talent team relied on Marketing to build a website experience to clearly convey that information and integrate with their hiring platform.

We've also collaborated to find ways to communicate with and engage our talent pool, especially those who may be a good fit at Lev but who don't align to an opening at that specific point in time. Marketing and the talent team collaborated to capture information from candidates interested in specific opportunities, communicated important updates via marketing automation to them, and hosted recruitment-focused webinars. The webinars in particular allow the talent team to provide an overview of Lev—our culture, benefits, and primary roles—to large groups of potential candidates while utilizing the skills and technology available to the Marketing team to execute the event, including promotion, registration, webinar hosting, and post-webinar follow-up.

Internal Communications

As a startup, you may be a "one-room" company at first and internal communications might be less of a priority. After all, with a dozen people sharing an office, it's easy to communicate, but as you grow, add locations, or work remotely, internal communications become more of a priority. Taking internal communications seriously and professionally from the outset will set you up for future success.

As marketers, we understand the value of sharing a message multiple times over various channels in order to make an impression on our target audience. Don't they say that a person needs to hear a message seven times before it sinks in? That approach is no different when it comes to communicating internally, and even then people might not get your message!

People register and retain information in a wide variety of ways. Some prefer in-person communication, some prefer email, some prefer Slack or Teams messages, and some people only look up at a poster in the kitchen. There's no right answer and no singular approach that will satisfy everyone's needs. This makes communicating one message across multiple channels a necessity, and an area where Marketing can make an impact.

In alignment with the People team, the CMO and marketing team can play a big role in advocating for employees as a key audience and strategically approaching internal communications in ways that will maximize engagement. From working with the CEO and other fellow executives on company-wide meetings to facilitating the sending of company-wide emails, Marketing can provide oversight in ensuring that we're communicating clearly and effectively with employees, no matter the message.

Employee Engagement

Everyone on the executive team has a role to play when it comes to growing and nurturing the company's culture and delivering a positive employee experience. For the CMO, however, the opportunities are particularly fun.

In the push to launch a product, feature, or offering, the emphasis is rightfully on getting the job done and checking all the boxes when it comes to external promotion. It can be easy, however, to overlook educating everyone internally about the work being done and celebrating the wins collectively. Building an internal launch plan into the overall marketing execution for a launch ensures that this piece of the puzzle doesn't get overlooked, and provides an opportunity for the company as a whole to be on the same page and take a moment to celebrate the achievement.

Marketing also has a role to play in supporting internal events. While we wouldn't advocate for Marketing taking full ownership of organizing employee events like employee appreciation activities or holiday gatherings—that is more appropriately the purview of the People team—there is value Marketing can add in providing creative ideas, producing swag, and supporting logistics planning.

Lastly, as the company transitions from startup to scaleup, your employee population will rise and bring with it the opportunity to partner with the People team in introducing affinity groups. From facilitating the sending of internal communications to designing text treatments or other design assets for the groups, to providing marketing expertise to help these affinity groups grow and expand their impact across the organization, there is a lot Marketing can do to add value to these groups and support them in their missions.

Chapter 45

Breaking Down Marketing's Functions

The three primary foundations of marketing—brand, sales, and culture—are not standalone, esoteric ideas but are actually supported by seven primary functions: Brand, Digital, Events, Content and Communications, Product Marketing, Operations, and Sales Development. Depending on the needs of the business and the size of the marketing organization, how you execute against these functions and organize your team may be wildly different. For a startup, one individual may cover multiple or all of these responsibilities, while a scaleup may have teams of contributors supporting each function.

No matter the size of your marketing team or the business itself, no matter whether you're a B2B, B2C, or nonprofit, no matter whether you have no budget or an unlimited budget, it is essential to understand the value each function brings to the overall marketing strategy and the ways in which they can intersect in order to get more out of each campaign and deliver maximum impact.

Brand

As we've discussed already, building and maintaining the company brand are the primary responsibility of Marketing. Your brand is not only the visual elements—the logo, color pallet, and supporting design elements—it's also the textual elements—the tone and personality you want to convey and the core messaging behind those elements. The combination of these is the driving force behind everything Marketing creates. It's not uncommon for people to think of the brand as one or two things, like the website, the logo, and the tagline and it's your job to continually remind people that the brand is more than that. The brand embodies your company values, culture, personality, go-to-market strategy, products, services, and more. Your brand will set you apart from competitors, it will attract customers to you, and it will help potential employees reach out to you. The brand is definitely more than a website and a logo!

We developed a deep appreciation for the power of a brand at Return Path and regardless of our role—whether an individual contributor or a manager—we considered ourselves to be the lead brand ambassadors throughout our time with the company. We adopted a simple mission: if anything had a Return Path logo on it, we were responsible for it. That means apparel, websites, event banners, invitations—anything with our logo was something that we were responsible for. With that directive, ensuring the consistent application of our brand standards became our primary goal.

Ideally, your brand should provide you, the marketing team, vendors, and employees "freedom within a framework." Simply put, what that means is that brand guidelines are supposed to provide a framework for the rest of the organization to operate within. The brand provides the basics—the colors, icons, language, logo, templates, etc.—and then enables the rest of the company to leverage those assets in order to deliver on their own objectives. They are empowered to take the framework the brand provides and mold it into their use case.

If you don't create "freedom within a framework," you'll constantly be battling the fast-moving renegades within your company, the people who believe that getting their content out into the market, or into the hands of customers quickly, is the critical task. They don't think about whether or not their content follows the brand guidelines or is consistent with other collateral. And, honestly, they have a reasonable point of view. But if you don't have a framework, you will spend your time both policing infractions of your brand and a lot of time recreating what's been done to fit it into your brand. Marketing will become the obstacle within the organization instead of the enabler, and Marketing will be blamed for all sorts of things in the market. Take it from us, the best thing you can do as a startup CMO is to create brand guidelines for the rest of the company, provide clear, repeatable training, and set them up for success for others to use them as they see fit.

It's important to note that consistency is key when it comes to your brand. Brand guidelines, or that framework to work within, ensure that throughout the customer lifecycle, the company is delivering a consistent, repeatable, and recognizable experience for your audience. The most-recognized brands have this consistency, and it not only outlasts the customer experience, but they are unique so you would never confuse Apple with McDonald's, for example. A random person should be able to look at any output of your company—a page on the website, a presentation from an account executive, an invoice from Finance, or help article from Customer Service—and immediately recognize it as belonging to your brand. This not only drives brand awareness, it also builds trust. And that trust is essential when it comes to identifying, winning, and growing your customer base.

That consistency should be evident not only in the visual style of your output, but in the voice as well. Tone, word choice, cadence; all those play a significant role in communicating a brand. Is your brand friendly and off-beat or is it more straightforward and buttoned-up? What is the tone of voice your audience will respond most positively to? Your brand voice should reflect that. Any marketer probably remembers (or is at least aware of) the famous "I'm a Mac. I'm a PC." commercials of the late 2000s. They're an excellent study in how brand voice and visuals can come together to communicate the desired message. You can even look at commercials for fast food chains like McDonald's, Burger King, and Arby's. Not only is their look unique from one another, the way they speak and the words they use are completely different as well, but aligned with each particular brand.

For a CMO, the brand marketing work doesn't stop once the brand has been established and the guidelines have been shared across the organization. Rather, that is just the first step of many. Think about your brand as being as alive as you or me. It's a living and breathing contributor to your company, and should be treated as such. The best brands receive constant attention and fine-tuning to keep pace with the nonstop change and evolution that naturally occur within any business.

At Return Path, the Brand and Digital Marketing team made it a habit to set aside time each year to evaluate every aspect of our brand, from our color palette to our core messaging. Typically, we took two to three days to focus exclusively on this work, making it clear to the rest of the company that we were "heads down" and purposefully avoiding the interruption of work on other projects. During that time, we would brainstorm, explore changes, make revisions, and fully execute any updates. Particularly when the company moved from three lines of business to one after a couple of divestitures, this time was invaluable in helping us determine how we would visualize and contextualize this shift. From reworking our color palette, redrafting our core messaging, and updating all of our company-wide templates and assets, the shift in the business was reflected and rolled out in our brand guidelines and assets.

No matter the level of output, make it a priority to evaluate your brand and associated assets on an annual basis to ensure they're doing everything you need them to do. The annual evaluation is a minimum—if your situation changes dramatically, if you acquire a company, divest a business unit, enter a new business, expand globally, or any other major event happens, you'll want to rethink your brand more often. You'll want to ask basic questions about your brand, like, are you still communicating your value props effectively? Are there new design trends that help you visually communicate your brand in a more interesting way? Are your brand guidelines working (or not working) once they're in the hands of the sales or service teams? As the CMO, you need to make sure you're creating space for you and your team on a regular basis to evaluate and refine your brand. You'll need to ensure that the brand continues to grow and improve in line with the company as a whole.

A final note on the brand: an individual brand marketer or brand team makes the greatest impact when they act in service of the rest of the company and a lot of times this means doing day-to-day tasks that are not necessarily visionary. You might be tasked with cleaning up PowerPoint presentations or creating another iteration of a one-pager for a specific use case. That isn't glamorous work but in doing that you're helping to put the company's best foot forward in as many instances as possible, and that provides value. As Matt used to say regularly when congratulating our design team, "Thank you for making us look so great to the world." The key for the CMO is to establish the ground rules for engaging those tasked with brand marketing (for example, establishing a project request system), enforcing those rules consistently, and delivering the requested work on time and above expectations. Doing this not only demonstrates the impact using the brand to its fullest can have (I've seen some incredible before and after decks in my time) but also builds relationships and fosters collaboration between Marketing and other functions in the organization. Seeing the marketing team actively make work better, deliver what is needed, and continually push its own boundaries influences how the rest of the company perceives the marketing organization as a whole.

Digital

Once you have a handle on your brand, the next step is translating it to digital marketing. Your digital marketing activities are where you take the combination of your visual brand (the logo, the colors, the design elements) and textual brand (the core messages, the tone of the writing, the personality) and bring them to life. That is where things really start to get fun.

For a startup, digital marketing provides the greatest opportunity for building brand awareness and driving demand for your product or service. Basic digital marketing activities include creating a website, establishing an email marketing program, generating a social media presence, and running digital advertising campaigns. Each of these is powerful on their own, more powerful when combined together, and fully maximized when collaborating with the other marketing functions.

Website

At Return Path, we often said one 100x25 pixel Call To Action (CTA) on our website was our most valuable marketing asset. That 100x25 pixel is about the size of a postage stamp—and it's our most valuable asset? Well, it was our "Request a Demo" button and it was responsible for more than 10% of our lead generation per year and was consistently our highest performer when it came to converting leads to bookings. That's the power of digital (and marketing, really): it's not an end in itself, it's an enabler, an accelerator, it's a sieve that culls the good from the bad, the serious from the inquisitive. Even something as minor as a button on a website can be a powerful tool.

Thinking more broadly than a small CTA, the website itself is the storefront of your company. It will be one of the first introductions a person has to your brand, your offering, and your mission, so it's critical that you have a firm understanding of what you need to communicate and what actions you want visitors to take.

A strict focus on the desired action you want visitors to take should drive every choice you make when it comes to your website's design, content, and layout. At Return Path (as with most good SaaS companies), that Request a Demo button was impactful because we intentionally directed visitors to it through visual cues, like the color of the button and its position on the page, and through limiting other possible actions they could take. For example, on gated landing pages—pages with a form the visitor must complete in order to get to a content piece like a whitepaper, video, or other branded asset—the only options were to A) submit the form, B) return to the home page, or C) click the Request a Demo button.

Your desired action may not always be to get the visitor to request a demo, or to only take one action. At DePauw University, I (Holly) managed alumni communications and one of my first undertakings was to streamline our website content in order to focus alumni action around specific goals—Go, Give, Help, and Connect. Each of those pillars had a specific action we wanted people to take, and all four were critical when it came to engaging our alumni base and keeping them connected to the institution. For each goal, there was one specific "best action" we wanted to drive alumni towards:

- Go = make a donation
- Give = register for an event
- Help = sign up to volunteer
- Connect = create an account in the online directory

By having a clear perspective on the action(s) you want your visitors to take, you can then build a website experience to create the opportunities and deliver the messaging to make that action more likely.

Email

Another foundational element of any digital marketing program is email marketing. We promise you that email is not dead, and won't be anytime soon. In fact, there is a significant amount of research that shows the incredible ROI an email program offers; as much as four times the ROI of other channels. The DMA, for example, reports that email returns $38 for every $1 spent*. With almost 4 billion daily email users, the ability to access our email on laptops, mobile devices, and even our watches, and the rise of 5G, access to the inbox has never been easier.

There are a few straightforward things you can do to ensure that your email marketing program is successful in engaging your audience:

- **Deliver meaningful content.** First and foremost, you need to be sending emails with content your audience will care about. Whether it's information about a product launch or your latest content release, make sure you understand the value it provides to the subscriber and describe it clearly and concisely.
- **Have a clear call to action.** When you overload an email with links and buttons, or forget to include one at all, you make it hard for the reader to know what to do. It can be tough, but try to limit your emails to one specific call to action. If you do have more than one CTA (in a newsletter, for example), lead with the action you most want the reader to take.
- **Know and follow regulations.** Regulations that govern privacy, like GDPR and CCPA, are strict and complex. Make sure you both understand how the rules apply to you and build up an email program that fits within the guideline. It will always be tempting to try and go around the rules, like sending an email to a group who isn't opted in, but it's rarely worth the potential risk and usually doesn't return worthwhile engagement.

*Source: https://blog.hubspot.com/marketing/email-marketing-stats

- **Make testing part of your process.** Consumer interests change as often as the seasons. Make sure you're consistently testing the various aspects of your emails to see what's working. From the subject line, to the button color, to the tone of voice, to the entire design, testing your options to see what works best will only make your emails more impactful.
- **Get creative with growing your list.** Growing your subscriber list should always be top of mind, and there's a lot of ways you can go about gaining opted-in readers. Tactics like email signature banners, one-to-one outreach to current customers, callouts on your website or blog, are a solid start. You can also make it more of an experience, like building the opt-in process into your onboarding campaign or incentivizing subscribers to get their friends to sign up.

Social

Social strategy can be a bit of a wild card. There's no debate that having a presence across various platforms is table stakes, but the ROI can be wildly different based on industry and target audience. Whether social engagement is driving sales or not, being active on social media will give you valuable consumer insight, and taking as much time to engage in social listening as you do in social posting will help you stay on top of trends, changing attitudes, and general sentiment.

At the most simplistic level, there are three things you should have a plan for in order to be successful in your approach to social media.

1. **Plan for being active.** Whether it's Facebook, Twitter, Instagram, or something else, you get what you give. Posting your own content and updates is important, but it's equally if not more important to engage with other profiles too. Whether they're customers, partners, or complementary brands, sharing, liking, and commenting can go a long way in building relationships with other individuals and companies that can broaden your reach.

2. **Plan your rules of engagement.** Trolls are a part of the game when it comes to social media. Whether it's an aggressive competitor, a disgruntled former customer or employee, or someone just looking to cause trouble, negativity will find its way into your feed. Our advice: never delete negative comments and remember you're not obligated to respond. There may come a time when the record needs to be corrected, but more often than not it's better to simply let the mean, false, or provocative comments go.
3. **Plan to make others aware.** At some point, you will engage with a current or future customer, and it will be important that you have a plan for how to make others aware of those interactions. Did a customer post something good or bad? Let their account rep know. Is there an issue? Know who you need to alert in IT, Engineering, or Customer Service. Understanding who needs to know when you stumble across potentially useful information will go a long way in making your social presence valuable.

At the end of the day, know what it is you can realistically get out of your social strategy. For B2B brands, sometimes just being present is enough and realistically it may never be a major lead driver. That's OK, but being there is still important and builds trust, legitimacy, and brand awareness. For B2C brands, social can play a much larger role. Don't get too caught up in trying to be what you're not, or trying to go viral. Represent your brand consistently and accurately above all else. Establishing that consistency and demonstrating your engagement with your customers (and future customers) can go a long way in building loyalty.

Advertising

The internet is a noisy, crowded place. Luckily, digital advertising provides an opportunity for marketers to make sure their brand stands out. From pay-per-click (PPC), to display, to search, to social, there is a lot of potential for promoting your product or services and reaching your target audience.

Digital advertising can be complex, especially when it comes to managing your budget, but there are a few basic concepts you can keep in mind to ensure that you're using those dollars effectively while your campaigns are running.

- **Be concise.** Almost every digital advertising option comes with a character limit. Know what it is and stick to it. If nothing else, writing for ads is a great way to tighten up all of your messaging because it has the most specific constraints. When you can make your case for engagement in as few words as possible, you're onto something good.
- **Have a clear call to action.** Every ad you're running is going to be driving the user to do something. Do you want them to give up their contact information to download a report? Do you want them to request a demo? Do you want them to register for an event? Know what the desired outcome is, make it clear in the ad, and make it clear on the landing page.
- **Stay on top of conversion.** Forget the seller's mantra, "ABC: always be closing." For marketers, it's "ABO: always be optimizing." Clicks are great, but conversions are what counts. It's important to understand which ads are driving clicks, but if you aren't seeing good conversion on the desired action, have a plan of action to make adjustments. Maybe your copy needs to be clearer, your layout needs to be rearranged, or you need to try an entirely new angle.
- **Try new things.** This is relevant for your messaging and ad design, but also for the types of digital advertising you're using. Maybe you're putting everything you've got into search advertising, but haven't given social advertising a try. Or you're running some successful PPC and display ads, but you've not really experimented with Account Based Marketing yet. Each approach offers its own value, and you won't know the perfect recipe for your success until you try out a few variations.

While we're on the topic, we should dip our toes into the wider pool of advertising possibilities, which of course no book or book section on marketing can ignore. In the world of B2C, digital advertising offers speed and flexibility, but there are many other avenues to take that provide the reach you may need. From billboards, to television, to radio, to newspapers or magazines, there are considerable channels to explore. As with any promotional activity, the most important thing to consider is whether the delivery method enables you to deliver your message in the most impactful, effective way possible. In the world of startups, traditional offline advertising is a rarity both due to its lack of addressability (although that is improving for most channels) and due to its sheer expense. Although we see the occasional proverbial Super Bowl dot-com ad, and a handful of billboards driving on the 101 between San Francisco and Silicon Valley, those don't represent enough of a startup CMO's advertising endeavors to warrant a lot of discussion here. If you find yourself being pulled into traditional advertising channels because your business is taking off explosively and you're a strong Direct to Consumer brand, you're likely beyond being in the startup phase and have much larger budgets and agency support to lean on!

Omni-Channel

One way to maximize the impact of your digital marketing efforts is to deliver the same message across as many channels as possible. What's important, though, is how you craft those messages to fit within the specific channel. As Marshall McLuhan said, "the medium is the message" and nowhere is this more practically applicable than in digital marketing today. Although McLuhan first wrote this concept down in 1964 in the advent of television, his concept is given more weight by the abundance of "mediums"—or communication channels—we have today. Think of it this way: if you were going to share one thought across all of your social channels, the output would be very different between Facebook, Twitter, Instagram, and LinkedIn. That's because each medium dictates the way messages are best delivered: Twitter favors short and sweet, Instagram is all about the visuals, Facebook is best suited for longer, more personal content, and LinkedIn is the home of concise, professional posts. Now apply that to the entirety of your digital marketing program, not just social channels.

If you can take one message and tell it on your website, through an email campaign, across your preferred social channels, and in digital ads, you're well on your way to an omni-channel marketing campaign and you've greatly maximized the output of that single message. Duplicated across a variety of value propositions, product offerings, or brand-building messages, you have a backlog of campaigns you can run to build up your output and get your content into the market.

The beautiful thing about digital marketing is that you have the capacity to continually make changes to optimize against your goals. Unlike approaches like print marketing, for example, where the message is locked in once it's out in the world, digital channels give marketers the opportunity to test, pivot, rewrite, and reimagine constantly. For a startup, this level of flexibility is a huge advantage, especially as you work to find your voice, build your brand, and establish the narrative that will help your company grow.

The Case for Grouping Brand and Digital Together

We are passionate supporters of the notion that brand and digital marketing go hand-in-hand in companies where the overwhelming majority of advertising efforts are digital (if, as we noted above, your brand has achieved some level of escape velocity and is pursuing a lot of offline advertising, then the previous sentence does not hold!). You can't have a strong digital marketing presence if you don't have a solid understanding of your brand. Your brand informs every digital move you make—how you speak, what content you create, and how everything fits together. Similarly, you can't have a strong brand without taking advantage of the opportunity digital marketing provides to constantly test, iterate, and refine your brand across a variety of channels. Your digital marketing efforts inform your brand evolution—how to strengthen the brand, how you adapt to new trends and technologies, and how you tell your story in new ways. Like we said, brand and digital go hand-in-hand.

Having both a brand and digital team is a luxury, most likely only attainable for a company more in scaleup mode, but thinking about the two functions together is absolutely achievable for a startup with limited marketing resources. At Lev, Holly is lucky enough to have one team member who is a powerhouse when it comes to both brand and digital marketing. Having both design and digital execution within one role has allowed us to move quickly in building our brand and our digital efforts simultaneously. As we build the website, we're refining our messaging; as we elevate our brand design, we're redesigning the website; as we test new channels, like digital advertising, we're finding new ways to tell the Lev story.

While a full-fledged brand and digital team may be a longer-term goal, the startup CMO can still make it a priority to find one or two people who can execute brand and digital marketing initiatives together to ensure that the constant growth and innovation that need to happen are adequately supported.

Chapter 46

Events

Nick started his marketing career in events and it's a marketing channel that we're particularly passionate about. We know "events" sounds old-school and traditional, especially after talking about technology and digital, but remember—marketing involves promoting, selling, and distributing products and services in ways that build the brand, generate leads, and solidify the company culture. Events are unique in that they promote all three of those objectives. If you do it well, event marketing can be the single most effective approach for achieving marketing goals. If you don't do it well, however, it can quickly eat away at your budget and time. It can tarnish your brand, reduce leads, and actually hurt your culture. Besides that, if you don't do event marketing well, you'll waste a good amount of productivity. Contrary to how most people view event marketing, the "event" isn't the most important thing to concentrate on. Effective event marketing requires equal investment in pre-event activity, event execution, and post-event follow-up. So, the event is the middle of a three-stage marketing model, and not the single most important thing.

We should note that COVID-19 has really thrown an interesting twist in events. Not being able to physically gather and having to cancel all marketing events is leaving a huge gap in many Marketing team's demand generation plans. At the same time that COVID-19 has impacted events, it has also spurred new innovative and creative approaches on how to pull off incredible virtual experiences. Webinars and virtual events are now the norm, and since individuals are stuck working at home, many are still hungry for these types of experiences. It's interesting to see how the pandemic has challenged marketers to think differently. Some of the innovations include sending a package a day prior to the event, including the itinerary, some swag (a tee shirt, some gifts), and even some food to keep people satisfied for that block of time. We expect events to continue to evolve and to push the limits.

Before getting into event planning, you need to know the "why" behind your event strategy. What is it you want to accomplish? Are you looking for net new leads or to nurture leads? Do you see event marketing as a way to advance existing opportunities or to retain and upsell clients? Or perhaps you want to generate awareness around a new product. Maybe it's a combination of more than one of these. Regardless, you'll need to know the "why" behind your event because this will allow you to choose the right event, set and measure the right goals, and put together the best overall strategy. Once you know the "why," you can then determine what event or combination of events will best achieve what your business needs.

Types of Events

Generally, events fall into one of two categories: third-party or self-hosted. Third-party events include trade shows, conferences, and partner events. They are hosted by a third-party and for the most part you are leveraging their audience. Sure, there are ways you can target and engage a segment of the audience, but you are largely paying a premium to be part of something that is already established. In third-party events you should expect to pay a sponsorship fee and receive in return various entitlements based on your investment. Third-party events can be great for getting access to an audience that you otherwise would not have access to. I said "can be" because the event itself might not lead to anything unless you do the pre-event planning and the post-event follow-up.

Self-hosted events are the events that you host and where you are responsible for ensuring you get the target audience to participate. You'll have more control over most aspects of self-hosted events, but also more responsibility. Self-hosted events can take just about any form you can think of, ranging from networking events, to hospitality events, educational events, webinars, customer events, user conferences, client advisory boards, and more.

Event Planning

Once you know the "why," or what you are looking to accomplish, you're ready to start your selection process and plan out your event strategy. Event planning is not a one-time task but more likely a series of tasks that should be planned out over a period of time. If your company is new to event marketing, you may want to carefully select one or two events to experiment with before building a more comprehensive plan. One of the best things you can do, even as an early-stage startup, is to map out your event calendar a year in advance. Sure, things will change. But as long as you allow for some flexibility, mapping out your event calendar a year in advance is a healthy exercise for several reasons.

First, mapping it out will allow you to build your event strategy in conjunction with other marketing and company initiatives. Things like product launches, major content releases, quarter end, other marketing campaigns, and internal events can be taken into consideration. There's nothing more ineffective than doing a marketing campaign weeks before or after a well-attended event. The right event in combination with a product launch can be a great way to boost excitement around the launch and also provides your event with fresh content.

Second, by mapping your event calendar, you'll be able to integrate your event strategy with seasonality. In a lot of businesses, the beginning of the year, end of the year, and midsummer are not optimal times for events, for obvious reasons. So, look to avoid those times and instead spread out events throughout the year to reduce strain for the teams who work on and leverage events.

Third, planning out a year ahead will really help with your demand generation planning. At Return Path, historical data gave us a pretty decent idea of the types of business results we could expect from most events. By itself, that data was just an interesting "good to know," but being able to map those events out along with other marketing initiatives provided us with a demand generation heat map. Now we had something powerful. We could see when we were light or heavy during certain periods and we could easily adjust our event schedule or other initiatives accordingly.

Partnering with Sales

No team in your company will be more interested in what marketing is doing with events than Sales. The relationship between Sales and Marketing is always an important one and it will become magnified during events. The sales team plays a crucial part in staffing and speaking at events. They'll also have ideas on products or services to feature and they'll want to contribute to the invite list and experience. Collaborating successfully with Sales can have a heavy influence on the level of success for an event. You'll want to work with Sales Leadership to know what types of events are needed, what goals they would like to achieve, and gather any ideas about event location. A thoughtful conversation with Sales Leadership will cover topics like quotas, existing pipeline, opportunities in flight, partnerships, staffing needs, geography, and bandwidth. These conversations happen during the event planning stage, not a few days before or after the event, and they inform your overall event strategy.

Another way to collaborate with Sales is to look for opportunities where you can leverage location and people. We were always big advocates on maximizing team travel time in-market when we're doing our event planning. So, if we were planning to travel to NYC in March for an event, we'd figure out what prospect or client meetings could be arranged in conjunction with that travel. Could we create a unique experience for a specific prospect or client by leveraging the event? Could we create a separate experience the day before or after an event and leverage a company thought leader or executive who will also be in attendance? Thinking carefully during the planning stages about the event, the people, the location, and what you want to accomplish will allow you to create pre- and post-mini-events that are more intimate and can build relationships with clients and potential clients.

Pre-Event

We referenced earlier that effective event marketing isn't only about the event itself. The work done before, during, and after are equally important. It's easy to lose sight of the fact that the pre- and post-event activities are equally important because for most people in your company and for event participants, the event itself will have a specific date, venue, and marketing message, so it's natural for them to think that the event is THE EVENT. It's best to constantly remind others that event marketing has pre- and post-activities that help to make an event a revenue generator. For third-party events, it's likely you'll get a pre-event attendee list and you should leverage this as soon as possible, although sometimes you may not get it until right before the event. Your goal when you receive the attendee list is to build unique experiences only available at the event, and comb through the attendee list to identify who in your target audience would benefit from these unique experiences. Ask your Sales or Sales Development team to reach out personally to the right individuals to engage with these attendees in advance. For example, at Return Path one unique experience we offered was complementary on-site consultation. Marketing would create an email template explaining the value, often including the normal dollar value of this now complementary consultation, and we'd ask the individual to reserve their time. The event team arranged a schedule, space, and staffing plan for these meetings to happen. We would be offering a unique opportunity for attendees to engage with us and nine times out of ten the attending individual would walk away with real value from the time spent with us.

After the complementary consultation, it made it a whole lot easier to continue discussing ways to work together. You should also share the attendee list with the internal team members attending the event in person so that they can arrange in-person meetings. And for any account owner, whether attending in person or not, we always made sure they knew who from their account base would be attending the event, and we provided account owners multiple ways to engage with their base. They could engage through one of our complementary consultations, attend our speaking session, stop by our booth to pick up something special, invite them to a VIP experience available to us as sponsors, or invite them to one of our own dinners or ancillary events we were planning in conjunction with the main event. The point is, you'll have an attendee list but it's not enough to know who's there so you can say hello, but to know who's there so you can engage with them in ways that lead to a deeper connection.

For self-hosted events, it's all about getting the right people to the event in the first place. Without the right people in attendance, you're wasting time, money, and effort. Marketing can promote an event to targeted lists through marketing emails (honoring individual preferences, of course), social, website, and other channels. However, the real value in an event happens through personalization. Sales and Sales Development typically know the value of events, and most of the time they will be chomping at the bit for the opportunity to engage. Regardless of the enthusiasm of Sales and Sales Development, Marketing still needs to fully participate and do whatever possible to help make the event a success. Provide a list of potential customers for Sales to target, provide templates with the marketing message you want, and have them reach out themselves to prospects. Don't forget that Marketing can engage with customers and prospects personally as well. Our field marketers got pretty good at building relationships with individuals in our industry and were a great channel for us to get the right people to attend our own events.

Event Execution

For event execution itself, we'd focus on two main elements. First, make sure that you provide an amazing attendee experience. Regardless of the type of event—virtual, in-person, conference, seminar, networking—think through every aspect of the attendee experience. How is your brand physically represented? Who or what is their first engagement point at the event? Is the content valuable? Are the time and place right? And, ultimately, is it a memorable experience? The second element is to be present. It sounds obvious, but how many times have you gone to an event and seen the hosts engaging with each other, and not with the attendees? Ensure that all team members attending the event, regardless of level or function, are representing the company at all times and engaging whenever possible with attendees. They should be fully present and that means talking to attendees, participating and engaging in content, being a thoughtful host, and contributing to the event. Teamwork at events is invaluable for a good attendee experience, and it's a good way to build on your company culture.

Post-Event

The final phase of event marketing is the post-event. This is where some of the most important work begins. As soon as possible, you should get your attendee data to your CRM so you can measure results. Follow up with event attendees as soon as possible, which can be as simple as saying thank you or sorry we missed you. Provide content or takeaways for those who attended as well as those who did not attend. Establish next steps. Make sure Sales Development and Sales teams know the details of the event for effective personal follow-up. Find a way to document and share notes from conversations that happened at the event. Keep the momentum from the event going. How do you keep that momentum? Share the good, the bad, and the ugly from the event. Share what went well and what you would change, share the "wins," and share the lessons learned. Events are not standalone and discrete but are things that will be repeated in the future and even if they're not repeated, the lessons learned will transfer to your next event. If you can capture people's ideas and thoughts about the event while it's still fresh, you'll be ahead of the game for your next event.

Using Data

Events are big investments for every company, regardless of stage of growth because they are a high investment in cash, time, and staff. Figuring out where and when to invest should be deeply rooted in business results and historical data. You'll need to measure the right things, over time and consistently, in order to have this data in the first place but common metrics like leads, MQLs, opportunities, pipeline, sourced bookings, influence, and ROI should definitely be tracked.

We found that starting with leads and then monitoring conversion to pipeline and to closed-won business were the most valuable indicators of event value, with sourced bookings winning overall. You can take this sourced number divided by the investment amount to provide ROI, which is equally important. I found influence was more valuable as a metric for impact on our customer base, which we measured separately. One additional reason to support reporting on event impact consistently over time is the fact that, depending on your sales cycle, you may not be able to see value from an event for some time. It could take one quarter, two quarters, maybe even a year or longer. Be patient, and don't get frustrated if you don't see ROI quickly. There are plenty of ways to nurture and move along event lists and conversations.

Chapter 47

Content and Communications

Content comes in many different shapes and sizes. From slide decks, to one-pagers, to customer stories, to blog posts, to podcast episodes, to webinars, to infographics, to press releases, to white papers, to ebooks … the list goes on and on. No matter the final output, though, the objective from a marketing perspective is always the same: to add value. Each bit of content, no matter how big or small, needs to have value for prospects, value for customers, and value for the business. Without that value, you're just wasting everyone's time.

So how does a content marketing strategy add value to your company? It actually checks all three boxes of your marketing goals: (1) it helps build your brand; (2) it drives demand for sales; and (3) it supports the company culture. Three out of three!

Marketing

- **Brand building.** Producing and promoting content that addresses your target audience and provides insight into challenges you know they face reinforce the value your brand can add. By using content to demonstrate that you understand your prospects and customers' needs, and can directly improve their lives for the better—either through your offering or related insight—you build brand awareness and make your offering more relevant to customers, which in turn creates respect, and loyalty.
- **Driving demand.** Your content also serves as a major vehicle for driving demand. On the one hand, Marketing can use content to attract potential customers to engage with your brand and eventually lead them down the path of

engaging with your sales team and committing to a purchase. On the other hand, Sales can actively utilize content to engage prospects they are attempting to engage or are already in conversation with. Content that helps to advance their conversations goes a long way in strengthening those relationships and moving prospects closer toward a decision.

- **Company culture.** The content creation process provides an opportunity to engage experts across the organization. By engaging people in all parts of the business, you'll be able to develop a greater diversity of topics and share a broader set of perspectives than if you just feature the same handful of people or departments. It also builds internal affinity to the marketing team, developing stronger relationships between Marketing and other parts of the organization by giving a platform for experts in the company to share their knowledge and fostering cross-collaboration. There are many people in the organization who have ideas and want to share them more broadly, so providing an opportunity for them to do that increases the diversity of ideas and has a greater chance of resonating with people outside your organization.

For any content marketing strategy, it's important to have a clear understanding of the goals you're trying to accomplish with your content. Do you need to demonstrate expertise or address common questions or concerns? Do you need to refine your brand voice or positioning? Having one or even a few specific ideas about the goals you want to accomplish with your content will help focus your efforts and allow you to prioritize ideas and requests.

Once you have those objectives in mind, the next step is to establish realistic expectations about what you can produce. For example, maybe you can commit to one blog post every week, a webinar every month, and a research report every quarter. Setting these goals up front then puts you in a position to outline a content production schedule to work against and communicate that internally in order to get buy-in and (if needed) content development support.

With your goals set and your content calendar roughly established, it's time to start writing. When it comes to content production, it's important to keep a few things in mind. First, everyone will have an opinion, they'll have another way to say something, or they'll have a last-minute change. At the end of the day, the CMO and/or content marketer are responsible for publishing what will serve the best interest of the brand and should be empowered to make the call and publish content when they think it's ready for prime time. That leads into the second point: sometimes done is better than perfect. Very few marketers are sitting on their hands waiting for something new to work on. Instead, they're often overloaded with tasks and moving quickly from one project to the next. With big goals and high expectations, obsessing over the nitpicks or minor changes that prevent a piece of content from going out the door won't help you achieve your goals. Plus, if your content is being distributed digitally, you have the greatest benefit of digital marketing at your disposal: you can make changes at any time!

Public Relations

While blogs, webinars, and ebooks are often top of the list when it comes to content creation, public relations (PR) also falls into the content realm. PR is a misunderstood and rapidly changing arrow in the marketing quiver. Years ago, traditional PR was about writing press releases and working—either directly or via a PR firm—the key journalists who covered your industry. As the media business and reporting in general have changed, PR still includes press releases but is much more often tied to a broader content program with multiple formats of content and multiple channels of distribution. But simply put, if advertising is the art of telling the world your story (via multiple channels), PR is the art of getting other people to tell that story for you.

There is still nothing better and more cost-effective for your marketing than to have a well-known and well-followed influencer tell their audience that your company, your brand, your product or service, is a must-have. Achieving that kind of third-party mention is almost free (at least relative to ad spend), and it carries a 10–100x multiple of credibility since it's not coming from you or your company directly. Sometimes those mentions can be generated by outbound PR work either from an internal team or from an external agency. That was certainly the case in years past when more traditional journalists were the most important and in some cases sole gatekeepers of publicity in whole industries. Today the effort can be a lot more diffuse, with micro-influencers like bloggers or vloggers providing their own mentions or endorsements, leading to more organic social media activity in the form of likes and retweets, which in turn generates other influencers to take notice, and, well, you get the idea.

A strong PR strategy can go an incredibly long way when it comes to brand building and demand generation. For example, at DePauw University, where Holly worked, PR served goals across almost every department—it highlighted notable alumni achievement, recognition received by the University, awards and accomplishments of the faculty, and important staffing changes, all aimed at attracting future students, recruiting talented professors, and energizing the alumni base.

Matt's experience running Marketing at MovieFone (777-FILM) in the 1990s was such that the company's entire trade customer promise—bringing addressable and measurable media to movie marketing for the first time—was given a massive boost by the presence of a major article in one of the company's key trade publications at the time, *The Hollywood Reporter*, that the MovieFone team nurtured for years and referred to as "the holy grail" article. That article generated an increase in sales for the company, but more important, it made other advertising vehicles, including relatively new but powerful internet giants like Yahoo, Interactive Corp, AOL, and Microsoft, take notice of MovieFone as a company, which was a significant driver of the company's blockbuster $600m sale to AOL less than a year after that article ran.

Content Creation

The biggest hurdle you might face when it comes to content marketing is making a case for its value and creating space for content development to happen. In the midst of brand building, event planning, sales development, and digital campaigns, carving out time to focus on content marketing will be challenging. A lot of times, content creation is the task on your to-do list that keeps getting pushed to the next day. However, content has the power to benefit your brand in unexpected ways so we'd suggest to keep in mind that "done is better than perfect."

At Lev, despite having a small team (just four people in marketing roles, including myself), having a person fully dedicated to content creation was a huge priority. By ramping up our content creation, we were able to address needs across the business while also building brand awareness within our primary lead source, Salesforce. We put a major focus on creating content that was vertical specific because we wanted to address consistent questions from prospects and partners about our expertise in specific industries. Thanks to our close collaboration with the sales organization, we knew they were running up against these questions a lot, so we were able to quickly identify key verticals to focus on and specific questions to address, and start knocking out blog posts, webinars, infographics, and ebooks to create a library of industry-specific content.

This library addressed the initial need of responding to common questions about our experience, but also provided two additional benefits. First, it created a backlog of content for the Sales team to utilize when engaging with new partners and prospects, preempting questions about our industry experience they might have. Second, the high volume of diverse content across verticals established Lev as a thought leader, building up our brand recognition and solidifying trust within Salesforce, a key business partner.

Gated Versus Ungated?

There are two different schools of thought when it comes to whether or not content should be gated. Gating content is a traditional approach to driving demand: a high-value piece of content, like a white paper, is put behind a form to collect contact information. The visitor trades their contact information for access to the content, opening the door for the sales team to engage with the individual. On the other hand, many companies are moving entirely away from the gated content model, choosing instead to make everything available without any barriers to entry. The benefit here is that because visitors are not being asked to trade their information to access content, they're given control over when to offer up that information through other means. For example, a person might voluntarily fill out a contact form or interact with a chatbot, not because they have to gain access to a white paper, but because they want to provide their information. With ungated content you can have a potentially stronger footing for the sales team when they engage with that prospect.

Chapter 48

Product Marketing

You can have a strong brand, unique corporate positioning, and innovative digital, content, and event strategies, but if you miss the mark on product marketing, the impact will be felt across marketing and your entire business. Product marketing is all about understanding where you fit in the market, the needs of your target audience, and how best to package and position your offering to generate opportunities. Without these foundational elements, every startup or scaleup company will struggle.

While cross-departmental collaboration plays a role in the majority of Marketing's activities, nowhere is that connectivity across the organization more pronounced than within product marketing. Product marketing is the most important marketing you'll do and it requires input and collaboration from nearly every department in a company. Effective product marketing requires intentional collaboration and engagement between Marketing, Product, Finance, Sales, and Customer Service in order to do the following:

- Deliver product offerings and innovation that meet market needs.
- Create client-facing product collateral to drive interest.
- Enable sales and service teams to sell and support product offerings.

So, the marketing of products touches nearly every part of the company and can be make or break for whether or not your company survives. Matt calls it one of a small number of "glue" functions inside the company because it holds so many things together.

Deliver

Understanding your market and the needs of your target audience is the cornerstone of product marketing. You can't effectively position your product if you don't understand the competitive landscape, if you don't know where you fit within that landscape, if you can't say how you are unique, or if you can't articulate how you solve the common pain points of your target audience. Your understanding of all of these components should be a key driver behind your initial offering and how you expand your product or services over time.

The marketplace can be pretty crowded, and product marketing plays an important role in keeping a pulse on the marketplace. Buyers' needs and interests are rarely static, so keeping up with what is happening around you will give you a leg up in remaining at the cutting edge. Similarly, other startups can enter the scene at any time, along with already established companies broadening their reach into new areas. Tasking a product marketer with staying up to speed on the competitive landscape allows both the Marketing and Sales organizations to maintain awareness of what is happening around them and be prepared when entering competitive situations.

Beyond marketing research and competitive intelligence, product marketing also plays an important role in shaping what is delivered to meet the needs of the market. Once you understand your audience, the needs you're addressing, and who you might be up against, using that knowledge to hone in on the right pricing and packing to address those needs within the scope of what is realistic for the targeted buyer is key. Product marketing can ensure that you are taking a competitive offering to the market, with either comparable or (ideally) superior offerings and competitive pricing in order to win business.

Create

Another important role that product marketing fills is the actual creation of assets that support the promotion of your service or offering. Product marketing should lead the charge when it comes to the messaging and positioning associated with your product. This messaging should be different from the messaging and positioning of the company overall, which established the foundation of all of your marketing collateral. Rather than competing with or replacing the corporate messaging, product messaging and positioning should build upon the foundation and provide specificity to dive deeper into the "what" and "how" of the product or solution you're providing.

That messaging then leads to the development of external promotional assets. These could include data sheets, PowerPoint slides, website content, event and webinar content, and demos. Each of these outputs provides a different way to build upon and articulate the product messaging. For example, PowerPoint slides and data sheets are ideal for highlighting value propositions, differentiation within the market, and pricing information. Webinars and demos are perfect for showing off how the product works and showing use cases potential buyers might encounter.

As these assets are developed and executed, you're then set up to launch promotional campaigns to get the assets into the hands of your target audience. From email campaigns to digital advertising to one-to-one outreach from sales development representatives, leveraging your resources across digital, content, and sales development functions will maximize the reach of your assets, rather than simply leaving them in the hands of the sales team for use.

Analyst Relations

Speaking of resources, leveraging industry analysts can really help boost your performance in the market, especially for B2B vendors. You're likely familiar with Gartner, Forrester, and some of the other large firms of the analyst world. Analysts focus on a relatively narrow segment of technology or industry, and they understand these segments better than anyone else. Many businesses rely on industry analysts to keep them informed of industry trends and to inform their buying decisions. Establishing relationships with industry analysts provides weight and feedback to your messaging and product development, and can help influence customer decision making. They also publish research on their segment that can have a real impact on your button line (we rode the Forrester Wave for years at ExactTarget). There are paid and unpaid ways to engage with industry analysts. Naturally, paying customers will get more attention. Regardless, it's worth considering, depending on your category, stage, and budget.

Enable

The last product marketing responsibility to address is the need for centralized enablement of the sales and service teams. Internal enablement can easily become an afterthought, especially in the rush to build a product, set the pricing, build the collateral, and activate the sales engine, but ensuring that time is taken to get the sales and service teams fully up to speed on the product or new offering is a huge component of product marketing.

For sales reps, in order to be successful in selling, they need to know a lot, including:

- what the product or offering is;
- what problem it is solving for the buyer;
- what pricing and packaging offerings are available;
- what discounts they're able to provide;
- what questions or concerns they need to anticipate.

If product marketing can provide clear and concise information to address each of these questions, along with the created promotional assets like data sheets, presentation slides, and demos, the sales team will be in a much stronger position to successfully sell. Without this critical enablement, the message being delivered could be inconsistent, incomplete, or, worst of all, inaccurate.

Similar to the need to enable the sales team, it is also important for product marketing to enable the services team. While they need to know all the same information as the sales team—especially for those in a position to cross-sell or upsell existing customers—they could also need technical training when it comes to troubleshooting or resolving issues related to the product or offering. Training requires close collaboration between Marketing and Product so that the right information and training are delivered to the services team prior to its rollout. Without comprehensive training the services organization is ill-prepared to address customer needs, questions, and concerns.

Chapter 49

Marketing Operations

Marketing is maturing and has become more complex over the years. The evolving marketing technology landscape, access to more data and channels, and increased value of reporting and marketing ROI have led to the creation of marketing operations as a role within marketing. This development, while subtle, is increasingly common for companies of all sizes and industries. For many companies, marketing operations are the backbone of the marketing organization and that individual or team enables Marketing to increase efficiency and drive results for the team, and ultimately, the full organization. It builds the foundation for your marketing strategy through metrics, process, technology, budgeting, and reporting. Individuals in marketing operations require different skills than a traditional marketer. They are still strategic and creative, but they also need to be analytical and technical, with solid project management experience. Because operations is deeply ingrained across all of marketing, we've touched on several elements already in this chapter. For this section, we'd like to explore four pillars of marketing operations: (1) technology; (2) process; (3) measurement; and (4) strategic planning.

Technology

We discussed the evolving and vast marketing technology landscape. Marketing Operations is responsible for managing the marketing team technology stack. What tools should be used and when is the right time to introduce them? Marketing operations manages the vendor relationships, selection process, implementation, and training. They keep track of the different technologies within marketing, and cross-departmentally, to look for opportunities to integrate or to eliminate redundancy. Because they also help manage the marketing budget, they will have a good idea of when technology spend is becoming disproportionate to overall marketing investment, and when adjustments need to be made.

Process

Efficient process management is always key to scaling and growing, and operations will own and streamline processes on behalf of the marketing organization. At a minimum, you can make a big impact managing processes around budgeting, invoicing, marketing automation, campaigns, and planning. With consistent processes in place and with either an individual or team maintaining a system for everyone to follow, you will be more effective and efficient.

Measurement

Establishing and tracking the key performance indicators for individual campaigns, as well as all of marketing investment, are some of the most important aspects of the marketing operation role. This role needs to ensure that there is alignment across the department on which metrics are tracked, the process for doing so, and sharing out and reporting of those metrics. The marketing operations role also will create and maintain dashboards, scorecards, decks, and reports. Good marketing operations teams not only report on metrics, but have a keen eye for interpreting the data and making recommendations for improvement.

Strategic Planning

Due to a deep understanding of technology, process, and measurement, marketing operations is in the perfect position to be key in the strategic planning for a marketing department. Marketing ops understands lead flow, conversion trends and they have deep visibility of historical campaign performance. Not only can they help manage the monthly, quarterly, and annual planning process, they can make valuable recommendations rooted in data to improve the success and efficiency of your marketing plan.

When to Start Marketing Operations

We've referred to Marketing Operations as a team, but it will likely start out as an individual, such as a marketing operations manager or a director of marketing operations. So, when do you know when this role is needed? For startups, your marketing leader will perform most of the marketing operations duties. However, scaling companies that are starting to see complexity in their technology and channel mixes may have an opportunity for improvement through a marketing operations function. Things like underutilized marketing automation, siloed analytics or conflicts between Sales and Marketing are red flags that indicate a person devoted to marketing operations is needed.

Chapter 50

Sales Development

For B2B companies, your sales development team can be the heartbeat of your organization. This team is made up of your inbound group, who is responsible for nurturing and converting warm inbound leads (marketing-driven), and your outbound group responsible for cold outbound prospecting. Sometimes, this is a hybrid function where the same people are responsible for both inbound and outbound. The sales development team is usually the first live interaction a potential buyer will have with your organization and it's crucial to the success of your revenue team. Although the sales development team members carry many titles like inbound sales representative (ISR), lead development representative (LDR), sales development representatives (SDR), business development representatives (BDR), account development representatives (ADR), or many others, their purpose is always the same. They will qualify a lead's likelihood of purchasing, determine whether the lead is worth pursuing, and pass ownership to the sales team at some point depending on the handoff process. Their goal is to create pipeline for your sales reps.

Sales or Marketing?

You may be asking yourself why Sales Development is under the Marketing section of this book, Sales Development usually sits under Sales, right? According to the Bridge Group, 65% of sales development teams do reside with the Sales Organization. But it's increasingly common for Sales Development to reside under Marketing, and inbound teams are roughly twice as likely to report to Marketing compared to hybrid or outbound groups. Most of our experience has been with both inbound and outbound teams reporting to Marketing, and it's a structure that has several advantages.

First, the inbound team is the beneficiary of one of Marketing's most precious resources: inbound leads. Having clear communication, both ways, between the inbound team and the marketing team on lead sources and response is super valuable here. The proactive "what's coming next" and "why it's important" type of information to the inbound team, and the feedback of "how it was received" and "how it can be improved" back to the marketing team are basic data but important to collect. While you can certainly get this information across teams, it's likely going to be more complete and easier to collect with inbound and Marketing working collaboratively.

Second, if there are dedicated inbound and outbound reps, consider putting them on the same team. Inbound and outbound reps are measured similarly, require similar training and motivation, and there is plenty of natural career pathing opportunity to leverage between the two groups. As an added benefit, it's easier to manage one team versus two, and the opportunity for cross-learning is significantly higher if in- and outbound reps are in one group.

Third, both inbound and outbound teams should always be leveraging marketing resources like upcoming events, content, campaigns, case studies, etc. While the same can be said for Sales, the main responsibility of Sales Development is qualifying prospects, and the best way to ensure this qualifying is impactful is to have both teams be super close to the marketing calendar.

Finally, as we've iterated over and over, the relationship between Sales and Marketing is really important. Sales Development is an essential bridge between Sales and Marketing and facilitates plenty of healthy collaboration. Sales Development can also serve as a fertile breeding ground for future sales reps, thus making the bridge between the two teams even stronger. It's worth noting that while good sales development reps often make good sales reps (a very common career path), it's not the only path for SDRs. We've seen good sales development reps be successfully promoted to just about every department across an organization including Sales, Marketing, Customer Success, Sales Enablement, Operations, you name it. Except Legal and Engineering! We haven't seen any SDRs go to Legal or Engineering ... yet. A high performing SDR knows how to put in the work, knows the value proposition better than anyone in the organization, knows how to collaborate across teams, and knows how to talk to and convert prospects. What's not to like?

Sales development can live successfully under either Sales or Marketing. Ultimately, ownership of Sales Development should come down to the most capable department, and you should ask the questions, who has the bandwidth, knowledge, and passion to lead this team most effectively?

Leadership

Your Sales Development leader is one role that you absolutely need to get right because the success of the team is dependent upon it. We were fortunate to work with Christy Weymouth, one of the best Sales Development leaders there is, at ExactTarget and Return Path. She embodies the best characteristics of any leader, and the Sales Development teams she builds are world class. Being an SDR is hard, really hard. Grinding it out all day, connecting with one out of 30 calls, and being rejected over and over again will take its toll on anyone. A good SDR leader makes this job exciting, rewarding, and even fun. It requires constant coaching, training, motivation, and celebrating success. A good SDR leader is someone who can communicate effectively across departments, especially Sales, and resolve conflict quickly. Perhaps most importantly, the SDR leader requires someone who can manage a never-ending rotation of individuals moving in, out, and up the organization while still hitting team targets. The onboarding experience for the SDR team needs to be a machine. You'll have to hire quickly, train effectively, and leverage all your resources including your existing team to get new hires to ramp as soon as possible. Then, you'll have to keep them engaged and learning while producing. The onboarding, training, coaching, and other tasks all need to happen during a short period of time before an SDR moves on to their next role. The average tenure of an SDR is about 1 ½ years, so it's one of the highest turnover departments of every company. Ideally, as noted above, your SDRs are moving on to other roles in the company!

Where to Start

Before you spend the time and resources to start up a sales development team, or before you just "put a person in there," you'll first need to figure out if it's even the right time to start a sales development team. You might not need sales development for some time if, for example, you have the capability to route leads directly to the sales team, or if you can get Sales to spend more time prospecting (which can be really healthy). Take a good look at your sales team and the sales process. How much time are they spending on creating opportunities versus working deals? Is the sales process conducive to a handoff without losing too much momentum? Is your messaging and value proposition solid enough to build a repeatable process and train a new and more junior team to execute? If the answer is yes, it's likely time to start a sales development team.

You'll then need to determine the reporting structure and nature of the role. We suggested that Sales Development can report to Sales or Marketing but remember it's bandwidth, knowledge, and passion that are most important. After you figure out the structure, you'll need to focus on the nature of the role. Here you'll want to consider generating pipeline through inbound lead qualification, outbound prospecting, or a combination of both. You may have enough inbound leads to start there. Or perhaps you'll want to start a dedicated outbound role as well? Or maybe it's a hybrid role that does both? We believe there is power in focus and repetition for this role, and we would suggest dedicated roles, but your business, sales team, and needs will dictate that.

Then, make sure you have alignment with the sales organization. What are the goals and expectations? Is your sales team generating meetings or qualified opportunities? What is the ratio of SDRs to sales reps and how are they aligned (by territory or segment, for example)? What is the handoff process? This one is important. Be sure to build out clear definitions for an official handoff. What is the process, and who is responsible for what if there is a no-show, or if an opportunity isn't quite ready? Shared and consistent processes, goals, and expectations early on will save a lot of headaches down the road. Remember, the success of each team will be dependent upon the other team.

You'll need to finalize compensation plans. This could be a whole chapter in itself, but a base salary plus variable plan based on both number of meetings or opportunities and quality by conversion to pipeline or close is common. Consider a way to incentivize team performance and align with sales success metrics as well. Compensation plans will change as the team and business evolve, so build with that in mind.

You'll need to build your technology to support the team including your CRM, sales productivity tools, call software, and collaboration tools. Define your KPIs. What is the behavior and channels that you expect to be most effective? And by "channels," we mean broadly everything: calls, connects, emails, LinkedIn Inmail, social. You'll likely need to experiment, so be ready to test and measure the effectiveness of each of these over time, and make adjustments based on what is working and what is not.

Building a highly effective sales development team is hard work, but can be so beneficial to your organization in many ways. Strive to build a solid foundation early with the right leadership, and with strong alignment between the company and sales leadership on expectations. With this foundation you will get on the right track. Remember that the work this team does is incredibly hard, and they are often under-appreciated. Invest in training and development, provide them the tools and resources they need to succeed, and celebrate their success. You never know: today's SDR may be your future SVP of Sales!

A Note on Working with Agencies

At some point in your career as a startup CMO, you will find yourself hiring and working with some form of agency: digital, advertising, PR, or some kind of hybrid agency. Agencies can be the biggest boost to your team's capabilities and capacity and drive transformational thinking and work for your marketing efforts. They can also be an enormous waste of time and money.

In days gone by, traditional ad agencies simply charged clients a percentage of their ad spend—and that spend was so large (TV commercials, etc.) that the charge covered all the agency's expenses. PR agencies almost never billed that way and stuck to more of a monthly retainer model. What both of these models have in common is that they were very hard for clients to decipher and trace back value. But as with everything in the last 20 years, all that has changed. If you're a startup working with an agency, you're almost certainly paying them on a "time and materials" basis—and possibly a retainer, but one that is closely tracked back to activities.

Whole books can be written (and have been written!) about working with agencies, so we'll boil down our thinking here to three simple rules we have to maximizing your agency relationship, regardless of the type of agency:

1. **Treat the agency like a close partner or employee.** The more time you invest in the relationship, the more you get out of it. If the people working on your account at an agency are invested in your company and your brand—if they see your quarterly all-hands meetings or your Board Books as if they were a trusted senior member of the team and therefore feel like a trusted senior member of the team—they will do higher quality, more relevant, and likely greater quantity of work for the same price.
2. **Use the agency to gain a fresh perspective.** Even if you follow rule 1 above, your agency team is not, in fact, employed by you full-time. Even long-term agency relationships rotate people on and off accounts from time to time. You and your permanent team are very close to your brand, your business, and your customers, but that closeness can occasionally be myopic. Listen closely to your agency, even if they are a little bit incorrect in their word choice or assessment of a specific situation. They are taking into account a much broader set of data about the world, with less bias, than you are. About everything.

3. **Use the agency for competencies you can't hire in-house.** Whether those competencies are in a "peak burst" (e.g., you are doing a website redesign which will require 3x your current in-house capacity but only for 3 months), or those competencies come in the form of expensive specialists that you don't need or can't afford on a full-time basis (e.g., brand refresh creative development or crisis management PR team), this flexible use of talent is a much more cost effective and rapid response way to go than hiring full-time talent. If all you're doing is using an agency as an "arms and legs" extension of your team, you are probably paying too much money for too little relative output.

Chapter 51

Marketing as a Partner/Collaborating with the Rest of the C-suite

We've talked a lot about marketing and collaboration. Marketing requires deep collaboration with all other functional leaders. Sales, Product, Customer Success, Finance, Business Development, HR—you name it and Marketing plays a key role. These departments and leaders understand the value of Marketing and they've experienced the benefits of having a deep partnership with Marketing. For example, customer marketing and partner marketing focus will only happen with deep collaboration with your Chief Customer Officer and Head of Partnerships. The customer and partner marketing manager positions themselves may come to fruition in the first place because these individuals identify the need. Individuals who perform these roles may report to Marketing, but spend a significant amount of time with the customer or partner team. This isn't going to work unless the leaders are on the same page, and are collaborating regularly and efficiently. In today's fast-changing and largely digital world, other functional areas realize that they need Marketing to be successful. And it works both ways: the marketing department is only as healthy as its relationships with other teams. Because Marketing is critical across the organization, the marketing leader will have first-hand knowledge and a deep understanding of priorities across the company. We get to see the "bigger picture" when it may not be so obvious for others to see. We also have the opportunity to extend the reach of an initiative across departmental aisles, to broaden the scope of a campaign, or to surface the success of an otherwise isolated idea, often resulting in a snowball effect. The marketing leader has a big seat at the executive table, and a unique perspective that can add value in multiple ways.

Also, remember that Marketing is a shared service and can sometimes make the biggest impact when acting in service to the rest of the company. No, we're not suggesting that we're order takers, but we can lead the charge of demonstrating what healthy collaboration looks like. And we're in the perfect position to do so. Today's marketers are creative but also analytical and methodical, we're both left-brained and right-brained. We're responsible for the company's brand, generating demand, and for helping to maintain a healthy culture. It's impossible to do this alone. As the CMO, you'll want to start establishing healthy relationships with leadership, continue building healthy relationships throughout the rest of the organization, and demonstrate how to collaborate broadly to your team so that they do the same. Finger pointing is not an option.

What I Look for in a Chief Marketing Officer

Brad Feld

Foundry Group partner, Techstars co-founder

Let's get this out of the way: I despise the word "marketing"— it's often the weakest link in a startup company. "Marketing" is vague and non-specific, often poorly executed and measured, and usually a colossal waste of money relative to the output. There are plenty of classical approaches that any marketing consultant would be happy to charge you lots of money to explain. However, practical marketing strategies have evolved in the past decade, especially as user-generated content, online marketing, and the idea of building a brand have become ubiquitous.

For a while, I asserted that marketing was all about "thought leadership," prompted by an email from my business partner, Chris Moody, who wrote:

> Don't do marketing. Focus on becoming a thought leader in your space. Talk every day with your customers, prospective customers, partners, and the world about why you do what you do and why you think it is important. The reality is you can only talk about what you do one

> or two times before people think "got it" and stop listening. But, if you talk about what you believe and point to countless examples that exemplify your beliefs, you can build real engagement with people who care/believe the same things.

Would I look for a Chief Marketing Officer who is a "thought leader?" Nope. That term, like "user-generated content," has become omnipresent and the meaning diluted. How many thought leaders are there anyway? While you can probably find one, they aren't going to manage your marketing organization at scale. Instead, look for someone with a point of view that can resist the marketing fad of the day. They should be driven by an obsessive focus on the customer and the product, rather than a marketing budget, initiative, or brand. If the person you are talking to about the role leads with phrases like "social media marketing" and "marketing spend," or talks about "refreshing your brand," be skeptical.

Instead of talking about marketing, a CMO should focus their energy on your customer and your product. They should be a key, integrated part of your leadership team, working side by side with your CRO and your CPO because many tactical marketing initiatives are regularly changing. Any CMO can manage a team executing them. The great CMOs understand why your customers care about what you do, and they talk to them more than anyone else.

Chapter 52

Building a Marketing Machine (Scaleup)

As your company transitions from startup to scaleup, your marketing organization should be in lockstep with that evolution. This is the time to build upon the foundation you've laid and level up the work being executed across the marketing team.

Growing Your Team

As your marketing output grows, so will your marketing team. While a startup marketing team may have a few individuals who wear many hats—for example, a single individual who handles both brand design and digital marketing execution—as the need for more marking work grows, it will be important to identify where responsibilities can be shared and built upon in order to make the team more successful. How to scale your marketing team will be very much driven by the needs of the company and requires the marketing leader to have a pulse on the work being done today as well as where marketing can make more impact in the future. Marketing team growth also depends on the opportunities to increase output with additional people and it depends on the aspirations of the individuals on the existing team. How would they like to spend their time? What are they passionate about and want to dive deeper into? What do they dislike and avoid? All of these elements—company needs and existing marketing team members' needs and skills—factor into the roles you add and how you structure the team.

When it comes to growing your team, it's important to keep in mind that the best talent can be found in unexpected places. One of the great benefits of marketing is that, quite honestly, it isn't rocket science. Most people are pretty well equipped to learn the basic functions of any marketing role. What really sets some marketers apart, however, are the intangible qualities that aren't necessarily honed in marketing-specific roles. Qualities like creative thinking, problem solving, willingness to work outside of title, interest in collaborating, and strong writing and communication skills are not taught in a marketing program, for example, but those are the qualities that make people effective in marketing roles. They're also the qualities that can be found in anybody. So, for example, if you're looking for a content marketer, you don't need to limit yourself to only people with past content marketing experience. Instead, journalists can be a great fit for these roles because of their ability to work against deadlines, communicate complex topics in a concise manner, and write in a variety of different styles or formats.

From personal experience, Nick's career started in video production in a role that was essentially project management. In that role, however, he managed project timelines and budgets, worked directly with customers, managed the workload of team members across multiple projects, and did a lot of storytelling. He didn't know it at the time, but it was all setting him up to be an effective marketer. Luckily, his future boss was able to see that even though he'd never been in "Marketing." He had the right skill set to lead a brand team, essentially launching his true marketing career.

Execute Across Channels

As your team and resources grow, your ability to execute campaigns across channels more seamlessly grows as well. The "omni-channel" experience—having a consistent brand and message "everywhere"—is what every marketing team aspires to create. With limited time, budget, and personnel, sometimes that fully realized vision can be a challenge to achieve As you scale, beginning to elevate the definition of a successful campaign allows you to push your team and your marketing output to new heights.

When you're thinking about a cross-channel campaign, every function of your Marketing team should have a role. More than that, each function should be in full collaboration with every other function, either owning, leading, or contributing to each activity within the campaign. Take a product launch campaign, for example. Your "Go Live" date would align with the chosen launch date of the new product. From that date, your marketing team can plan out everything to happen leading up to, on the day of, and after that date, in order to launch a fully cross-channel campaign. Launch Day may include a new webpage, email marketing campaigns triggered to customers, prospects, and partners; it might include the launch of new blog content and/or thought leadership content, promotional messages across social channels, coordinated outreach to prospects from the sales development team; it might also include a training webinar for current customers, and an in-person event in a key market. All of these activities would then be tracked back to a centralized dashboard with KPIs to measure engagement and impact across audiences and channels (Figure 52.1).

At Return Path, while we took an omni-channel approach to every product launch, our most exciting, collaborative, and far-reaching campaigns were often our annual brand campaigns. We weren't promoting a specific new product or offering, but we were focused on driving brand awareness and evolving our brand at least once per year. These campaigns required close collaboration across all functions of the marketing team, coordinating content launches, one-to-one and one-to-many outreach, in-person and virtual events, and new digital experiences. These campaigns allowed each function within Marketing to shine independently while being supported by the work of all the other functions, and they truly demonstrated that the whole is more than the sum of its parts.

Not only were these brand campaigns exciting challenges for our marketing team, they had a significant impact on Marketing's overarching goal of driving pipeline and bookings. Each year our data showed that these campaigns generated a substantial number of new leads, influenced revenue within existing business, and consistently increased brand awareness. In fact, our 2016

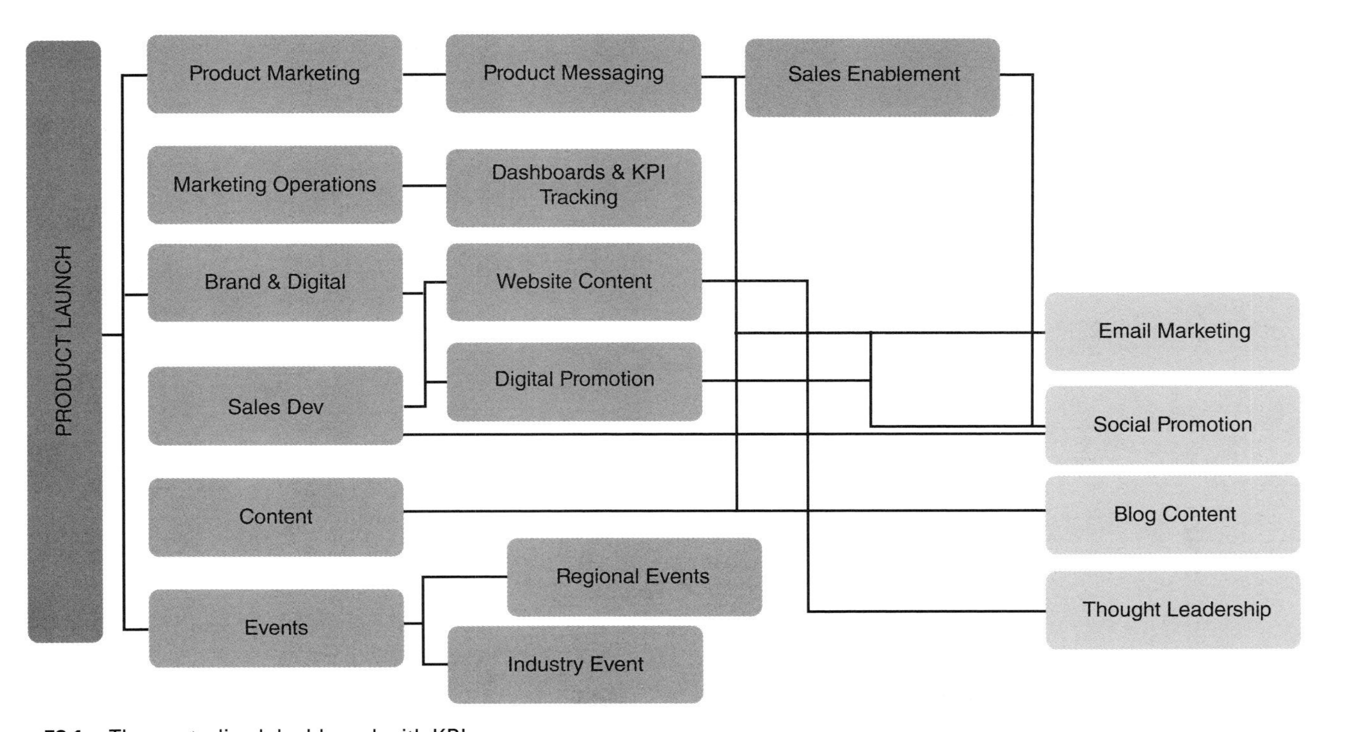

Figure 52.1 The centralized dashboard with KPIs.

campaign delivered a 10% increase in brand awareness over our 2015 campaign.

Broaden Your Reach

If you've scaled Marketing in lockstep with your company, you've added significant talent to your team, you're firing on all cylinders with your campaign launches, and you're seeing quantifiable results between marketing and revenue. What comes next? Global expansion. If your scaleup is already moving at the pace to be making significant marketing investment and seeing that investment pay off through innovative, successful marketing campaigns, chances are that global expansion is already a topic of conversation at the executive level.

When it comes to breaking into new markets, localization is key. And we don't just mean translating your English content into French, Spanish, German, or other languages. Each market is unique, with their own cultures, ways of doing business, and processes for evaluating purchases. It's critical for your business and your marketing team to have a presence in-market to influence what needs to shift from your existing US-centric approach to be successful in a new region. Here are a few ways a marketer in your target regions can positively impact global expansion.

- **Accurate translation.** Google Translate certainly won't cut it, and even a stellar translation agency can still fall short. It's hugely important to have a resource in-market who can ensure that all the hard work you've put into developing a particular brand voice makes sense in an entirely different language. A qualified local person can also help ensure that your brand and message are applied consistently across campaigns and channels, in the same way you'd expect consistency across your English language work.
- **Local examples.** Most global markets don't just want to hear about the great work you've done in North America. You may have to start there as an entry point, but it will be key to get local examples as soon as possible. Local examples will help significantly in gaining legitimacy in the region and

demonstrate your awareness of their particular challenges, beyond just their business size, industry, or business challenge.

- **Cultural insight.** An event that works for a French audience won't necessarily work for a Brazilian one, and vice versa. Not only is it important to have a local resource who can provide insight into what works or what doesn't, it's equally important to *trust* that person to do what they know to be right. Forcing an approach onto a region because it worked in the US or in another country won't necessarily breed repeatable success. Instead, it could work against you and highlight a lack of understanding of them, their country, and their culture.

Final Thoughts

From organizing a marketing team, to tactical execution, to collaborating across the business, we've covered a lot of ground. Whether you're a marketer, a marketing leader, a current CMO, an aspiring CMO, or even in an entirely different role but suspect marketing might be for you, hopefully you've found something valuable here to take away with you. Marketing, when it's done right and taken seriously, can be a game changer for a business, especially a startup. We've witnessed first-hand at multiple companies what Marketing can do to shape a company, drive revenue, and impact how the business evolves.

So, what is marketing, really? Marketing is everything. Marketing is the big ideas and the small actions. The art of the possible and the science of the attainable. The pretty pictures and the cold, hard facts. It can and should impact every part of your business. Sometimes that influence is big, and sometimes it's small, but there is always a role marketing can play to advocate for the customer, bring different perspectives together, drive the narrative of the business, and both define and deliver the promise of your company.

That makes marketing an incredibly exciting place to be. One of the true joys of marketing is that the possibilities are endless. There are countless best practices to follow and proven ways to get the job done, but also a million sideroads you can take or small changes you can make to do something more efficiently, inject more creativity, or reimagine it entirely. With an unending stream of new technologies, creative talent, and innovative experiences out in the world, marketers are the beneficiaries of a world full of inspiration.

If we can leave you with one final piece of advice, it's this: Marketing should be fun. If it's not, you're doing something wrong. Yes, there are goals (sometimes big goals) to hit, like generating opportunities, pipeline, and revenue. These are critical to the business and the primary objective. But at its core, marketing is about creating connections.

We are real people creating real relationships with other real people. The human element is undeniable in everything we do. Lucky for us, no one's lives are in danger if an email goes to the wrong list or a research report falls flat. We can fix, update, replace, or remove our mistakes in pretty short order. Without that pressure, we're able to create an environment within our marketing teams to foster freedom and creativity. Marketers will always do their best work when they're having fun. If you can create a space where the goals are clear but the pathway there is open to debate, where all input is welcome and no ideas are bad ideas, and where everyone can bring the best they have to offer to the table while continuing to improve in areas for growth, you'll have created the perfect environment for marketing magic to happen.

Chapter 53

CEO-to-CEO Advice About the Marketing Role

Matt Blumberg

What comes before a full-fledged CMO? In most startups, there is at least a medium-sized and quite busy marketing department with multiple mid-level leaders well before there is a seasoned leader at the helm. One of those leaders may be a VP of Marketing—depending on the nature of the company, it is likely someone with a specialized area of focus within Marketing (brand, digital, event, etc.) who has some working knowledge of the other areas.

Signs It's Time to Hire Your First CMO

You know it's time to hire a CMO when:

- You wake up in the middle of the night concerned that no one in your company but you knows how to orchestrate a successful product launch.
- You are spending too much of your own time managing smaller pieces of marketing because your marketing leader isn't experienced enough across all of the function's many sub-disciplines.
- Your Board asks you how you would spend an extra $2m in marketing if you had it—or what you'd cut if you had to reduce your marketing spend by 50%, and you don't have a great answer and aren't sure how to get to one.

When a Fractional CMO Might Be Enough

A fractional CMO may be the way to go, if you have a generalist marketing manager or director who has strategic inclinations but not enough experience operating as a strategic executive and who just needs a little more supervision in order to "level up." Or if you have a series of more junior leaders of marketing sub-functions, none of whom is experienced enough to coordinate activities across groups but none of whom require a full-time leader.

What Does Great Look Like in a CMO?

Ideal startup CMOs do four things particularly well:

1. They understand that the marketing budget starts with drivers and business results and works backwards in a modular way to spend, not the other way around. They understand what the business needs to achieve—the sales plan—then what the funnel looks like. Then they know what marketing levers they can pull to both optimize the funnel and make sure the funnel is full. And they build the plan in a modular way so that if the budget needs to be trimmed, they can ask the right questions and easily trim it.
2. They make spend decisions on data, not on a hunch or because "that's what's always worked." Even in traditional B2C businesses that make heavy use of traditional non-addressable media like print, outdoor, and TV—today, everything can be tested and measured to some degree. A strong CMO is one who starts every answer with "Let's look at the data."
3. They behave like a CEO in terms of being able to orchestrate the different pieces and parts of their organization. Just as a CEO has to manage a litany of disparate functions, so too do CMOs have to manage a litany of disparate channels. Gone are the days when CMOs were either "brand or direct" or "online or offline." Today, the average CMO has

to be able to manage 20+ different channels. The level of complexity and number of points of failure for the job have exploded. A great CMO handles this with the fluidity that the CEO handles moving from a sales pipeline meeting to a product roadmapping exercise.

4. They spend time in-market and in-product. Given all the responsibilities around multi-channel orchestration, systems, budgeting, and execution in general, it can be very easy for a CMO to operate 100% from behind the desk. The great ones want—need—to be out in the field, attending sales calls, partner meetings, events, serving as executive sponsor on some key accounts, in general, collecting primary data on the company's products and brand.

Signs Your CMO Isn't Scaling

CMOs who aren't scaling well past the startup stage are the ones who typically:

- Treat all tasks like French fries (see the Part Four textbox). Wait, what? Marketing can easily be a service center as opposed to a strategic function. I don't think that's ideal, but that may be how a company decides to run it. You're never too full to eat one more French fry. You may be too full to order a plate of fries, but not one more individual, tasty, crunchy, salty, fry. Marketing at its worst can be the same. There's always one more task to be done on the long list of tasks. CMOs who focus on task execution (eating the next fry) and can't pull up to think about whether they're doing the right thing (should they be ordering another plate of fries?) are simply not scaling.
- Report on activity as opposed to outcome. This is related to my prior point. When all the world is a task list, then report-outs are just volumes of tasks. I'm not sure why Marketing ended up like this, but it's frequently the only function in the company that spends time producing beautiful reports on all the stuff they do. It probably comes from years

of working with agencies who report like that to justify client spend. Regardless, can you imagine seeing reports on activity instead of outcomes from other departments? The report from the CFO that talks about how many collections calls the team made as opposed to reporting on bad debt—no, thank you. The report from the CRO talking about how many meetings a rep had with no mention of pipeline or closes—seriously? CMOs who can't link activity to outcome with a focus on outcome are not scaling with the job.

- Spend disproportionate amounts of time on creative or agency work. That's the glamorous and fun part of marketing, for sure. Having made TV commercials as a head of marketing when I was at MovieFone, I can attest to that. But even if you're a big B2C marketer with a lot of agency and creative spend, while you should be supervising that work, spending all your time on it is a sign that you're not interested in all the other, well, French fries.

How I Engage with the CMO

A few ways I've typically spent the most time or gotten the most value out of CMOs over the years are:

- With the rest of my go-to-market executives as a group, not in a silo. This is even more important than it is with respect to other GTM roles like Sales, Account Management, and Partnerships. While of course I have always had one-to-one meetings with my CMO, I find that the most valuable conversations are the ones with the GTM group as a group, talking about shared objectives and the underlying drivers and coordination points to get there.
- Thinking sessions where we take time away from the day-to-day to do deep dives on strategic topics like the company's positioning, voice, or brand. Sometimes I like to do these in the context of reading a relevant marketing book or business journal article, sometimes not. I find that the most creative thinking and ideas—and even the quantitative part of marketing involves a lot of creativity—happen in some of these longer form, unstructured conversations.

- Before they become the CMO. For years, we went through CMOs at Return Path at the same clip as other companies—one to two years—and we had a pattern of hiring them in from the outside. Over time, though, we realized that we would be much better served by having more continuity in Marketing by grooming people and promoting them from within. The last few CMOs we had at Return Path were all promoted into the role—so I got to know them pretty extensively ahead of time and was not only thrilled to give them a shot at the top job but I was in a great place to understand their strengths and weaknesses coming into the role so I could most effectively mentor them. The same could be said of other functional departments, but Marketing is more acute, based on the average tenure of CMOs.

Part Five

Sales

Anita Absey

Chief Revenue Officer

Revenue to a company is fuel to a vehicle; it's air to an organism. It's the lifeblood of any successful company. Can you get by without revenue? For a while, but sooner or later you have to figure out how to generate revenue and, more importantly, figure out why you can't. The time to do all this figuring is in the very early stages of your startup, like week 1.

You should immediately dismiss the notion that "if you build it, people will love it and buy it." It's difficult to dismiss that kind of thinking because a lot of founding teams have spent months, years, and even decades developing a product; they have put all their savings into the startup; they have tapped their friends, relatives, and possibly angel and venture capital investors. I appreciate that effort and risk, but falling in love with your technology, product, or service doesn't matter. What matters is whether or not you have something people want to buy.

I'll give a personal example. A while ago I went to buy a car and I found one that I absolutely loved. I loved the feel of it, I loved driving it, I loved the whole experience of it. When I got back to the dealership, the salesperson opened the hood and started talking to me about the horsepower, the torque, and other performance features. He lost me, and he lost the sale. Why? He didn't understand what I valued about the car and his focus on the car (the product) and its features were important to him but not to me.

As a startup, you need to understand how potential customers value your product. Does it save them time? Is it a better solution to something they need? Is it more cost-effective? The way to go about this is to get out in the world and talk to customers. In the very early stages you'll most likely explore and validate your product with the CEO and Head of Product. But to scale, you'll need to develop processes, systems, and hire and develop a team. In this Part I share ways for you to build a great Sales Organization.

Chapter 54

In the Beginning: From Prospect to Customer

There's a framework that Matt borrowed from one of our original investors at Return Path, Greg Sands. Greg always talked about the evolution of an enterprise selling process as going from "selling on whiteboard" (a largely custom and conceptual sale) to "selling with PowerPoint" (a sale that requires creativity and tailored pitches) to "selling with PDF" (a standard sale that can be taught quickly to an inexperienced sales rep and used with a high degree of predictability to all customers). Part of what that means is that, as a startup, your goal is not to deliver a polished, final PDF to customers, something that is buttoned down and refined, but to start with a mindset of discovery. As the Chief Revenue Officer, along with your CEO and maybe Head of Product, you should go to a prospect's office and literally use a whiteboard, drawing things out, drawing charts and frameworks and circles and arrows and exclamation points, while you try to understand your potential customer's problem and then go back to the office and work on building the solution they need. You're creating this with the client because you don't have a deck yet, much less a PDF. You've got your initial product, but you're still very much in discovering and positioning mode. You are an evangelist. You don't know what resonates in your prospect's mind. It can be very interactive and engaging with you selling on a whiteboard and using that very intimate moment to try to develop the right story for your product.

Then, as you evolve and grow, chances are you are going to have a sales deck and a pitch because you won't be discovering what the customer needs, but you'll have very refined (and tested) ideas about their needs, maybe even customer segments. But a caution here is that what could and often does happen is that your deck, your pitch, gets modified along the way—for every single pitch. So, if you have four salespeople, each of them has a different version on their laptop, and there's probably no central organizing body yet that has thought about what the tone and tenor of the brand should be. This stage is called "Selling with PowerPoint" because good, clever, senior, business development-oriented salespeople are most successful by creating custom pitches for each client based on their learned history of what has and has not worked in other places. You're lucky if sales reps share their knowledge and slides across the group. You are still miles away from being a sales machine.

What you want to strive for is a level of sophistication and market understanding that enables you to get to a PDF presentation. A PDF is something that's complete, that can't be modified or altered, and it ensures that everyone's speaking the same language. At this point you have the kind of consistency and message and positioning that enable you to be repeatable and scalable. Obviously selling by whiteboard—and even PowerPoint—is sufficient to a point in time but if you're thinking about unleashing your product on a massive scale, then you have to get to the point where you have a very smooth presentation and message that you know resonates with the audience.

Once you have the consistent, polished PDF, you might think that all you have to do as your company scales is to use that and tweak it a bit here or there for particular customers. That could work, but sometimes the best way to sell is not to use materials. Sometimes, if salespeople rely on that magic piece of collateral or the better deck, they actually miss the point: you want your salespeople to connect with the customer, to discover with the customer, and you want them to understand the customer's problems. A great salesperson needs to constantly hone their questioning and discovery skills and that's what you ought to look for when hiring and building your sales team.

A great salesperson also needs to be able to ask questions that allows them to understand when a "no" is really a "no." There's a common belief among salespeople that "a quick 'no' can often be better than a long 'yes'." Although there are a lot of people who do believe that, I reject that thinking. A quick "no" is an easy way out for a salesperson, especially if they're dealing with a qualified prospect. But I would say that you should only accept "no" if you feel as though you've gotten an honest response to important discovery questions, after you've had a chance to paint a picture of what success looks like.

A salesperson needs to be able to discern whether a response by a potential customer is a brush-off—or is it legitimate? Or are they getting a response like "we don't have time," or "we don't have the budget," or "we don't have the need"? Getting to an honest, legitimate "no" quickly is better than a long protracted "no," but it's more nuanced than what most people think and it requires probing and thoughtful questions from the salesperson. It also depends on the product or solution sales cycle but, as long as you're in control of the process and you can accurately forecast your pipeline, then a long "no" is fine. A warning here, though: if your sales cycle turns out to be 18 months, then even if you have 25 "yeses," you better find a way to get companies where you have higher appeal because you'll starve waiting for those to come in.

You might be thinking that you can get your sales team from whiteboard to PDF quickly, that it's a matter of understanding the process and then executing it. But the reality is that there is no quick way to get from whiteboard to PDF and it's not a linear process. You can't put into your business plan that you'll spend the first six months selling with whiteboard, the next six months selling with PowerPoint, and the next six months selling with PDF. It's much more nuanced, there's a lot of trial and error, a lot of experimentation, and a lot of thinking and rethinking based on customer ideas and feedback. At Return Path, for example, it probably took us somewhere between five to ten years before we got from whiteboard to PDF and it was only after refining our approach and materials that we were able to build a sales machine.

Chapter 55

Hiring the Right People

In the beginning, as you're discovering more about your product and solution, you'll go on calls with the CEO and others at your company and eventually, you'll do that yourself. Once you validate your model, you might think that you're ready to hire, but I would set a reasonable stretch goal in the beginning, like 25 customers using the product within three months, or even ten customers. You are the first salesperson, of course, and you won't be able to hire additional people until you really understand your products and customers. So, while you might believe that you can hire a salesperson quickly, I would err on caution and make sure you're very clear on your product, solution, competitive alternatives, pricing, and customer benefits before hiring your first salesperson.

Once you hit that target, you'll be ready to hire salespeople to help you, but who do you hire, and how do you figure out if they will be good at sales, especially if you're in a new market? You might think that hiring someone fresh out of college is the way to go, that a young person will grow with the business. Certainly, there are a lot of talented people with college degrees from first-rate universities, and I appreciate what it takes to get into a highly competitive university, the intelligence and work ethic required. But for a startup, it's more important to find people who understand the nuance of selling, who can listen, who can understand other people, and who can tell a story, a narrative. The best salespeople often are the people who have faced and overcome adversity or failure—maybe financially, maybe in their relationships, maybe in a previous job—because those people will have higher empathy for others, and that's what you need to be successful in Sales.

In terms of hiring salespeople, it's difficult to create an interview process that hits on all the important points, but there are a few points that are critical to get a handle on early when you interview candidates. I'm certainly interested in a person's expertise and experience. So, I want to know whether or not you successfully made quota, carried a bag, and in what verticals you've been successful. I want to hear about your success stories but I also want to know what you've learned from failure. I don't want to just focus on the art of selling, but I also want to know if someone is focused on the math of selling because you need both—you need the empathy and listening skills, and you need to be process-oriented and deliberate and organized in your approach to building a pipeline and moving things through the pipeline and forecasting.

It's not easy to ask questions that will direct you to those answers. But it's easy to have a conversation about things like, What is most important to you? What do you think is the most important part of the sales process? How do you manage yourself so that you can manage the process? What kinds of questions do you ask customers or prospects about business as part of a discovery?

These are very simple broad questions that if someone is really trying to be successful or has been successful in sales, they'll be able to answer, effortlessly. I say that with confidence because I've seen it when it works. And I can know in five minutes if it doesn't work.

You might be tempted to create a fake sales situation for a candidate, thinking that you'll get a better sense of their ability than by asking them a series of questions. It's very difficult to create a realistic sales situation since there's so much nuance and listening involved. It's artificial. But I have done that and can recommend two approaches that go beyond interviewing. One way to approach it is to test whether the sales candidate has done enough work researching your organization to be able to craft some potential discovery questions. So, one way that I stress test a candidate is by asking them to imagine that I'm a potential customer of Return Path (or whatever company I'm hiring for). How would they begin the conversation with me? In doing this, I can learn if they have the right discovery questions. Have they done their homework?

Another way to vet candidates beyond interviewing is to say to the person, "I have someone equally as qualified as you for this role. Please tell me why I should select you." And that very often flusters people but if you can't sell yourself with confidence, then you're probably not going to be able to sell a solution.

One final thing on hiring. I think it's important to have people from different functions in the company interview the salesperson because a salesperson needs to work well with others—the product team, the marketing team, the technology team, the finance team. You can be sure that if a salesperson is going to be demanding of these groups, which inevitably they are, then there absolutely has to be a natural sense that this person is a collaborator and not a demander and that they will bring ideas back into the organization. So, getting perspectives from different departments is critical. I wouldn't hire someone who ticks all the sales boxes but is not well received by others in the company.

Hire with Head or Heart?

In hiring salespeople, sometimes you go with your heart, not your head. Of course, you consider someone's experience and expertise and prior success—but it may be their passion, their heart, and their genuine enthusiasm that will triumph, even if they may not have the polish or demeanor you hope for. One of the best hires I ever made was very successful by sheer force of will, good humor, creativity, and persistence ... despite being a bit rough around the edges. He looked more like a disheveled radical than a business person, but he was one of the best salespersons I ever hired.

Sometimes, despite all of your assessments, interviews, and reference checks, you make a big mistake, and that is time-consuming and costly. My worst experience was hiring someone who ticked all the right boxes, and in hindsight, maybe too well, and too smoothly. We found out within a few months of hiring them that they actually had taken another job at another company at the same time and were attempting to work for both while "calling in sick" or "working remotely" half of the time. We immediately terminated them but lost time spent in onboarding, training, and quota attainment. My only counsel here is when you make a mistake, correct it as soon as possible, whether it is in an unusual situation like this, or whether someone is clearly not succeeding in the early days.

Chapter 56

Profile of Successful Salespeople

There are a lot of misperceptions about salespeople—and no wonder, there are so many types of sales approaches, and we've all experienced being "sold" that it's easy to think of sales as "convincing people to buy something they really don't want." Nothing could be further from the truth. A great salesperson has to lead with empathy and have a real interest in helping others solve their problems. This is difficult for some people to understand, especially people with a product that solves problems, but even though a successful salesperson is a solutions seller, they're not doing that from their point of view, but from the customer's point of view. A great salesperson can paint a picture of a future state and bring the prospect along on the journey by asking questions and helping them to understand impact, value, and outcomes.

I've found that a great salesperson is not a talker, but a listener, and they listen for understanding and comprehension. Yes, a great salesperson deeply understands the product, but they're not focused on features and functions, like the car salesperson I referenced earlier. Instead, they're focused on the benefit and advantages to the prospect. But first and foremost, they have excellent listening skills and even more broadly, the ability to understand visual cues from customer interactions. Does the salesperson notice that the potential customer is getting bored, is not paying attention, or is anxious? Those are cues that a good salesperson will recognize and rather than keep talking, maybe wrap up a conversation.

Sales is a process and discipline, as well as an art. A good salesperson will understand and use the "sales math" that gets you from first call to cash. That means they'll know how many phone calls they'll have to make each day to get a customer to respond; how many responses they'll need to get a customer to agree to a meeting; how many meetings they'll need to get a customer to say "yes." A great salesperson will always have this calculation in mind, they'll know where they are in the process, and they'll be able to lean in, give more effort to the right areas, so that they can hit their weekly, monthly, or quarterly targets.

There was a good hiring model we adopted at Return Path at the suggestion of a private equity investor. We looked for sales candidates early in their career, who were successful at selling commodity products, and who were likely not from top tier universities. The thought was that if they could sell a commodity, they would be able to connect with prospects, and could be taught how to sell a more complex solution. We also believed that they would fight harder to win, as they may not have had the educational pedigree that made it easy for them to land a plum role. There was an assessment that we did as part of the interview process, and if the candidate passed that (a must), and did well in the interview process, then they were likely a good hire. This does presume though, that you have a sales trainer and training program to help them develop solution-selling skills. This proved to be a good way to "grow our own" team, and took some of the risk out of the onboarding process and improved the time-to-value period.

If you can create a model of the qualities needed to be a successful salesperson in your organization, that is the surest way to hire the right people. If you can't (yet) create that model, then I suggest searching for people who have high integrity, a zest for success personally and professionally, perseverance, and persistence. These are critical traits that I look for in potential candidates and salespeople with these traits can be successful in most situations.

One word of caution: the characteristics that define a successful salesperson can be very different than what is needed in a Sales Manager or leader. A successful salesperson has to be an excellent communicator and listener, self-motivated (and self-interested to some extent), and inspired by the chase and the win. The sales manager should have those skills but also the ability to coach and mentor, to manage performance, and provide strategic guidance. Those additional qualities are critical in the successful sales manager. A leader will set the tone and tenor for the organization, embodying the vision and mission for the team, and inspiring high standards of performance and accountability.

Chapter 57

Some Myth Busting

Here are some common misperceptions about salespeople:

- They are only motivated by money.
- They sell primarily based on relationships, dinners, drinks, and false promises. (I say never let the relationship get in the way of revenue.)
- A successful salesperson is the lone wolf.
- Successful salespeople are extroverts who have to be in control of the conversation.
- The successful salesperson talks endlessly about product features and functions.
- The successful salesperson has a deep, extensive, rolodex of contacts and customers.

All of these ideas listed above are not true of great sales organizations—although they are true of some people, some cultures, and some organizations. But for great sales organizations, they're misperceptions, they're ideas that people have about salespeople. As a CRO, if you believe these ideas, then you'll recruit based on them, and that's what you'll populate your Sales team with. And then you'll naturally talk about salespeople as extroverts motivated by money and needing to control the conversation. That's not a recipe for success, that's not conducive to a successful sales culture. A successful sales culture, which I talk about in a bit, is one of sharing ideas with others on the team, mentoring others to build a stronger team, listening to the customer rather than telling them about products and features, and building a relationship of trust with the customer through empathy.

It's likely within a startup that other members of the executive team will have these same misperceptions of salespeople and one of the first things you can do is to educate them on what a successful sales team is. If someone on your executive teams says something like "Well, salespeople are only motivated by money," you'll have to be direct: "No, that's a misperception and I'm building a sales team culture that will out-perform any team focused only on financial motivations." The time to set the stage for the culture you want to build is Day 1, not 18 months into your startup, and in doing that, you'll set your company up for growth.

Chapter 58

Compensating Sales Team Members

Of all the myths surrounding salespeople, the one that is most widely accepted is that salespeople are wholly driven by money, which is not necessarily true. But compensation is a big part of building a sales team and it's one of the areas you'll have to think about early in your startup and revisit over time as you scale. I've seen two things occur in startups that you have to watch out for. One is that you hire someone who's willing, and enjoys the chase. You know they're motivated and they'll be aggressive but there's so much uncertainty that it doesn't make sense to have commissions. So, you pay that person an inflated salary, noting for them that part of their salary will eventually be turned into an incentive compensation. The challenge there is that it's very hard to dial back money once you put it out there. And even if you're explicit about it, even if you have it written down, the salesperson is likely to view any reduction in salary as a negative. So that is not my preferred form.

My preferred form of incentive in a startup is to give someone an aggressive salary, something that is at or above market. You want (and need) to get the best salespeople in your company, so pay for it. Don't try to go in on the cheap. That never works. Pay for the talent you think is going to help you be successful, based on market rates, and then pay a high flat commission rate on every single dollar that comes in the door. So, my recommendation is always the same for a pure startup: when you're just getting out there, $0–10 million in revenues, pay everyone a commission on every single dollar.

$10 million might be too high for many companies, but pick your number so that it's appropriate for your finances, cash position, speed of the sales cycle, or other factors unique to you. But sales incentive compensation is critically important. It's one way to absolutely influence the behavior of someone on your team. As you get a little more mature, I think a leveraged plan where you pay a high variable rate and a lower base salary works well. I always like to base compensation on a quarterly target because it's easier for people to have something in their line of vision. And as the CRO, it allows you to adjust for seasonality as appropriate. So, if my quota is a million dollars a year, I'm overwhelmed, but if I have to hit $250k this quarter, I know exactly (if I know my sales math) how many calls I have to make and how many meetings I have to set up to get to close, based on the conversion rate. That's an easier number to think about!

With a quarterly plan you'd want to pay people increasing amounts based on them getting closer to their target. Maybe you'd have tiers or bands of $0–25,000, $25,000–50,000 or whatever is appropriate, but you'd pay them a higher commission rate as they go through those tiers. So, they're really rewarded for getting that business and for pulling things into the current quarter. Once your salespeople meet 100% of their target, it's appropriate to offer additional incentives based on the incremental dollars that come in, above goal.

I recommend having incentive plans with no caps, so that you do not diminish the earning potential of anyone on the team. This rewards attainment, and emphasizes the fact that the company is willing to pay for high performance, regardless of how high the commission payout may be (See also Chapter 18).

I would caution against applying incentives that encourage reps to pull deals into a quarter at a discount, for example. This diminishes your value proposition, and sets the expectation in the market that you will play on price to close business.

There are obviously a lot of ways to compensate and motivate salespeople, but I've found that having a clear plan early in a startup and recognizing the potential for growth will lead to better results and also reduce turnover, which can be a huge problem in a sales organization.

Incentive Compensation Revisited

Evolving Your Compensation Plans

When you're expanding, you're beyond the startup phase and you'll probably need to revisit your incentive compensation plans. Like other elements of the sales organization, you'll outgrow what worked in the early days and it's far better to be proactive rather than reactive. This is especially true for incentive compensation since it's a key motivational driver within the sales organization.

As a startup and as you're scaling, I believe that a flat percentage for every dollar of bookings is what works best. As you expand, you'll want to leverage the incentive plan with escalating percentages as a salesperson gets closer to their annual number. I suggest quarterly targets because it's easier for a salesperson to determine the number of calls, the number of meetings, etc. that they have to do to hit a quarterly target. And, yes, the clock goes back to zero after each quarter so that also motivates the salesperson to be sure that they're constantly building their pipeline. You want them not only to be working in the current quarter, but also to be thinking about how to keep building the pipeline as they're closing deals. That takes a lot of discipline and a lot of coaching from you, the CRO. A good compensation plan will help people figure out how to do that, how to pull things in from the quarter ahead so they can get the higher percentage. They'll want to make sure their pipeline is always robust.

Once you're beyond startup, the processes around onboarding become much more important. You'll have a differentiated market, buyer segments, personas, and verticals, so your sales team members will vary based on experience and sophistication. You'll need to hire more team members and as you do that, since your company and markets are more complex, you'll have to onboard new people. The length of your onboarding depends on how quickly a person can become fluent in asking the questions, following the sales methodology, and assimilating your sales philosophy. Obviously, you want that time to be as compressed as possible so their time to value is 60 days rather than 120 days.

For that first quarter, rather than have a salesperson starve, you might consider paying them on achieving other goals, like completing any number of trainings, going on X sales calls, and beginning to develop some pipeline. Your onboarding should have as its goal a series of buildable skills so that new hires don't lose their zeal to be successful in subsequent quarters. A non-recoverable draw for the first 90 days is possible too, especially if you have a complex business. The risk is that you lose that money, but if a person doesn't hit these early goals, you're probably better off letting them go after 90 days. Hopefully you made a better hiring decision so that doesn't occur. If it does happen, you may want to revisit your hiring practices.

Incentive compensation plans, like all of the elements in the sales organization, have to evolve as you scale and expand. If you don't constantly upgrade them, you'll find yourself flatfooted in the market and scrambling to catch up.

Chapter 59

Pipeline

Two of the biggest sales challenges in a startup involve pipeline and forecasting because there are so many unknowns. Pipeline and forecasting are different processes and require different skills and you should develop models and ideas about them separately. I know a lot of companies that make both pipeline and forecasting so complex with lots of details, metrics, and analytics, but they still can't get a strong pipeline, nor can they develop accurate forecasts. Creating a pipeline is not something to be done within the silo of Sales, it takes collaboration with Marketing, with Business Development if you have that function, with Product, and with Finance. With so many people involved, it's important to have a system of record, be it Salesforce.com or HubSpot or something else, but you absolutely need to have a place that's a central repository of information.

In the beginning you need to begin collecting data to understand exactly what the dynamics of your pipeline are. Even if that information is incorrect to start, trying to map out a pipeline framework to your buyer's journey is critical. What data is important? Information from your buyers about their decision-making processes, about timing, about how they evaluate products is a good place to start. You need to think about your activity but also how that activity influences the buyer, what it is that's affecting their journey. Of course, you can assign probabilities to each of those stages and you'll begin to develop some well-worn pipeline management paths. But in the early days it's important to at least start with a framework, because if you don't know where you're going, everything you do will be inefficient and ineffective.

The other questions I would ask prospects in addition to decisions, timing, and evaluation, are almost mentor-like questions and if you've developed a relationship with the customer, they will help you by answering them. You should ask questions like, "What will help us get to the next step? What things should I be considering?" It takes a lot of boldness and courage to ask those questions and the buyer might not even know the answers, but you form that together.

As an early stage startup, you're working with so little, but it doesn't take very long to develop a philosophy, a theory, about what's the right entry and exit criteria, to develop several paths that, with your probabilities, tell you something important about how well you'll close deals. And if you use data historically as you go forward, you'll be able to see how a deal closed, how many days were in each stage. So, when something comes up, like this deal was lost in the fourth stage, you'll be able to figure out if you should have spent more time in the early discovery stages.

As a startup, you may have only one customer type but as you grow, you'll find that you have segments, you have different buyer journeys, and maybe different experiences with your product based on region, vertical, or some other factors. But you absolutely have to start understanding your pipeline, you have to understand things like days and stage and time to close so you can improve your pipeline management and, ultimately, your forecasting. Probably the most potent combination of art and science that exists in the whole revenue process is understanding how to manage your pipeline and how you can tie that to accurate forecasting.

The major differentiator between the startup that is struggling and the startup that is poised to scale is pipeline management and forecasting.

Chapter 60

Scaling the Sales Organization

I would not suggest scaling (from a revenue perspective) until you have a consistent viewpoint of the buyers of your product or service. If you're still at the whiteboard stage, no matter how passionate and optimistic you are about the future, it will be difficult to scale. You'll need to be at the PowerPoint stage if you want to scale. You should have answers to questions on the buyer's problems, you should have a good understanding of the competitive landscape and where you fit, you should have a philosophy on your value, and you should have a methodology that allows you to know where you're at in the pipeline and what's forecast in the future. Without a solid understanding of these basic elements, scaling is just a pipe dream.

In the very, very early stages of a company, when you haven't stratified to the point where you have enterprise sales or small and medium-sized business (SMB) sales, you'll have all those sales in one person or possibly in one "type" of sales rep—people who are jacks of all trades. If you have a person who is a go-getter and they can tell the story as an entrepreneur, they will likely have the ability to close large enterprise deals and to speak to mid-market. As the CRO, you should make sure that they can close business in all those segments to feed your enterprise. You'll probably find that smaller sales happen quicker than enterprise sales, but you should validate that with your business. The worst thing, the last thing you want, is an enterprise-level effort for a transactional sale. So, you have to be sure to balance all of that but also carefully map retention rates by segment.

As you start to scale, you'll have a lot of things in place that you worked on as a startup, for example, you ought to have a pretty good idea about the type of person that's best suited to be in your organization. You should have your hiring profile done so that you can just turn on that machine and always be hiring. You never want to fall behind in sales power because if someone walks out the door, even though you have a pipeline, you don't want to be surprised with a departure and then have to say, "We just lost that quota." You want to be sure you have a deep enough bench to allow for any kind of voluntary or involuntary exits from the organization and you can accomplish this in two ways. You can either have more junior people who can be promoted to carry more quota, or you can have more quota in your existing pipeline. Either way, you don't want a sudden departure to shove the sales organization off the rails. You'll also have to be thoughtful about career-pathing as you scale.

I currently work with ethics compliance software and we've found different tiers of the market, different appeal of the product to small and medium-sized business, and different appeal than what we see at the enterprise level. If you can match the market segment by the type of salesperson, you can create career paths that provide growth opportunities for them and reduce turnover for you. So, you might have someone early in their career focus on the SMB mid-market because they may not have developed the sophistication and expertise to appropriately handle a larger enterprise sale that may go across many divisions and even into the C-level. And an entry-level salesperson with no prior experience I'd start out as a sales development rep (SDR) doing inside sales. Sometimes an SDR is housed within Marketing and sometimes Sales, but that's a good place for inexperienced people to start to learn the sales methodology, process, math, and approach.

To really be able to scale, you'll need to have a profile of the people you want to hire, and a career path for them to follow. Without those, you're likely to spend a lot of time looking for people, in part because you don't know what you're looking for and in part because you'll have a lot of turnover in the sales organization.

Don't Blindly Promote Your Best Salesperson

There is often a classic mistake made by sales leaders (and this happened to me), where you promote your best salesperson into a sales management role, and it turns into an epic fail. In my case, the salesperson insisted that management was the next best, and only desired, step in their career. At the risk of having them leave the organization, I put them in a sales management role, despite my misgivings.

Within months, we lost some quality members from their team—they were just not a good manager, coach, or mentor. Eventually, (finally) we made the decision to let them go as well, so the outcome was bad for them, for team morale, and for the company.

My guidance here would be to craft an interesting career path for people who stay in an independent contributor role. This could mean having them take on more strategic enterprise accounts as their career progresses, or creating a Global Account Management role that spans geographies, and provides an opportunity to lead and collaborate with teams internally to successfully penetrate and manage global accounts with multiple subsidiaries. The important point is to match impact with expertise, and to reward and recognize those who choose sales, and not management, as a career path.

At the end of the day, have the corporate courage to "just say no," when someone is not a good fit for a role, and move on. The long-term benefits will far outweigh any short-term pain.

Chapter 61

Scaling Your Team Through Culture

Your team will scale naturally as you grow and by that I mean, you'll hire people based on what works for your company, and based on the growth rate of the company. When I think of scaling your team, I think of creating the type of sales culture that will allow your team to scale—not so much just hiring more people, but hiring people who make a big impact in the company in the way that fits with your company values. If you build the right sales culture first, then you can spend your time proactively building the pipeline rather than reactively solving internal problems. You create this impactful sales organization through your management, your systems, and your culture. The sales culture is the essence of a good sales organization and that means having an organization where everyone is holding themselves and one another to really high standards of performance and accountability.

There are a lot of companies that subscribe to a productive competition mindset as a way to motivate salespeople, to focus them on getting their quota, making their numbers. That works to some extent, but it's more effective to reward performance through recognition. For example, it's possible to create standard awards for recognition and even crowdsourcing awards among the team so that you are recognizing performance, like biggest deal to close, breakthrough in a new industry, whatever it might be. This is different than making it competitive among the sales team members.

There are some sales organizations that create a President's Club, with trips to island paradise resorts for the winners, but that ends up fostering competition among the team, it's elitist and it doesn't play well to the company as a whole. If you have a President's Club for Sales, what about Engineering or HR? So as an executive, I don't like it. There are other organizations that pay spiffs, where a salesperson gets an additional amount of money if they facilitate a particular product. Yes, you can drive some top-line growth that way but you're not helping the sales team to focus on the buyer's problem, they're just getting some extra money for selling something the product team rolled out.

Avoid Special Treatment

At Return Path, we created something called "Performance Aspire," which recognized (with relevant data) our highest performers from across all departments in the organization, and could include up to 100 people from around the globe. We held a single big event that included recognition, but also coaching, mentoring, and information-sharing opportunities. The feedback from across the organization was positive, and something that people "aspired" to attain as part of their time with Return Path (See also Chapter 34).

But, as CRO you may have situations where you recognize that you're having a down quarter or you feel as if people just need to feel some level of enthusiasm, need to be jazzed up for whatever reason. A way to motivate the sales team if there are across-the-board problems is to provide a public spiff and tie it to the pipeline. Give everyone some type of reward as you *approach* your goals—don't wait until you hit them. You could call it a sales contest, for example, who can close the most accounts? Who can make the most calls today, or have the most meetings, or create the most pipeline this week? Hopefully, the work that you're doing as part of your CRO day-to-day workload management would eliminate the need for that, but it can be helpful, especially when markets are impacting your entire organization.

So, there is a role for awards, for recognition, and for spiffs in a sales organization, but those tools are really managing by exception. Instead, if you want to build a great sales organization, focus on creating a culture of transparency, of sharing, and learning. The best salespeople can learn from one another rather than compete with one another. They can learn from their managers, customers, and prospects, and they can share that with their colleagues. To build a great sales organization, create an environment of ongoing learning and continuous process improvement, not a cut-throat environment, but one that's highly, highly accountable for performance, highly recognized and rewarded for performance, or one that also inspires a sense of common purpose and collaboration.

As the CRO, creating this type of culture starts with you, starts with how you interact with others in your organization. I learned an important lesson early on in my sales management career, from someone who had been a peer, and was now one of my direct reports. As I was consistently asking Ronda whether she had included certain information, or talking points for an important presentation, to Fidelity Investments, and making endless suggestions for change, she suggested that I trust her to do her job, and that she wanted this opportunity as much as I did.

I realized then that the most effective way to manage and develop people is to work through them to make them successful. It does not matter how much I know, or how well I can do something—it matters that I use that to help others become successful. Work through others to ensure their success, and my own.

In managing a sales organization you'll certainly look at the numbers, but it's not just about the numbers. You don't want a brilliant jerk in the organization, a star performer who treats other people without a level of respect. Similarly, someone who's really nice but is not a performer is not someone to continue with the organization either.

Lone Wolf or Team Player

I noted earlier that there is a myth that the "Lone Wolf" is often the most successful salesperson. Actually, that may often be true in terms of putting numbers on the board, but is often not true when considering the rest of the organization. For example, here's a situation I had that's quite common for the CRO: I had an extremely capable performer who consistently met and exceeded quota expectations, but was roundly disliked by many of the people with whom they had to collaborate. They showed a lack of respect for other people's time and responsibilities, often ran over people to get their items to the top of an agenda (be it in Product or Marketing), and failed to acknowledge or show appreciation for the support they were getting from others that helped make them successful. They clearly were not a good culture fit, and that was corrosive. Despite multiple conversations about working well with others to get things done, we eventually had to exit them from the organization. For me, WHAT you do and HOW you do it are equally important. Just hitting your numbers is not enough if you go about it in a disruptive way.

When a CRO has to provide feedback, I'd suggest that the feedback be directed toward helping a salesperson understand your philosophy because that's where the breakdown for most sales behaviors happen. For example, if I do provide feedback, it's typically around helping the salesperson to look "outside in" and not to fall into the trap of talking about our product. That stems from my philosophy that we conduct sales with empathy and understanding of the buyer. It's a constant for me to remind our sales team to be looking at our company through the customer's eyes. I'll sometimes use software tools that allow us to listen to a sales rep's call and help us analyze and understand how much talking the salesperson did versus the prospect. That provides analytic insight to the salesperson and to the manager and helps each strive for continuous improvement.

That type of feedback can really help a salesperson refine their communications and help them understand if they are asking questions and listening as opposed to just talking. I want our salespeople talking about customer problems and how we solve them. I want our sales team to get the buyer to the point where they're actually thinking that they're controlling the sale, even though the salesperson is controlling it through their questions positioned in the right way. That's the feedback I would provide, given our sales philosophy.

So, a powerful sales culture is one that has at its root an environment of constant coaching, mentoring, support, and encouragement with standards, not an environment of fear. A great sales organization will not be punitive but will have high standards and high emotional IQ as well. And if you can create a safe environment, where people can openly discuss, disagree, and share, then you'll have an environment where feedback is welcome, where feedback is recognized as a gift. When you get that, you can scale as quickly as the market will allow.

There are other sales cultures that are effective, however, and the choice is up to you. You can be successful, for example, by running a boiler room that churns and burns out people and customers. You'll have a lot of conflict and, most likely, low morale. It's not a fun way to spend the day for you or your sales organization, it's not sustainable, it doesn't really develop people, and it's not sophisticated. The approach I advocate, what I believe a great sales organization will embrace, is the opposite of that in all respects. I choose to invest in people, to encourage them, to create a space for sharing and learning. But, different industries have different characteristics, and ultimately the choice is yours.

Chapter 62

Scaling Sales Process and Methodology

When you're scaling your sales organization, you should have a very good handle on your sales processes and methodology. It's wise to embrace some type of sales methodology, to have something that enables you to have a framework you work from. You should have a common language around the buyer's journey, and how that aligns with the stages in the pipeline. Otherwise, it's going to be very hard to develop a consistent point of view across the organization about what works and what doesn't work. The sales methodology will help guide the way you question and move the sale along, and the concept of the buyer's journey will enable you to understand how the buyer is looking at you and how that aligns with what must be true for you to come in and out of a pipeline stage.

In the very early stages, when you're selling on whiteboard and you're just trying to get business, developing a sales methodology will be difficult because there are too many unknowns. But as you close in on a few million dollars in sales, when you really need to think about scaling, that's when a methodology is important. You'll be able to ask questions like, why is something getting stuck in evaluation? Or worse, why is something stuck in negotiation when it has sped through the evaluation stage? Clearly you didn't do your homework early on, you didn't think about the methodology and asking the kinds of questions that would have helped you understand the problem, and be able to understand the gap between the problem and the solution you fill with your product or technology.

As you scale, your methodology will expand, it will evolve to include more scenarios that impact both the sales organization and the rest of the company. In the early stages, for example, you might sell to every customer that you can but when you're scaling, you will have figured out the effort and payoff of certain customers and you'll be able to resist selling to everybody and only sell to the most profitable customers. In the beginning you might have the philosophy that "the customer is always right." I don't believe that the customer is always right and I've been in the situation where a salesperson is dealing with a very, very demanding customer. In that instance, it's helpful to coach the salesperson to ask questions to discover those demands, to make sure that they're having a conversation with the customer and letting them know that we are here as a partner to provide a solution. Our approach is to do that in an environment of mutual respect and understanding.

In a really difficult situation, if a customer undermines our salesperson, or shows a lack of respect or takes advantage of someone in the company, then that's not a positive relationship and I would happily walk away from that if it's not tenable. People are eminently reasonable and if you try to understand where they're coming from, and what's motivating their demands, potentially you can find a solution. But not always.

At Return Path we did have one customer who was downright rude and threatening to one of our salespeople, and we fired the customer. I made a call to Matt, and he responded with "There's no place for this, our people aren't going to get treated like that. The negative to our culture isn't worth the positive of that one contract." So, your sales methodology is more than a selling process, it also informs what you don't want to do, what customers you don't want to have.

You can also run into a situation, anytime in your growth, where your methodology compels you to do something you'd rather not do. Very early on at Return Path, we chased a high-profile, fast-growing phenomenon and we had a salesperson on it, but because it was an important multi-million dollar deal we also brought in Matt to help close it. At one point during negotiations they demanded (and we reluctantly agreed to give them) most favored nation pricing which is something I loathe because it's just a race to the bottom. We carved out a couple of exceptions that had been longstanding in the US and as we were negotiating the final deal, someone from the Finance team came to me and said, "Anita, I'm sorry. I made a mistake when I gave you the carve-outs for the agreement. I forgot to include company XYZ."

I said, "OK, that's fine. No problem. Thank you for owning up to the mistake." I went to Matt and told him we made a mistake and the result would be $250,000 to the good for the client (meaning $250,000 to the bad for us). I said to Matt, "I think that we should live by our values and we should do the right thing." He responded, "Absolutely. Of course." And with that I went to our customer and I said, "John, we made an error in your favor and I want you to know this will be the hallmark of how we work together. One of our credos is to do the right thing and when we found out about the $250,000 mistake, we felt we had to honor our agreement." He literally signed the agreement that day and instantly became a huge advocate for us and a reference customer based on how we handled the situation.

Your methodology, as I see it, is at first about customer discovery, understanding their journey and pain points and creating a roadmap of questions and stages for your pipeline that can be measured, understood, and managed. But a good methodology, if it evolves, does more than that and it can help the sales team and the organization define who they are and what they want to do, and not do, as the company scales. A great methodology will help you get from whiteboard to PowerPoint to PDF, and that's the key to scaling.

Chapter 63

Scaling the Operating System

Your operating system, like your methodology, will evolve over time as you scale. In the early stages you may find yourself doing far more than sales, and you may find yourself chasing a lot of things that don't play out, that take up a lot of time, and that leave you with little to show for your efforts. That's an artifact of being a startup, with all the unknowns. But as you generate some consistent revenues, as you build a methodology that can scale with you, you will be well-served by developing and abiding by an operating system.

In any sales organization, and especially if teams are distributed, it's critically important to create an environment that's safe for people, where they feel that they can be mentored and learn from each other. As the CRO, a 15-minute standup in the morning can be helpful in just getting everyone oriented to the day, it can help the team understand what's going on in the community, and it helps keep them accountable. As the leader, you would want to have one-to-one conversations once a week with each individual salesperson and ideally, a meeting once a week with the full team.

Meetings, if not managed properly, can quickly lose their value and it's the leader's responsibility to ensure that, if you're bringing everyone to a meeting once a week, there's value for the individuals and the company. So, I don't want to have a weekly meeting where each salesperson goes through their pipeline. I can read; I know what's in your pipeline. Tell me about the opportunities you have, the challenges you're facing, where you're stuck, what you've learned, what you heard about the competition. Talk about the dynamics of the business, both for the benefit of your team, and for my benefit, as we all learn together about how our product is being received in-market.

That being said, it is critical to have weekly or bi-weekly pipeline and forecast reviews with your team, and then with management, that includes an assessment of "safe case" and "best case" scenarios for opportunities, so that you have visibility into the likely outcome for a month and/or quarter. This level of scrutiny improves discipline around pipeline management, and forecast integrity. It also is a tool that helps assess the relative performance of team members, and can offer coaching moments, especially if opportunities consistently "slip."

It's also important to have strategic meetings as well, at least quarterly. Thinking about the past quarter, what worked, what didn't? What would we do differently? What do we add, what do we subtract? And during the quarterly meeting we'd also roadmap for the next quarter. What are our goals? And how will we get there? Those kinds of questions are good to share to be sure that we're both tactically and strategically engaged as a means of being successful and as a means of learning from one another.

As the CRO, you want to create that fiber within the organization that connects the salespeople, so you want to bring everything from the outside in and bring everything inside to the salesperson. I would also encourage you to create as part of your operating system an invitation for others in the organization to interact with Sales. For example, I want my Head of Product to know exactly what our prospects and customers are saying as a means of improvement and would invite that person to get on a sales call with someone from my team. That's an important part of both team engagement and engagement with the market.

The other component of an operating system that's important is performance management. It's a red flag if you have a sales organization that is consistently missing its numbers or is unable to deliver. You might not find out that you have a problem until you're six to nine months into a startup. But if you don't have a robust performance management program where you're very deliberate about the kind of coaching you need to get people up to performance, that is a potential hazard. You have to know, especially in a startup, both the speedway and the guardrails to get things going.

The specific performance management plan is obviously dependent on your sales cycle, and the time it takes to get from lead to close. Generally, I would give a rep one quarter to ramp, and begin to build pipeline. Assuming they get the training, coaching, and clear guidelines on the action needed to get to goal, you can begin monitoring performance on all aspects of selling, not just attainment. If a rep cannot get to 70% attainment for two quarters in a row, and they have received consistent feedback and support, I recommend an exit plan. The rep should be aware of performance expectations at the outset, so that they can do what it takes to perform and succeed.

Chapter 64

Marketing Alignment

No relationship for the CRO is more important than the one you have with the CMO. Marketing can help you get from the PowerPoint stage to the PDF stage, especially if Marketing has SDRs reporting to them and if Marketing is paying close attention to what works. If you're a CRO without a very good personal and professional relationship with your CMO, you're setting yourself up for failure. Notice that I didn't say, a very good professional relationship only, but a personal relationship, too. That's important, especially as you're scaling because there will be mistakes, miscues, miscommunication, and a host of other issues that strain a professional relationship, but a personal relationship can overcome these. But developing a strong relationship with Marketing is critical in getting to the PDF stage for the sales organization.

If you're new to the CRO role, or if you're starting with a new company as a CRO, your first order of business is to develop a relationship of mutual respect and understanding with the Head of Marketing, and together develop a commitment for Sales and Marketing to work together. If you don't do that, you'll have problems scaling. For example, while I was at Return Path, we missed our quarter, missed our targets and goals, which is a yellow flag. And then Marketing stood up and said everything's green—we made our goals, we made our quarter. That's a problem! If Sales is yellow and Marketing is green, that's a huge indicator that there's misalignment because Marketing and Sales should be tied to revenue. After I gave that feedback, however, we never had that issue again.

The CRO and Head of Marketing need to make sure there's alignment around goals and objectives, around messaging and positioning, around a concept of continuous improvement and experimentation. The level of visibility, interaction, and alignment around positioning messaging, strategy is critical. Beyond that, Sales and Marketing need alignment around shared goals and objectives.

One of the best ways to strengthen the relationship between Sales and Marketing, to accelerate that alignment, is through structure. You do this by having the SDR and inside sales report to Marketing and by doing that you'll have shared goals and objectives for both the pipeline and lead generation. We had this structure at Return Path and Voxy, where the inside Sales team reported to Marketing so that Marketing deeply shared our pipeline creation goals. After all, Marketing manages the team that delivered potentially half of the pipeline.

Not only were we aligned on the pipeline, but also we were aligned in our go-to-market messaging. Our outbound campaigns to support the Sales team had to be completely aligned with Marketing; they had to direct their marketing machine to build the pipeline. I'd also suggest that the CRO look for opportunities to bring Marketing into the sales process. Have them listen in and participate on sales calls, or have them staff the booth at a conference or trade show along with your sales team. That way you're learning from one another and creating content that can be used for webinars, blog posts, LinkedIn posts, or any other social media. Through these opportunities you'll build the market knowledge so that Marketing deeply understands the market, from both a Sales and a Marketing perspective.

There are several different ways to structure the role of the SDR, based in part on how your company is structured regionally, and also personal preference. One way is to have the SDRs support all of the salespeople. If you have a weak performer, a single sales rep is not saddled with that person and if you have a strong performer, that talent will be distributed more broadly to all the sales reps. I'd also suggest that SDRs be compensated for both the number of meetings and percentage of opportunities that close. You want the SDRs to have the opportunity to benefit from working with a variety of salespeople and you want the SDRs to learn about Sales, so that they can understand possible career paths for themselves. Other CROs take a SWAT team approach, but I prefer an environment where we can learn from shared best practices, as opposed to creating little silos. If you manage the sales organization in secret, without full candor (like what happens when things are in silos), then it has the potential for problems.

Although Marketing and Sales each have their individual goals, they also have shared goals around pipeline creation, around lead generation, and by having SDRs report to Marketing, those shared objectives keep them on the same page. It enables you to celebrate successes and to learn from one another's misfires, campaigns that didn't work, or messaging that didn't work. Having that openness and communication and shared goals helps you get to the point where you can collaborate for the most impact and success. And that collaboration is fundamental to being able to scale.

What I Look for in a Startup Chief Revenue Officer/Head of Sales

Greg Sands, Founder, Managing Director at Costanoa Ventures

About once a week I hear an entrepreneur during a financing presentation tell me that the product is "almost ready" for general availability. They'll follow that up with, "Now all we need to do is hire a VP of Sales." I typically respond by asking them some tough, pointed questions about their sales process because usually entrepreneurs at this early stage haven't had enough repetitions to truly understand the ideal customer journey and making it result in sales. Hiring a VP of Sales at this time would not be the right person to hire first, because you don't want to scale up a process you don't truly understand yet. It would be a grave and costly mistake for the entrepreneur and for the person hired.

So, who should startups hire first? Sadly, there is no one answer other than "it depends." It depends on the stage of the company and the maturity of the process. The key thing in the first phase is to figure that out. The simple truth is that most startups at the early stages do not need a highly experienced sales leader. The first "sales team" is founder-led by the person who brings a deep knowledge of the product to market, who has the impressive title of founder/CEO, and hopefully, who has the charisma to help people overcome their reluctance to buy something new.

At some point, the CEO needs to build a sales team, and the temptation is to hire a VP of Sales first. I've rarely seen this work. A more proven approach is to hire a smart, hungry sales apprentice. Former CEO of MobileIron, Bob Tinker, refers to this person as a Davy Crockett, "someone who can find the path through the woods." The job at this point is not merely to go out and try to find revenue, but to help develop a process that can be replicated in the future. To do that, the first sales hire has to be smart, creative, and aggressive. They need to have more ability than traditional reps to understand the product and where it fits, to find new use cases, to invent new paths into the market. And they need to be strategic enough to learn from the experiments and figure out what is potentially repeatable. Your first sales hire doesn't need to be particularly senior; in fact, a senior leader isn't usually hungry enough to get this job done.

Once the sale process is more defined, which often means you've got a couple of "Davy Crocketts" out selling, *then* you need a VP of Sales. This person needs the leadership and judgment to hire and build a team (including figuring out the right rep profile), the decisiveness to lay out a basic framework (quotas, territory assignment, mix of SDRs, reps, and systems engineers, early sales training), and the adaptability to step in and change when one part of the formula isn't working well enough. And at this stage, the company still requires someone who is willing to be out in the field, working deals with reps to close the quarter successfully.

When the company is ready to step on the gas and scale, you'll need to hire a Chief Revenue Officer (CRO). For a SaaS company, this is typically at $15–20m in ARR, more than 15 sales reps, and a fairly repeatable sales process. At this point, the leader needs to be a systems thinker, a process builder, and a planner. This job is really more like a field general in command of significant resources that need to be positioned and coordinated to achieve scale and efficiency. They need to have the judgment to hire good managers and the sales capacity to close the best ones, the ability to work together with the rest of the executive team, and they need to know how much process is appropriate for the current scale of the company. As the company is ready to add new geographies (e.g., EMEA, APAC) or new channels and partners, the CRO should have the experience, judgment, and network to be able to initiate those efforts with a strong initial plan and team.

But there is much more that is highly variable depending upon the stage of the company and, even more importantly, its sales maturity. As above, the answer to what you need in a CRO/VP of Sales/first sales hire is, "It depends." But hopefully, I've given you some ideas about "on what exactly, does it depend?" and how to determine what the right role scope and the right capabilities are for your company right now.

Chapter 65

Market Assessment and Alignment

As you're scaling, you should have a lot of enthusiasm around a new product, around the market you want to penetrate, and around the new business that you'll generate. But that enthusiasm can also be a trap unless you are able to be honest about your prospects. I don't mean your prospects as potential customers, but your prospects in the market. I like the framework of "What must be true?" as a way to assess new markets, as a framework to understand if you can expand and move to adjacent markets.

Being truly honest in determining and understanding the factors that could potentially affect your growth is a necessary first step and you should be rigorous (and conservative) about answering questions like the total addressable market and other areas. For example, what privacy or legal restrictions may you have, or may you face that you haven't initially anticipated? What does the competitive landscape look like? Do you have an upstart competitor that potentially sees opportunity and is willing to disrupt? Have there been any factors that affected the overall vitality of your offering and that might impact your pricing? Is this sale repeatable enough to a large enough market that we can move to selling with PDF, or are we forever going to be stuck at the PowerPoint stage?

Scaling involves reevaluating the factors that made you successful as a startup and you need to revisit those factors to be sure that you can expand either the distribution channels or geographies. If you just go in blindly without being very conscious of what factors must be true for you to be successful, you have the potential to trip up and either miss additional opportunity or over-estimate what your real total addressable market (TAM) and opportunity are or create an organizational mess in pursuit of growth.

It is important to note that not all growth has to be organic. At Return Path, we made several critical acquisitions in the early days (Delivery Assurance Systems and Bonded Sender), that gave us the product and technology platforms that helped us grow and scale the business. That was incredibly positive.

However, somewhere between $50–75m in revenue, we began to understand that our target market wasn't large enough to support the kind of long-term growth we'd need to go public, one of our goals at the time. So, we thought about adjacent spaces we could potentially step into leveraging our core assets to try to expand our reach. And that decision, to expand our reach, led to all sorts of ideas and experiments and new products. That expansion led Scott Weiss, one of our independent Board members, to comment that we had become "the world's smallest conglomerate." We ended up with disparate businesses that had slightly different audiences with slightly different market messages with slightly different product requirements. The resources we needed to keep all of those going was a real drain on the organization. We ended up selling off all of those businesses.

I ran one of those businesses, Consumer Insight, for Return Path at the end of my tenure there. It was a great, great idea because we could use the same data we were using for our core business and repurpose it for a different customer segment. The problem was we could never invest enough to truly leverage the opportunity across the Return Path customer base. We isolated and identified that market research companies and hedge funds had the sophistication and the appetite to collect these huge data feeds, so we chased them. But once we got past the low hanging fruit of the 10 market research companies and the 25 hedge funds who could use this data, where are you? The investment to make our data truly consumable, and packaged for a broader customer base of mainstream was beyond our means.

We didn't waste time and money (like I said, we sold the business), but we had engineering resources that arguably could have been deployed elsewhere to improve other products. If you're going to chase an opportunity, if you're thinking of expanding to adjacent markets, make sure that you're really honest about what the actual potential is with what you have today. The best way to begin that exercise is to create a list of answers to the question, "What must be true?"

Chapter 66

Expanding Distribution Channels

My experience has been primarily in B2B businesses and in other companies it will differ, but in my experience when you're starting out, you start out as a direct seller. As a direct seller, you really understand the dynamics of the market, what resonates with customers, and you create your own positioning, messaging, branding, pricing, and packaging, that gives you a very refined and comprehensive understanding of what's important in order to be successful in selling.

Soon thereafter, it's really important if you want to expand the reach of your message to align with partners, to expand and develop your channel. By aligning with the right partners for your product, you'll amplify your partner's products, helping them sell more or helping them reduce customer churn. Channel business is not really about generating revenues for the partner from your product—it's great if it happens—but the real advantages are that you get access to a channel for your product that you wouldn't otherwise have, and you're helping someone else be successful.

If you can find partners where your product is additive to theirs, they'll be highly motivated to work with you. The incentive can't just be to sell more of your product, it has to be making their business better. If you can find those partners, I've seen it be incredibly successful, but not without conflict.

At Voxy, I had a small direct sales team of about 10 people but we could never hire fast enough to be able to materially grow our business. I introduced the concept of trying to work with channel partners and although they resisted in the beginning, we eventually got it working. The company's product was a technology platform that helps people around the world learn how to speak English. The companies we partnered with were ones where our technology enabled them to provide more value to their customers.

For instance, we partnered with employment agencies, who have huge numbers of people that they need to fill all kinds of roles and they need that employee population to learn how to speak English. By offering our solution to their employers, it was able to help its employee population speak English, so that was tremendously valuable to them.

Another channel included nonprofits who had as their mission educating a refugee population. A nonprofit like the Red Cross could see how providing our product as part of their suite that they were giving to their populations, employees or otherwise, would be enhanced. Within two years we went from zero and increased our sales by 20% in channel business, which is a big leap when you're a startup.

The right channel partner helps get leverage in the market and also helps you take advantage of branding with well-known companies. You may never have heard of Voxy, but you probably have heard of the Red Cross, of UNICEF, or other refugee agencies. For a small company looking to expand, a channel partner can give your brand cachet whether it's private labeled, or whether it's something that's powered by your brand. But it's a critically important growth strategy.

You may run into resistance with your CEO or CFO in approaching a channel partner like the ones I mentioned above, and the way I try to couch it with a CFO or even a CEO is to acknowledge that this is a lower margin sales strategy but the top line growth should far exceed the worry about the margin. And that has typically proven itself to be the case with regards to brand value. And your brand value can add cachet to your partner. Our company was very contemporary and innovative in the English language learning space, so in some cases the brand was the cachet and our partners could say, "powered by ...," signaling that they were working with a top company in the space. Others preferred their own brand, which is fine.

The channel strategy I'm advocating is not without work on your part. It's not the case that you can find a partner where your product is additive to theirs and then sit back and ship out orders. You'll have to work with the partner marketing team, helping them bring your product to market, establishing messaging and potentially marketing through them so that you can offer webinars and conferences that help educate their customer population.

It also helps to have a person or team that manages channel partners and is responsible for training the partner and making sure they're fluent in your product. Your team should also provide partner support and traditional customer service. But the partner should be responsible for delivering service and value to their customers.

I will add that this is not easy. There will be inevitable conflict between your channel and direct teams, unless you can establish clear rules of engagement (always lean into the partner), and commission that rewards both channel and direct teams in the event of conflict. This will promote trust with the partner, and encourage your direct rep to take the commission and move on to close more deals in their territory. You do have to be sure that your direct reps remain productive, and do not only benefit from deals sold through a partner.

Expanding your reach through channel partners is a lower margin business and if you have a really aggressive partner, they can take up to 50% of the bookings for themselves. But if they're marketing and they're servicing and they're selling, I'll take that as long as the top line growth justifies giving up that much margin.

Chapter 67

Geographic Expansion

When you're ready for geographic expansion, you'll use the market assessment framework you've already created for scaling up. The litmus test that you have to pass to know whether you're ready to expand to new geographies starts with a list of answers to the question, "What must be true?" What must be true for you to be successful globally?

With Voxy, the English language learning software, our first "What must be true?" was that the government had to support that their constituents learn English. Without that driver, it would not make sense to expand to a new geography. Similarly, with LRN, the ethics and compliance software company, our "What must be true?" was that ethics and compliance training had to be mandated by the U.S. Department of Justice, or by the zeal of companies to protect corporate brands.

There won't be just one "What has to be true?" factor, but many factors that you'll want to consider to be successful in a particular region. If you have a list of ten factors, they don't all have to be weighted equally and there might be five things that would be sufficient to get you to 85% of where you need to be. You'll have to experiment to be sure that you're not letting great be the enemy of very good. Nothing is going to be perfect at any given time so you'll have to be sensible about how you weigh and prioritize what must be true to be successful.

Some countries seem like ripe opportunities, just given their population, but you'll have to dig deeper to understand whether your business can take hold there. For instance, in China, it's very difficult to do business because there are so many entrenched competitors. Let that be your guide as well. People said to us at Voxy, "China's the largest population in the world demanding to speak English, why don't you expand there?" But our competitors had hundreds of people on the ground and we were just a 75-person company. Despite the huge demand and appetite to learn English, and despite our superior software, it didn't make sense. But Vietnam was attractive, so we went there.

The long and short of it is that you'll need to be honest with yourself about what must be true for your product and what must be true for your company to adapt and fit into a new culture with all that it entails. In most countries you'll need to have native people as part of your team to really be successful. Even if you get some inbound leads from regions you hadn't expected, you still need to be much more deliberate and collaborative with the executive team to expand globally. Because a lot must be true.

When contemplating geographic expansion, you cannot just chase dollars without understanding the complexities of entering a market, doing business locally, and accurately assessing overall market potential. There are some geographies that are just difficult, regardless of what you think the upside is, and even if your "What must be true?" is true. We succumbed to this temptation at Return Path as we started responding to inquiries to sell our solutions into Russia. We spent too much time and travel trying to wrangle opportunities, only to realize that we were deluded by the prospect of success, without truly understanding the long-term viability of the market. It ended up being a waste of our energy, time, and resources.

Your decision to expand cannot be at the behest of a lone wolf salesperson saying, "Hey, I can close this big deal in China." You need to have the people in the home office ready and able to support global expansion enthusiastically and embrace a total growth strategy. Otherwise, if you try to go it alone with your sales organization, you'll do it in fits and starts and you likely won't have the kind of resources you need to nurture and grow business outside of the United States. You'll have to market in a new environment, hire local people to sell the product and provide service, you'll need to have commitments from the product team because they may need to adjust your solution based on local needs.

As you're figuring out your business model and as you're seeing that you will be able to scale successfully, then you should absolutely be thinking about international expansion. In many of the companies I've worked at, international business accounted for 50% of our bookings. But you'll need to be nimble and flexible and it's critical that you do it in a way that's scalable and repeatable. You'll need the back office and framework for getting business done to be nailed down, noting that there may be little nuances to pay attention to that will help you be more successful.

Chapter 68

Pricing and Packaging

Pricing

Pricing and packaging are two areas where many startups trip up, either because they don't do discovery well in the beginning or they don't understand the competitive landscape. If you're selling into an existing market but maybe you have a new mousetrap, you've got a slicker product and more innovation, or deliver the solution or the technology faster, the best approach is to price the way people buy it. Don't try to break the model. If it's a per seat model, then don't try to change that structure, try to fit within it so that you're competitively priced. Unless you can deliver materially more value that enables you to get a higher price, then your best strategy is to be competitively priced.

It's always a matter of testing and having a working theory of what can work so that you're not just going in asking someone, "Well, what would you pay for this?" Because their response is likely to be, "Well, you know what, I'm going to give you a low price of what I would pay for it." But if you can create scenarios that help someone understand how the value and price equation works, then you have something that they can hold on to and you can help them understand as you're pricing, what value you're delivering at the same time.

If you're going to a brand new space, then it might require a little bit more experimentation to determine pricing. The strategy here is to align your pricing model with how the customer uses your product. If they use your product for every mail campaign they send out, for example, then you might want to charge by campaigns, and not by number of emails. But it's important to understand the cost structure that your customer has and what value they get from your product, what revenue they're collecting, so that you fit into that in a meaningful way and that your price point is not so high. It's very much experimentation when you're in a brand new market but with some working theories you can begin to discover, hopefully not through trial and error but through experimentation, what works. This approach is particularly helpful in the beginning, before you've scaled.

As you scale, price becomes incredibly important. I am viscerally opposed to discounting as a means of getting business. If you're priced too high and you keep losing deals based on that, then contemplate changing your pricing structure or creating a lower-end offering that is differentiated from the full price offering. But if you discount just to get business, you're demeaning the value of what you bring to the market. And this is even more true as you expand in the market because you need to understand what the market will bear, not how low you can price your product. Finally, this will be incredibly disruptive to your business in markets where customers talk to one another. Being as consistent as possible with pricing, and pricing practices, will help you maintain the price/value ratio in market.

In scaling up, your competitive intelligence is paramount and you'll have to constantly be looking around to see what the competition is doing. Ask yourself, is there a way I can do this more simply? Is there an easier way for a prospect to work with me and a customer to sign a contract? I don't necessarily mean you have to undo what you've done from the beginning, but you have to constantly be wary of changes in the market, you have to know how budgets may be impacted at your customer that could potentially affect how you price and how you package your product.

Packaging

In terms of packaging, there are several ways to think about it. Some companies like to do the menu approach where every single thing has a price. When you do that, you may be giving the impression that you're nickel and diming somebody. The opposite of that is bundling things together, like when the cable company offers "packages," and whether or not you want something, you have to pay for it. There's got to be a balance between those two, but I think in packaging again, it's best to listen to your prospect, listen to your customer. Just because something is bright and shiny to you as the seller may not actually be important to the buyer. I would be so much happier with my cable company if they didn't make me pay for a local sports channel, which I never use. Even if it's $1 a month, it's just irritating and I feel as if I'm being taken advantage of. But if you can be sensitive to what's important to your customer and prospect and you include their ideas, you'll have a better sense on how to package things.

But try not to provide a laundry list of pricing that you share with them because that is not a good customer experience and you should try to offer a holistic bundle of services or products that make most sense to your buyer.

Regardless of what you want to do, you may have to match the market you're in. In B2B and SaaS companies there's a fee structure based on the number of users. That's a seat model and it's very common.

But one thing you can control on pricing is how much flexibility you give to your salespeople. I'm in favor of giving them a lot of flexibility within ranges that we define. If a salesperson wants to discount by 10%, that's fine if that's within your realm because some customers always want a discount. But if it's more than that, maybe you want to have a more stringent set of internal approvals. Also, when I say that it's fine for a salesperson to offer a discount, I don't want them to discount on the last day of the quarter just to hit their quota. That type of discounting diminishes you as a brand, as a company, and as a provider. I would rather walk away than cut prices by 30% to close a deal.

I understand that the market is viciously competitive, and oftentimes you do have to use price as a means of leverage, it's just not the way I prefer to operate in the market. I prefer that we use the concept of value and impact and outcomes, and that we really paint a picture for the prospect on how we can help their business. And then that value price equation is much easier to accept and agree on.

I also understand that salespeople are often met, invariably up front, with the question, "What's your price?" That question often comes early in the sales process, before you've had a chance to discover the needs and problems of the customer and before you've had a chance to ask questions and demonstrate value. In this case, my response is, "I'm happy to give you a price. In return, I need to get some information from you. I need to be able to ask you questions so that I can give you a range of what's relevant."

Similarly, if you're asked for a proposal early in the selling process, then my response is always yes, but in return I'd like to get a meeting with the key buyers to understand your decision process or understand your issues better. That's a very common response to a question that comes very early in the process, typically by customers who are very price-conscious. It's not a deal breaker, but you should work to get something in return for your efforts. It's a give and get strategy, very simple but important.

As you scale, pricing is critically important and the process of discovering your pricing and packaging is very much that: a process of discovery.

Final Thoughts

In Sales, nothing is ever "in the bag." I cringe when I hear an overly optimistic salesperson display that level of bravado and confidence in the absence of actual data. A "good feeling" does not equate to success. And I often tell a salesperson that I don't care about their feelings, I care about the facts (in the most supportive way, of course). Another saying here, that may or not be appropriate, is that "it ain't over till the fat lady sings." When I joined Return Path, the original, departing VP Sales, who was a friend and colleague, assured me that $6m was "in the bag," for the year, although we only had just over $3m in sales at the time. I believed him, because I thought he had great "gut" instincts, and it was part of the reason I joined Return Path ... although certainly not the primary one. Needless to say, the cumulative sales for that particular product line (which we later sold) never totaled $6m after several years. Lesson learned. "In the bag" became a funny euphemism for something that might never actually happen.

Although the particulars of your own Chief Revenue Officer role will differ from mine, and while you'll undoubtedly have a very different career path than me, I can say with complete confidence that if you want to build a great sales organization, the approaches I outline in this Part will certainly get you there. If you're fortunate in your career you'll get to work with extraordinary people, totally aligned to turn your startup into a thriving enterprise—but it's not necessary. If you understand the importance of having scalable and repeatable processes, of working with direct and channel sales teams, of looking carefully at efficiency and productivity metrics, and of managing a business to the bottom line, you'll be able to make a big difference in your company. And the sales organization will pull the rest of the company forward.

The easiest, fastest, and most gratifying way to build a great sales organization is to cultivate a culture that's people-focused, one that provides opportunities for coaching, mentorship, and career growth for your employees and, yes, one where successful salespeople are amply and properly recognized and rewarded financially. If you work with and through others to execute on the vision, engage teams in the vision, collaborate for success, hold each other to high standards of performance and accountability, you'll have the foundation for building a great organization.

So, let me end with a few thoughts on what "great" looks like.

- Great is running an organization with a common sense of purpose, shared commitment, and enthusiasm.
- Great is creating a team culture of learning, sharing, and mentoring others.
- Great is representing a product or solution you believe in, and helping your prospects and customers solve business problems.
- Great is providing an environment where people can be successful, and advance in their careers.
- Great is being part of an organization that collaborates for success and holds itself and its people to high standards of performance and accountability.
- Great is being easy to work with, internally and externally.

The Wisdom of Anita

Matt Blumberg

I had the pleasure of working with Anita as a leader in Sales, Service, Customer Success, Marketing, and GM for almost two decades. Over that time, I gained so much from Anita's experience—but even more from her ability to concisely wrap up thoughts into a key phrase. Here are a few of Anita's greatest hits:

- **God gave you two ears and one mouth for a reason.** It's so important to listen to your customers in sales pitches instead of just running through your pitch deck and spouting out features and benefits.
- **Hope is not a strategy.** Too many sales leaders throw "Hail Mary" passes at the end of too many quarters. If you're methodical about sales math and understand your conversion metrics and work collaboratively to make your sales funnel and sales power line up, you'll hit your numbers more times than not.
- **Feedback is a gift, whether you want it or not.** Brilliant advice for anyone looking to improve in sales, or looking to advance their career from sales, into sales management, ultimately into sales leadership.
- **You can't fall off the floor.** Taking on a new sales management challenge, whether you're doing so at a fresh startup or in a scaleup looking to improve its sales leadership, or in a turnaround situation, is always easiest when you remember that sometimes, you can only improve on what's gone before you.
- **We need to go to the gym on that.** Sales training is not a three-hour seminar you attend once. It's something you need to work on every single day with your manager or sales trainer. When you've mastered one technique, there's another one to move onto.
- **May all your months have 35 days.** For any sales department ever … enough said!

Chapter 69

CEO-to-CEO Advice About the Sales Role

Matt Blumberg

What comes before a full-fledged CRO? In most startups, the founder is the first salesperson. As startups scale, they add sales reps or maybe some form of a sales manager once there are more than a couple of reps. In the journey Anita writes about—from "selling on whiteboard" to "selling with PowerPoint" to "selling with PDF"—all of this is in the whiteboard stage and beginning to make the transition to PowerPoint.

Signs It's Time to Hire Your First CRO

You know it's time to hire a CRO when:

- You wake up in the middle of the night concerned about HOW you're going to make this quarter's number—not just WHETHER or not you'll make it (since you should know that as much as anyone), but that you aren't clear what the levers are, or what the pipeline/forecast details are, to get there.
- You are spending too much of your own time managing individual deals and pricing, or teaching individual reps how to get jobs done.
- Your Board asks you if you're ready to step on the gas and scale your revenue engine (e.g., move from PowerPoint to PDF), and you don't have a great answer and aren't sure how to get to one.

When a Fractional CRO Might Be Enough

A fractional CRO may be the way to go if

- Even at small volume where you wouldn't yet be ready for a full-time CRO, your sales operation is very complex or to a very senior buyer, and a more junior sales team needs a fair amount of deal support from above.
- You're entering a new adjacent segment (e.g., mid-market going to enterprise), and you need a seasoned professional to help translate sales process from one segment to the other while keeping the initial segment running smoothly.
- You are not sure what kind of sales leader you need long-term and full-time because you're not at enough scale yet, but you want to try out a specific type of revenue leader to see if that type works (e.g., sales only, sales + customer success, manager of hunters, builder of a high velocity sales engine).

What Does Great Look Like in a CRO?

Ideal startup CROs do five things particularly well:

1. They know when to turn up the volume, and when not to. Thinking through our metaphor/framework for enterprise sales that Anita writes about in this Part—from whiteboard to PowerPoint to PDF—great CROs know when they aren't yet in PDF mode. In the early days when your organization is selling on whiteboard or figuring out the transition to PowerPoint, adding sales reps like crazy, even if there is a lot of opportunity to pursue, is inefficient and unlikely to be successful because it still depends on the success of individual hunters. Only when the organization has made the true transition to PDF can a sales machine scale rapidly.
2. They give credit to others first when things go well and look inward first when things go poorly. Great CROs are the first ones to thank their fellow executives in Marketing, in Product, in Finance, for collaboration and successes. They are

also the first ones to thank their team publicly for a good quarter. When they miss a quarter, the first thing they do is figure out why the Sales team blew it, as opposed to blaming the product or marketing.

3. They are maniacally focused on building a conveyor belt–style pipeline for sales talent so they don't lose momentum when a rep quits or gets fired. "Quota just walked out the door" is never an excuse in a well-tuned sales machine where multiple layers of reps are consistently trained, managed, and groomed for the next level of selling.
4. They say no to over-paying and over-promoting. The second-worst thing a sales leader can do is get compensation wrong by paying reps too much base, or have too much commission in easily-repeatable form. Reps who are over-paid get "fat and happy," when what you want is for them to be "lean and hungry." The worst thing a CRO can do is promote a superstar sales rep with no management aptitude or training into a sales manager role. That will not only lead to the person getting fired, it has the potential to poison a whole sales team. Great CROs know how to say no to the misguided request for a promotion.
5. They don't believe in the "magic rolodex" (yes, I realize that term is a bit dated!). Unless you are hiring a sales rep who literally just finished selling a competitive solution to the same target customer set, sales reps who claim they come with a built-in book of business can only deliver on that promise 1% of the time. It's alluring—but it just doesn't work out that way. Great CROs know how to ferret that out.

Signs Your CRO Isn't Scaling

CROs who aren't scaling well past the startup stage are the ones who typically:

- Gravitate to being an individual contributor sales rep and focus on closing big deals instead of mentoring sales managers and sales reps to do that work on their own. To be clear, sometimes the role of a sales leader (or a CEO) is to swoop in and help close a big deal. But CROs who can't shake their addiction to closing deals almost never build enough of that muscle into their organization and end up creating an unhealthy dependency on them.

- Get sales commission plans out in March or April. While it's true that, in a lot of businesses, it's very difficult to get sales commission plans out until after the year starts, getting them out after late February is a sign that your CRO doesn't have enough of a grip on numbers, isn't partnering effectively with Finance, doesn't care enough about their people, or isn't good at prioritizing the important over the urgent when needed. "They'll all be fine, they know I'll take care of them, the plan is a lot like last year's" isn't good enough for the best reps who are constantly doing sales math in their heads.
- Regularly deliver surprises at the end of the quarter—both good and bad surprises. This is a sign of a scaling problem—either your CRO doesn't have a good grip on the pipeline and in particular on larger deals; they are bad at managing expectations; or both!

How I Engage with the CRO

A few ways I've typically spent the most time or gotten the most value out of CROs over the years are:

- During travel time or in and around events. Particularly if you're a B2B company that engages with clients during the sales process, you'll probably find yourself at a lot of client meetings and events, either internal or external. Your CRO will be there, too, which gives you a great opportunity to spend large blocks of time together in transit. One thing we're learning during the work-at-home pandemic is just how much time we save by not traveling. So, when life resumes to normal, why waste time in an Uber or on a plane when you can have a deep strategic conversation or even a personal/social one with one of your senior executives?

Because that early morning time in the hotel gym or late night drink in the lobby bar before heading up to bed could be the time you have a breakthrough with some tricky client-facing issue.

- In a Weekly Forecast meeting. Jeff Epstein, former CFO of Oracle, was one of my long-time Board members at Return Path, and helped us architect a new core business process once our sales team got large and mature and geographically disparate enough that it was hard for us to have a solid forecast. By our CFO and me personally engaging in a Weekly Forecast meeting every week, we forced the discipline of a good roll-up of all regions and business units, and the CRO and all sales managers attended and knew that we were paying attention to the numbers and trends and asking tough questions. The result was that our CRO did a pre-meeting the prior day with all teams and units to prepare, and that in and of itself had a cascading effect through the organization of adding discipline, rigor, and accuracy to the forecast. It also made me a lot more empathetic to my CRO's issues with respect to the sales leadership team.
- Ad hoc, either internally or in-market. My most successful Heads of Sales have been good at winding me up and pointing me at things as needed, whether that means getting on a plane or Zoom to help close a deal or save a client, or doing a one-to-one mentoring session with a key employee.

Part Six

Business/Corporate Development

Ken Takahashi

Chief Business Development Officer

To Create a CBDO or Not?

If you're an early stage founder—one of a few people sitting in a shared space trying to figure out how you're going to create an actual business where customers will pay you money—the last thing you're probably thinking about is whether or not one of you needs to take on the role of Chief Business Development Officer (CBDO or a similar title) or if you should hire a dedicated person for the role. Even if you've scaled a bit (or a lot), you might think that the CBDO role isn't necessary, that it's a role reserved for Fortune 500 companies, not startups. I'm not saying that you're wrong in thinking that way because there isn't a "one-size-fits-all" approach for when you need a CBDO—if ever. But to help you decide, I'll share how the CBDO role can add value and sustainable growth for companies, regardless of industry, product, or size.

First, though, let me tell you that the CBDO role is different from other roles in a company because it's less clear when and if you actually need that role on a dedicated basis. If you're bursting at the seams with growth, it's pretty obvious that you need more people for Sales or customer service reps, or more people involved in HR or onboarding. That's not true for the CBDO role and while some companies have successfully merged business development, corporate development, and strategy into one role, others have one person for each of those roles. In this Part I've decided to consolidate all these roles into one, the CBDO. Because the CBDO role is often one that emerges through consolidation, there's not a traditional profile of the type of person who takes on that role, and there's not one approach to filling the role, either. You could fill the CBDO role by building it around specific people or a group of people or you could look for one person to take on the role. Some people who become a CBDO are data-driven and numbers-oriented and they often come from investment banking; others tend to be former management consultants. Others (like me) have more general customer-facing operational backgrounds. My personal timeline afforded me a wide array of experiences that built the foundation throughout my career. I will cherish my early years in pre-sales consulting as it allowed me to interact with dozens of businesses every week and be at the intersection of market need and solution feasibility. I then spent a few years in an industry relations role that allowed me to work with peers in our space, including competitors, industry associations, as well as state and federal officials and agencies that were looking to introduce legislation that could impact our business. One of my first responsibilities at Return Path was to grow our business outside the US, which had me interacting with all industry stakeholders from a local country perspective. I had a similar experience when leading a new line-of-business as its general manager. Having had the opportunity to interact with so many different stakeholders in my career gave me a first-hand account from so many different perspectives. Irrespective of the source type, the best CBDOs are independent thinkers and they take a broad, generalist approach to the company and the ecosystem in which their company operates. They are the person who provides a candid, thoughtful temperature check on the market and they're the one who often provides a contrarian perspective.

One common trait of a good CBDO, regardless of background, is that they are people who deeply value trust and they develop trusted relationships both internally and external to their company. Without trust, there is nothing. If you don't earn the trust of your colleagues, they won't prioritize the work you are doing, and you won't get very far. If you don't earn the trust of the ecosystem—of the other vendors, competitors, communities, customers, regulators, and funders—you don't have a chance of getting any deals done. In the end, as CBDO, you are an ambassador for the business.

The CBDO Is an Ambassador

Every position or role at a company is an ambassador, people just don't think of their role that way. I learned three important lessons very early on in my work career that highlight the importance of an ambassador perspective. The first lesson I learned was during my first W-2 job at my local supermarket, Waldbaum's store #242. I spent all four years of high school working there, starting first as a cashier, then stocking shelves, then working in the dairy/frozen department. I learned at Waldbaum's that relationships are everything, especially in small circles. Before the days of the self-checkout scanner, we hand-keyed everything at the register and I prided myself on two things: speed and accuracy. I didn't like long lines and I didn't like my cashier's drawer being over or under at the end of my shift. But I learned over time that certain customers needed different approaches. I started to recognize which customers needed to see the exact price on every item that I rang up while others only wanted to exit as quickly as possible. It was always nice to see the same customers line up on my register, and I made sure to cater to them the way they preferred. What I realize now is that, as a cashier, I was the face of the store, I was the last person a customer would interact with, and how I treated them had a big impact on their opinion of the store.

My second lesson learned stemmed from helping my dad. He moved to the US in his thirties for Mikimoto America and eventually became their COO. One summer, before my internship started at an advertising agency, my father asked if I could step in for a month as their wholesale team was a little light. It was a busy summer filled with numerous client events around New York City. Observing how he spoke and engaged with customers, and commuting time together, taught me that it's not *what* you get done that matters. What truly matters is *how* you got it done. Getting a deal done on your terms but burning the relationship is no win at all. That summer, and on multiple occasions, retail jewelers from across the country had commented to me how much they valued my dad's partnership. Even today, years after he retired, my dad continues to receive gift baskets, photo greeting cards, and hand-written letters from former customers.

The third lesson I learned from making partner calls to AOL during my time at DoubleClick. Back in the early 2000s, our clients were sending a considerable amount of email to their system and the two teams worked together to ensure there was continuity for warranted email. I would go to the Dulles offices quite often to review performance and explore new programs to ensure even more consistency. I learned that focusing your attention strictly on the primary decision-maker is wrong and that to be successful you have to build a relationship with the other people who interact with the decision-maker. From the minute you are greeted at the reception desk or while you're talking to a junior person on the team, you need to remember that they are all influencers for that person, and an extension of that person, really. Once, while visiting his Dulles campus, my partner told me (about his assistant), "Libby controls my day so much that if she wanted, I would walk off a cliff at 2:30 p.m. and not even know it." Fortunately, I had developed a good strong relationship with Libby, and she always made me a priority on his calendar when I visited.

All three of those lessons—understanding customers, how you get things done, and building strong relationships with everyone you contact—are key skills for the CBDO.

Unlike go-to-market teams, as CBDO, you aren't selling your product or services, you're selling or representing the business, its people, and the company's ability to grow and succeed. At some point in your growth your CEO can't be at every meeting, event, conference, or negotiation. That's where the CBDO steps in, that's where you need to be able to deliver a pitch deck, lead a client renewal, soothe an angry customer, host a client event/dinner. You are a senior executive representing the company. Internally, the CBDO brings a different lens, an external one, to the management team. While the majority of the company is focused on internal matters, like moving a single deal or contract across the line, your role is to provide some perspective of the entire ecosystem, to be able to see the "market" because everyone else is heads down on their work.

Much like ambassadors on the world diplomatic stage, if you aren't trusted internally or externally, your life becomes near impossible. But your role as CBDO is more than an ambassador and you'll also be highly involved in partnerships, go-to-market and product strategy, mergers and acquisitions, and general investment priorities. The ambassador mindset is the best way to go about being successful and making a big impact, both internally and externally.

Chapter 70

How to Make the Biggest Impact as a CBDO

The CBDO team is one of the smallest—if not the smallest—in the C-suite. As a startup it will probably be just you and even as you scale to $100 million in revenue or more, you'll most likely only have a team of one or two other people. If business development and partnerships are key to your growth, you may have as many as five or ten people on the team, but in general the CBDO team is small. Sales, Marketing, People, IT, Finance, Legal—all those functions will grow in headcount, budget, and complexity, while the CBDO team will remain small. But the impact you can make as a CBDO is massive and I've organized my thoughts into the key tasks or roles that you'll want to undertake to ensure that you make that big impact, right away.

I've already mentioned that the key traits of a CBDO are trust, relationship-focused, generalist thinking, and an ambassador mindset. In addition to those traits, a CBDO will need to have one clear objective that underpins everything that they do. That objective is to increase the valuation of the company. Yes, everyone in your company has a belief that what they're doing will increase the valuation of the company, but they also have a number of imperatives that are focused on the here and now: they are responsible for weekly updates, quarterly reports, annual reports, budgeting cycles … the list is endless of things people *have* to do in a company.

The CBDO role is different. It's like solving a puzzle without all the pieces and solving a puzzle where the pieces you do have are blurry. The information you gather to inform your decisions are often nuanced, gleaned from conversations with customers, partners, peers, or back channels with competitors. You'll probably never have the hard, quantifiable data to justify your ideas, your hunches, or your intuition. That's why trust is so important internally because a lot of what you're doing as CBDO is providing counterintuitive information that can only be substantiated by your word.

One of the key roles you'll have that's focused on the short term is to develop and work closely with go-to-market (GTM) partnerships to increase sales growth. I use the term "sales growth" broadly here to mean more leads, shorter sales cycles (or quicker "no's"), integrated solutions, selling your solutions via other companies or creating product/service partnerships to broaden your company's value to clients. You can also create reseller/referral partnerships that leverage your client penetration so that you're selling other solutions to clients. This often comes with scale and/or owning a specific part of the market. It's also worth noting that not all partnerships need to have a revenue focus. As we launched Bolster, we knew that building our talent network directly was not a recipe for success. We made a decision to invest in a partner strategy that aligned our business with the myriad of professional organizations with their own member base. Finding the right balance in value (how we can help their members' career trajectory as well as how we help their organization grow), may have an indirect impact on revenue. All of these tactics can be accomplished within a startup since you're adding value to others in the ecosystem—other complementary providers as well as to clients.

In the longer term, you can work to ensure that your company is poised, prepared, and ready to engage in a buy-side merger or acquisition, or a sell-side merger or acquisition. As the CBDO, you're ideally situated to understand the broader context of the ecosystem, and you'll have relationships with potential acquirers or companies that may be interested in buying parts of your company. If you've been operating with trust and a fair amount of transparency, you'll be able to get the alignment internally and externally to make a buy-side or sell-side deal work.

The other tasks that involve the CBDO and increase a company's valuation are fairly unique and might not ever happen for your company. If something comes up that requires a lot of coordination across the company, the CBDO might be the right person to lead that initiative. At Return Path we made several acquisitions, and integrating people, products, and solutions into the company required significant coordination with all functions—Legal, Finance, IT, People. Every aspect of the company is affected in a post-merger/acquisition integration and the CBDO is often the best person to lead that integration. While M&As often focus on additive efforts, sometimes the right move is to divest part of your business. At Return Path, there were a few times where this happened, and it allowed us to focus on and invest in the parts of the business that we felt best fit our long-term growth strategy. While on paper, divesting the business might seem rather straightforward, the reality is, and even more true for the longer your business operates, it can be a tangled web. Customer agreements are typically with the company and not the business unit. Technology resources, including people, usually span across business units. Most likely, you don't have dedicated sales and service teams by business unit. In the end, asking the team who runs the day-to-day operations being impacted to figure out a way to split the business may not be the best idea. For one, they have the business to run. Two, intentional or not, there is a tendency to make a decision on what's best for the person/team vs. the company as a whole.

Another situation that may or may not happen with your company is geographic expansion. I cover this in detail later in this Part, but to preview here, the CBDO plays a key role in managing geographic expansion and to be successful you not only have to coordinate internally with all the stakeholders, but you also have to coordinate and collaborate with companies that are external. If you're expanding outside of your home country, the amount of work can be enormous and takes both a physical and mental toll.

Chapter 71

Building Your Influence Internally

To succeed as a CBDO, you'll need to work effectively with all functional executives to set long-term goals, based on your knowledge and insight of the ecosystem, and you'll have to help them understand the connection between their long-term goals and how they influence short-term decisions. These conversations are not always easy since your CBDO perspective often equates to being the counterpoint or counterargument in many conversations/decisions. And at a minimum, your voice is one that will advocate a change from the status quo. For example, at Return Path, working with others on the Executive Committee (EC), I spent most of my time within the EC's GTM subcommittee, which included the Head of Sales and Head of Marketing, and I had to ensure that our short-term decisions did not negatively impact our long-term outlook. Sales is often compensated on quarterly goals (making the numbers) and there is no penalty to those quarterly commissions if the deal is detrimental down the road. Channel conflict is a topic I will explore more deeply later on but I will cover a bit of it here. There's a tension between how salespeople get compensated and what is best for the company valuation in the long term and I've had situations where we negotiated a "winning" deal behind the back of a partner, but in doing that we ended up creating a large roadblock for quarters and years ahead. Short-term win, long-term consequences. So, having those tough point-counterpoint conversations, having a healthy relationship with others on the executive team that allows you to have difficult conversations, are critical to a highly functioning management team (and in this case, a subcommittee that influenced all revenue decisions).

Another way that the CBDO interacts internally with others is by bringing the ecosystem lens to the entire management team. Aside from some of the GTM team interacting at a micro level with the industry (e.g., a rep talking to a prospect or a service member talking to a client), as CBDO you will have access to strategic people across the entire ecosystem and you need to find a way to relay the important bits of information back to the organization in a digestible format. This type of competitive intelligence oftentimes can only be gathered and shared by you. Sometimes you're getting more clarification on news that the rest of the market might not have access to; sometimes you're getting information early before it goes public, and sometimes it's information shared with only you.

But, in sharing information about the ecosystem, there must be an understanding that the information you receive and share with the executive team has to be kept within the team. If not, if what you share somehow gets leaked, you will lose your credibility and trusted ambassador status within the ecosystem. At Return Path, we used a "First Team" approach that worked really well for this type of information. The First Team idea is that, while all senior executives have functional leadership responsibilities across the company, your first commitment is not to your direct reports, but to the executive team. And no one can breach that trust. No one. Without that trust I wouldn't feel comfortable sharing sensitive and confidential information with the team and if I couldn't do that, the impact that the CBDO can have on the company is constrained. That would be a shame.

As one of the few generalists at a company (your CEO is probably the only other one), you may be tapped to incubate, develop, and lead projects or initiatives that would otherwise fall through the cracks, be unmanaged, or underdeveloped. Some people refer to these initiatives as "special projects," but that's a term I really dislike. We had a running joke at DoubleClick that anyone with the VP of Special Projects title was likely going to leave in the next six months. We were almost always right. For initiatives that are important to really have a chance of gaining traction and succeeding, they can't be labeled "special projects," as if they're something that people can easily dismiss or push off to the side. You won't be able to make much headway if people in your company think of an initiative as a side project or a special project, especially if what you really need is coordination across the company and executive sponsorship.

Here's an example of taking on an initiative that was important but didn't fit within any existing functional area. It also is one of my fondest roles in what turned out to be Return Path's first and only security product offering. Our core product was an email marketing solution, not a security solution, and this initiative was led by George Bilbrey, my colleague and Return Path President at the time, who needed to find a permanent home for the security solution. To give some context here, the email industry has always been littered with technical challenges associated with the openness of SMTP, which is the protocol used to send outgoing messages. One of the vexing issues was spoofing—when someone claims the identity of another entity. Our clients viewed it as a brand protection issue because they didn't want malicious emails being delivered to inboxes around the world, hijacking their brand. We drove a lot of the development toward a security solution by working with industry groups, and we decided we had a product for broader adoption. George knew that it needed a permanent home and someone to shepherd the product concept into an actual business. I recall the conversation he and I had, along with Matt, discussing if we should get someone with general management experience that we had to teach the Return Path business and the industry or an industry long-timer who with some targeted support, could fill out the general manager role. After some discussion, we all agreed the latter scenario fit our business as teaching someone the nuances of the Return Path business was far more complicated than investing in a few development areas for me.

After accepting the role, we pulled together a small team across the organization, and I temporarily relocated my family from the East Coast to Colorado because that's where the majority of the team was located. In less than two years we grew the business from a few clients to a $7+ million run rate business. I found a phenomenal successor, Rob Holmes, to run the business, and I returned to my traditional CBDO role. While we were creating a market for the security solution, we noticed that purchasing decisions and personas within our ecosystem changed—significantly. Now the driver for purchasing email solutions was being driven not by brand management, but by information security. This shift made us realize that we really didn't want to target a completely different buyer and have to create a whole new go-to-market team. It wasn't a big enough opportunity to pivot to, and to keep it properly going, serviced, and managed was not something we wanted to undertake. While it was not overnight, I had to shift from a "selling the product" mindset to a "selling the business" mindset. As noted in the earlier divestiture section, having someone who knew some of the intricate details of the business was a definite advantage, as was, for many reasons, not having the GM, Rob, lead the process. This allowed Matt and me to work on the entire process quickly and we successfully divested the business to Proofpoint—and at the time of this writing, our security solution continues to be a great asset to their overall offering.

To really build your influence internally, I think these are the first and critical tasks:

- Develop strong, trusted relationships with your executive team so that you can help them see the long-term view and opportunities of the company.
- Develop strong and trusted relationships within the ecosystem so that you can gather competitive intelligence to share with your team.
- Be ready to step in and take an active role on projects and initiatives that require collaboration and execution across the company.

Doing these things will undoubtedly increase the valuation of your company, which is your chief contribution.

Chapter 72

Building Your Influence Externally

Unlike other roles in the company, the CBDO has the opportunity to be a "boundary spanner," to make an impact internally and externally in the ecosystem and I've suggested that having the mindset of the company's ambassador to the market/ecosystem is a way to generate more opportunities. By earning the trust of all players in the ecosystem (competitors, partners, and other organizations that sell to the same customer persona), you should have personal access to industry CEOs, CBDO counterparts, bankers, and others. The back-channel relationships you cultivate will provide opportunities, perspectives, and a different level of guidance to your colleagues across the organization.

Being the company's ambassador to the outside world is all about trust—both learning to trust others and being trustworthy to them. As the saying goes, trust takes a long time to develop but can disintegrate in a second. The additional challenge here for you is that it's not just about your actions, which you can control, but it's also about the actions of the entire company, which you can't always control. There's nothing that will destroy your credibility as a trustworthy person more than standing on the proverbial podium saying one thing while the rest of the company is doing the exact opposite of what you are saying. Being a trusted company ambassador requires trust and alignment across the organization and if you have high levels of trust, it can help you sidestep some thorny situations.

What I Look for in a Startup Business Development and Partnerships Leader

Greg Sands, Founder, Managing Director at Costanoa Ventures

If you're the founder of a startup and want to sell through partnerships, look for quality, not quantity. Most often, your chief business development (CBD) lead should seek out at most one big partner who can sell your product and another significant one who can deliver leads. As a result, the ideal person in that role is someone who has not merely worked on partnerships before but has also set up one or more highly productive, commercial arrangements before. Your CBD lead needs to be experienced because you'll need to cultivate deep understanding with one partner, not superficial understanding with a multitude of partners.

The most important thing to remember is that you want to build a close working relationship with your ideal partner or partners so you typically can't go too broad. To ensure that happens, your business development lead should first help you find a partner who meets all of the following criteria:

- The partner is selling to a similar customer and type of company as you.
- The partner is reaching the right buyer profile in that company.
- Your product is a useful add-on to your partner's, that expands its value proposition.
- The partner is able to help you develop additional leads and/or sales.

To build a really successful partnership, your CBD lead will need to understand how to meet the partner's objectives—and often the specific objectives of their sales team (such as quota retirement). If the partner is actually going to be selling your product, your CBD lead should understand corporate objectives, how their reps get compensated, and how to resolve potential channel conflict. Anyone can do a *Barney* press release ("I love you. You love me") but if you aren't helping your partners make money, you won't sell more product. Joint events, seminars, and webinars can be productive but you need to measure them like any other marketing program. That's where having a business development person with a great track record of creating and maintaining a productive partnership is essential.

Make sure you're focused on quality, not quantity, when it comes to partnerships.

For example, we overcame a bad situation because we had high levels of trust internally and externally with partners in the ecosystem. Toward the last few years before we sold Return Path, we finally felt like we hit our go-to-market stride, both from a direct sales point of view as well as a channel point of view. Without question, our partnership with Salesforce Marketing Cloud (formerly ExactTarget) was our most important. We had deep ties with not only the product team but also the various sales organizations (and there were many!) around the world and the entire executive team. We got to this point through a lot of hard work across the leadership team. We obviously had a working relationship but it went far beyond that and our trust was founded on caring for the person, understanding their emotions, their challenges, and their successes. We coined the term "Hearts and Minds," which referred to winning the hearts and minds of everyone at our partner company, and we broadened it to be part of our everyday experience—with everyone. We weren't perfect in dealing with this partner and every now and then, there would be a rogue salesperson who would take the "easy way out" to close a deal and go around the partnership on a deal. We, and not just me, but our sales leadership cemented clear rules of engagement about how our reps were to handle deals that impacted our partner's customers. Months after cementing the rules of engagement across all of our sales teams and reps, one particular rep in the US made a decision to circumvent the rules of engagement and closed a deal that directly impacted (in a negative way) our partner.

As soon as we heard about this rogue deal, we had an internal call between three leaders: the CRO, the Head of US Sales, and me, all hopped on a quick video call to figure out our response. Before I had a chance to talk about short-term wins at the cost of long-term success and growth, the Head of US Sales said that there was no other choice but to terminate the rep in question for blatantly violating the rules of engagement that were put in place to preserve the long-term viability of this most important partnership. Talk about alignment! The CRO and I agreed and we made the difficult but necessary people changes in short order. Had we not acted quickly and decisively, how could we look our partner in the eye and present ourselves as trustworthy and honest in our dealings? They had invested heavily in this partnership as well, with product integration, sales enablement, joint marketing, joint events, and messaging stating their commitment to the long-term viability of the partnership.

You might think that we had some explaining to do with the other members of the executive team, but they were 100% aligned with us. Why? Because we had a high degree of trust and because we followed through on this clear lack of trust (by the sales rep). Our partners at Salesforce never made a big deal out of it, like they did in the past. Why? Because we never gave them the opportunity to. While the ideal situation would have been no conflict at all, we provided the best situation considering the hand being dealt to us—honoring our commitment. It's one thing to build trust, but to maintain it you have to follow through on difficult decisions.

I learned several important lessons from this episode. In my past lives and iterations of partner management and alignment in my career, there wasn't a high level of alignment on partner strategy and there wasn't really agreement on avoiding channel conflict. In some of the companies I've been involved with, partnerships are viewed as a "nice to have" or a "side dish" to the sales efforts. Investments in partner efforts are treated as secondary efforts and the thinking in these companies was that you take care of your own first. I'm not sure whether partner relationships are less interesting because of the revenue share or because it only adds a small portion of the total revenue. But I do know that it often gets less commitment or attention. I challenge that perspective and note that no partnership, no initiative will ever have a chance to grow unless you treat it as primary.

The other lesson I learned from the rogue sales rep incident is the importance of having a solid relationship and strong alignment with others in your company. It's easy to align with your Head of Sales when deals are closing and everything is great. It is very difficult to align with your Head of Sales when sales reps view the partnership as an obstacle to getting a deal done or it requires extra steps in order to get a deal done. Rep mentality can only change if sales leadership is pushing the rules and not just leaving this role to the CBDO team.

The same strong relationship and alignment are true with your Head of Marketing. It's very difficult to get marketing resources (people and budget) to be diverted from their own precious budget and spent on something they might view as secondary. I'll be more blunt: they're not going to divert anything to something they see as secondary. So, having clear alignment with your Head of Marketing, just like you need with your Head of Sales, is vital to place marketing bets on today in addition to tomorrow.

Chapter 73

Where Internal and External Meet: Your Relationship with Your CEO

Another way to build your influence externally is to work closely with your CEO so that you can be a surrogate, an alter-ego, a fill-in for them. Your CEO cannot be on every call or own every ecosystem relationship, even if that's what they desperately want to do. It isn't going to happen because of time constraints, because of travel, because of double-booking, or because of a myriad of other reasons. But you, as the CBDO, have a fantastic opportunity to represent the entire company, to represent the CEO, in those instances. You will be able to connect to your counterparts and to CEOs of competitors and you'll be able to develop relationships with channel partners and other industry players. This part of the CBDO role is similar to what we see on the political stage, where ambassadors represent their country abroad or where a surrogate speaks on behalf of a candidate. It's no different in corporate life and the CBDO role has the same level of responsibility within an ecosystem as an ambassador has in a country. Believe me, if you haven't developed a level of trust with your CEO, they would rather let something drop than send a person who doesn't represent the company and them well. So, there has to be a high level of trust and understanding between you and your CEO.

While we weren't successful at achieving our objectives with 100% of our partnerships during the Return Path years (some our fault and some out of our control), we certainly left it all on the field. Matt and I would have countless meetings with CEOs of potential partner companies, and we'd meet with GMs/MDs from marketing cloud platform companies who were in our partner ecosystem around the world. Matt had many personal relationships with many of these leaders and when he trusted me to take on those relationships, that was something I did not take lightly. Because if a CEO of a company in your ecosystem doesn't trust you, good luck moving the partnership anywhere! I spent time ensuring that these meetings were about mutual benefit, realistic in nature, and driven by trust.

Chapter 74

Influence Meets Operating System

As a generalist on the executive team the CBDO holds a unique position in being able to understand not only the day-to-day demands of the organization, but also the long-term trends in the ecosystem. Because of your dual perspective, you're in a position to build and instill a leadership environment from across the entire organization that is coordinated and collaborative in setting the strategic direction of the business. One strategic area that comes to mind involves the build vs. partner vs. buy dynamic. The choices between building a product yourself, partnering with others, or acquiring a target in the market is a constant, recurring issue in all phases of growth, from startup to exit. But I've seen many companies tend to tackle these issues in silos, which leads to all sorts of problems, not least of which is that it slows you down. For example, Product wants to build what they think is the most exciting product possible based on customer discovery while the BD team wants to partner extensively so they can put the largest number of logos on their partner pages. And Corporate Development might think that pursuing the most accretive acquisition target in the market is the preferred strategy.

The reality is that strategic decisions of this nature need to be a singular conversation. That's where the management team, with the help of others, needs to step up and align on what "solutions" they want to bring to market. Only after stakeholders are aligned on what their top priority is and what resources are available to them, can a discussion be made about build, partner, or buy. But alignment on this strategic focus is not a sure thing and all sorts of factors come into play from both internal and external data points. Internally, the most pressing question to answer is, do you have the skill set to create this product? And even if you do, does developing the product take the company in a different direction? Is it the best and highest use of your development efforts right now? Externally, the questions are not about skills but about long-term sustainability. Do the potential partners compete with you? Are they similar in size or scale as you? What other benefits might they bring? In addition, are any of these players looking to make moves in the market?

Nearly every company I have worked for or consulted tackles this important strategic decision in silos, and each perspective within that silo is valid, each will have data to support their argument, and each will be passionate about their approach. From each of their perspectives, that might be the right decision. But therein lies the problem. The best decision has to take all these views and get alignment on the approach that works best for the whole company, and not just for one function. The CBDO often plays that role since they focus on the entire company and not a specific function. At Return Path, the CBDO role had me leading the partnerships as well as M&A efforts (two of the three—partner and buy). I worked with the Product org to round off the mix (build) and we made sure that all three teams were aware of the solutions we were committing to, who in the market was already there, and if we wanted to take an inorganic path or not. With your perspective of both the internal and external situations, you have the ability to help coordinate and align the company to a holistic, long-term approach to the decision.

Chapter 75

Develop External Trust for the Company

Another group of stakeholders that you'll have to build influence with are the ecosystem participants and, here again, trust is everything. If clients, prospects, vendors, or bankers don't trust you, they won't buy from you—why would they? If the ecosystem doesn't trust you, making a deal—even a small one—is an uphill battle. Partnerships and acquisitions have trust as the foundation followed by strategy and financial considerations, not the other way around. You need to be trusted to be invited to the table, to be able to interact with, and even associate with others in the ecosystem.

This point about being trusted in the ecosystem was driven home to me (again) during an M&A deal. Before launching Bolster (in early 2020), I was advising a company on an M&A strategy, and there was a competitor shopping themselves to the market. The entire management team I was advising was bullish about acquiring their competitor. The company I was advising's private equity firm was leading the deal process, and it looked like we were going to lose this deal to a different PE firm looking to get into the space. While we had a few due diligence sessions with the other side that was led by our PE firm, the video calls were very mechanical, and we were unable to showcase our leadership, management, approach, and culture effectively. I paused the process and suggested we do a deep dive session for each functional area with no private equity or bank representation—just the actual leaders and people operating the business on both sides. A half-dozen sessions later, followed by the CEO and me meeting with their

Founder/Chairman and their CEO, we received the verbal agreement and handshake that they wanted to go with us. While we knew of each other in the ecosystem, they were pleasantly surprised to find that we were practically identical in how we operated our business and our approach to corporate culture. We were not the highest bidder for the acquisition, which goes to show that you can win the hearts and minds of a staunch competitor even without meeting them in person (noting the global pandemic in 2020) and with the understanding that people want to do business (or in this case, merge companies) with people they can trust and relate to.

Chapter 76

Build Your Influence in Strategy

As I noted earlier, having clear alignment on the most important decisions and investments the company will make is paramount to a company's success. Without realizing it, companies often spend the majority of their energy on very tactical or transactional decisions. Those are the "fires" that crop up from time to time, but spending time on those only, and not the long-term strategic positioning of the company in the ecosystem can lead to lousy results. Strategy is not an accumulation of tactics or operational decisions, it should dictate tactics! Management teams must set aside enough time and remove themselves from their day-to-day responsibilities and become corporate strategists on their own business. Naturally, each executive will have a different perspective on the market, but the CBDO can provide a critical perspective of the outside ecosystem to the conversation. At the end of the day, everyone on the executive team is responsible for thinking strategically, and they shouldn't defer that responsibility to more tenured or seasoned leaders. In addition to bringing in the outside perspective, as CBDO, you have the opportunity to quarterback the process, bring in outside resources, and coordinate a deep market analysis. From there you'll need to work closely with functional executives to ensure the decided strategy is communicated crisply and concisely across the organization and all downstream goals and decisions align. Externally, ensure a clear message is communicated and there is no misunderstanding, and avoid possible conflicts with key partners in the ecosystem.

Chapter 77

Building Your Influence in Business Development

Outbound

Some businesses generate 100% of their sales from their own sales efforts. Other businesses have no individual salespeople at all and rely solely on channels to sell for them (think Intel). In technology markets other than hardware, it's likely that a company uses a combination of the two and the choice of channel (internal or external) depends on whether your solution is a primary platform versus an add-on solution. At the end of the day, it is imperative that companies understand the market that they are selling to and who else in the ecosystem is selling to the same customers. This information—what's going on in the ecosystem—is something that a CBDO needs to have at their fingertips because, frankly, you're the only one in the company who will be able to get reliable data. If you (your company) decide to pursue sales through outside organizations, then the worst thing you can do is let this decision be made by a silo within your company. Instead, make sure the entire organization is aware of this possibility and that they know of the potential consequences.

At least three functional areas within your company will be impacted by external sales organizations selling your solution. First, the sales team needs to understand that prospects will now have more than one way to procure their solutions. They need to know the Rules of Engagement when both you and your partner are selling to the same target. I mentioned earlier that even with cemented rules of engagement, it's always possible that you'll have a rogue rep go outside the rules and you'll have to have a clear understanding with the sales team on how you'll handle that situation, if it comes up. In the case I mentioned earlier, we severed our ties and let the rogue rep go, but not all cases are so clear-cut. Second, the marketing team needs to understand that they must provide marketing efforts, resources, and people to the new partner. And at the very least, Marketing needs to know that they cannot change messaging or any promotions without notifying your partners. And third, Finance and Accounting need to be in the loop because the new revenue stream will most likely come with varying revenue share rates (partner-dependent), sales compensation changes, and other accounting/finance challenges, etc. Changing processes on the fly is a horrible experience for the accounting team, sales reps, and others. In the end, the CBDO drives the implementation of external channel partnerships and it's your job to help the functional areas with the greatest impact (Sales, Marketing, and Accounting) avoid mix-ups and frustration.

Even if all the planning goes well and you have alignment between Sales, Marketing, and Accounting on your channel partner, you should expect things to slip through the cracks. I believe that partnership launches are every bit as important as product launches and you should treat them with the same effort and care. The stakes matter in a partnership launch. Having partner companies and their reps start with a bad taste in their mouth is only going to make this partnership harder.

One way to get things off to a great start is by having well-thought-out goals. I've seen so many partnerships that are nothing more than a logo on a website and a contract that gets stuffed in a drawer. I subscribe to the notion of less being more. For growth companies, having a smaller number of meaningful partnerships is far easier to manage with limited resources. With that, it is critical to set partnership goals that are aligned both internally and externally. And it doesn't have to be revenue-focused goals. It can be a joint case study, x number of sales in a geography, etc. These aligned goals (and goal types) will probably change over the course of the relationship, and they should. At Return Path's peak, there was no question the most productive partnership we had was with Salesforce Marketing Cloud (previously ExactTarget). We had clear alignment between both organizations' sales leadership as well as a clear understanding with each finance team around financial metrics. Our quarterly meetings around YTD metrics, new product rollouts, and a focus on underperforming areas were standard—top-to-bottom alignment. This took years of hard work and evolution on both sides, but it was definitely worth it, and I'd do it again.

Inbound

On the flip side of seeking external partnerships, other organizations can leverage your go-to-market capabilities while you augment your own offering and sell outside products and solutions to your customers. Of course, you'll need to make sure there is alignment across the executive team that this is a key solution because, like any partnership, it takes effort and resources to make it work. The executive team is more likely to consider a partnership if they have already vetted the build vs. buy options and the last option available is a partnership.

Similar to the scenario in which you seek out partners to sell your solution, the same level of work, collaboration, and leadership needs to happen if you're going to make any traction on selling someone else's solution. The entire organization needs to be made aware of this partnership and any and all concerns must be addressed before launch. You'll have to work closely with Sales, Marketing, and Finance as before, but you'll also have to create alignment with Product since you're adding products to the equation.

For the sales team, you're likely to get questions like:

- What am I selling?
- Why would a client buy another company's product through us?
- Better pricing? Better terms?
- Integration with our offering?
- Or is it the same as the client buying from the source company?
- How does this impact my compensation?

- Will I be compensated on the gross amount or the net amount our company recognizes?
- Does this provide quota relief?

The product team will ask how to position a new offering among the existing products, and they'll wonder if there is integration required. And Finance and Legal will have questions like, our paper or theirs? What's the revenue/collection flow? How do we handle bad debt?

I look back at two inbound (technology) partnerships that, on paper, should have had the same chance for success but had two very different results. One of the companies was 8 seconds out of Europe and the other was BriteVerify (and as fate would have it, BriteVerify was acquired by Validity before Validity acquired Return Path). Both companies provided adjacent solutions to Return Path. Both wanted to leverage Return Path's large client base to drive their go-to-market engine. But only one of them was successful. What made the difference between a successful partnership versus a failed one? For brevity, I will note the top three here. One, simplicity. With BriteVerify, we ensured we had clear and simple messaging on the solution and why we were offering it. Every rep knew it and never had to refer to any support materials. Two, a single go-to person on our side who knew everything about the partnership. Whether it was a sales rep, a support person, someone in Accounting, we made a point, after making mistakes over the years, of ensuring that this was not running off the sides of a few people's desks but was a primary responsibility of one person. Lastly, channel conflict—or the lack thereof. I will go much deeper about channel conflict in an upcoming Chapter but it is worth noting that much of that learning impacted our decisions here. We ensured enough revenue share to provide reps full comp (off the gross) and quota relief. From a numbers standpoint, there was no difference if they sold our product or BriteVerify's. We also made very clear rules of engagement with BriteVerify. While it may have seemed onerous to them at the onset of the partnership, they realized any ambiguity or friction at the rep level would reverberate across Return Path and scuttle the partnership.

To the extent possible, thinking through the consequences of an external partnership from all the potentially impacted functions—Sales, Marketing, Product, Finance, Legal—before you start advocating for a deal is your best approach. And as I mentioned earlier, the "less is more" thinking pays off here because each partnership is unique, each has different people, culture, products, and importance and to reap the rewards, you have to be actively involved in the deal from start to hand-off. Ideally you'd want a single point of contact to own this relationship for internal purposes, to coordinate the myriad of questions and operations around the partnership. While there will need to be support from the entire organization, someone within your company needs to be responsible and accountable for each partnership. Selling a solution you don't control has inherent risks and challenges that need to be managed. And for a truly successful partnership, this structure and associated commitment should be mirrored on the other side.

Chapter 78

When Things Go Wrong in a Partnership ... and They Will

No partnership is perfect. And by "perfect," I mean something that once started just naturally evolves and blossoms into a fruitful win-win-win for you, the partner, and customers, free from operational hiccups. Think of a partnership like planting seeds in spring. If you don't water those seeds, if you don't tend to them, if you don't fertilize them, they might still grow, but they'll never be as healthy as the seeds you do take care of. Let me be clear, in a partnership, things WILL go wrong. It's how much you prepared for it and how you and the rest of your organization handle it that determine whether or not a partnership will succeed.

Your first partnership will be tormented by a zillion paper cuts and possibly one big one: your sales reps. While a lot of issues are small, just remember that from the sales reps' perspective, if the effort outweighs the benefit, no one will support it. People often say that salespeople are coin operated. While I don't like that analogy, it does highlight the reality that commissioned employees make 99% of their decisions on their ability to earn. If a sales rep finds it easier to sell product A versus product B, guess which one they will sell?

The other situation you're likely to face is when a company and their partner have the ability to sell the same solution to the same audience. Enter channel conflict. A set of rules (Rules of Engagement) must be clearly outlined and adopted by your company and your partner in the channel. Early in our Return Path days, we struggled with channel conflict. There was little alignment between our partner team and the sales organization. In fact, there was a divide within the organization where there was clear animosity between the two teams. The challenge we had that aggravated the conflict was that many of our partners offered some of the solutions we provided as a managed service. Without going into the details, they were over-selling their capabilities and/or under-valuing what we were providing, and that not only led to channel conflict, but caused a rift internally for us.

What I learned from that experience is that, to minimize channel and internal conflict, you need to think about and address the following issues. The sales reps need coaching or help to understand that the partnership is additive to what they're already doing. Your key goal as CBDO is to help everyone understand the situation. Why do we have this partnership? It could be that the partner is a bigger and more influential organization than we are. It could be that their solution is an anchor solution and ours is an add-on. It could be that we've calculated that giving up margin for volume is in line with our growth strategy and profitability goals. But if sales reps don't understand why you're partnering, then there will always be resistance. So, treat the sales reps as critical to making this work and bring them into your thinking early on in the planning stages of the partnership—don't just set up the deal and expect them to follow.

There's a saying that you need a paved runway for smooth operations. And by that I mean, if closing deals with a partner are littered with hurdles and exceptions, no sales rep will want to do it. The entire sales process (marketing, lead generation, pre-sales support, onboarding, billing/collection, sales ops, etc.) must be mapped and ironed out as thoroughly as you would with a direct sales process. This effort requires commitment and involvement from the entire organization, not just the CBDO.

You'll also need clear, explicit compensation details within the Rules of Engagement. Without a doubt, every salesperson is calculating their sales math—a deal's impact on their commission check and quota figures for accelerants. The number one way you can scuttle a partnership is when sales reps realize that they are going to get pennies on the dollar for closing a deal with a partner versus closing it directly. In almost every situation, I strive for 100% compensation and quota relief. If it is our product in question, it removes the desire/temptation for reps to block out a partner on a deal. And if it is an outside solution we're selling, it removes the prioritization around that product. Partners have always balked at my desire to compensate my sales reps at 100% (likely more around the revenue share I was negotiating) but they soon understood that this partnership would go nowhere unless sales reps on both sides are incentivized to sell. Again, the "less is more" idea applies here. It's a lot better to have a small number of highly effective partnerships to manage versus a laundry list of players that causes confusion across the organization.

Finally, to reduce channel conflict and to really make a partnership work, you'll have to have complete trust with and belief in your counterparts. During my DoubleClick days, we made a decision to divest our web analytics platform to an industry leader because we realized that the product wasn't core for us and there were more robust solutions out there. With our divestiture to Omniture (now Adobe), we also made sure those clients had a great home and that we could continue to sell their platform to our clients. I was spending a good amount of time in Utah with my counterparts there—John Mellor and Aaron Watson. With the acquisition and partnership framework being spun up rather quickly, we had our fair share of hiccups along the way. What got us through that period was a mutual understanding and trust that both sides wanted to grow the partnership and would commit whatever we needed to get us there. While many partnerships would revert to the words in the actual agreement, we often used the phrase, "spirit of this partnership" when we had to make tough decisions. The "spirit of a partnership" will always garner a sense of mutual commitment and growth in it. Plus, it keeps legal out of the negotiations until they're actually needed, which makes things much smoother (and faster!).

Chapter 79

Geographic Expansion

Geographic expansion comes across as a glamorous aspect of business. The idea of international travel, business dinners, meeting new people, and learning about new cultures are things that a lot of people want in their career. The reality is that it's a grind. The planning phase is mostly quantitative and model-/assumption-driven and the flag planting can be tedious. There are countless red-eye flights, meetings all day (including breakfast, lunch, and dinner meetings to make the most of your time away from home) and nights spent either doing your regular "day" job or sleepless with jetlag. Don't get me wrong, I made great friends and I've been able to go to places I wouldn't have gone to otherwise because of geographic expansion, but the work you put in, the physical and emotional toll, the time away from your family, hobbies, and friends, is significant.

The reality is that not all companies need to expand into new geographies. Markets are fragmented, they have their own unique nuances and very well may have significant regulatory requirements you need to be cognizant of, and the Internet is global, no matter where you are. And for those companies that should expand their footprint, there isn't a magic metric that tells you when you should actually make that investment. The decision to enter a new geography is one that requires coordination and alignment from the entire management team. It will require a significant amount of market research and preparation (both model-driven and in-market discovery) during the consideration phase.

At a minimum you will need to dig into areas and be able to answer questions like, Is there a need/market? How well positioned are we? What resources do we have there? What is the degree of difficulty? Which markets do we pursue, and in what order? Do we have allies?

Geographic expansion is not a decision to be made in a vacuum. While early research can be done in small groups, the consequences of a decision need to be thought through carefully with senior leadership. The reality is that even a simple decision like "sell to European clients" can trigger a series of operational realities that really require commitment from the entire management team. By selling to Europe, the expectation from European clients is that you will take the proceeds from this sale and provide more resources there (and eventually enter it), even if expectations were set otherwise. If you don't meet these expectations, the client will likely churn. As more European clients come onboard, the stakes become higher and higher. So now the company is on the path of needing more sales and service reps to maintain enough presence in the market, and with more sales and service reps, you'll likely need a regional marketing person. But it doesn't stop there because with a larger group of employees, you'll need even more service people and then a local HR generalist, and so on and so on.

This happens at the same time that the rest of the organization is being stretched, having to keep up with local European payroll and labor laws, finance teams creating stop gap operations, product teams being asked for region-specific features, etc. Sure, some companies can stomach it if the growth curve is manageable. But for fast-growing companies, it can be quite painful. That doesn't take into account all of the product development and other projects that didn't get done because we were supporting the European efforts. In short, the CBDO's job is to paint a picture (or a timeline, really) of what entering a market really means, what the commitments from around the executive team have to be and making sure that you have a firm "yes" to the commitment. Anything short of that will result in frustration and misalignment across the management team. It will also be costly.

Flag planting, or actually getting something going in a different location, is the end result of all your pre-work. But it only happens once alignment has been achieved and commitments are made. Now comes the execution. Do you hire local talent right away? Do you assemble a SWAT team and send them on assignment for a period of time? In any scenario, executives and senior leaders from the organization will need to allocate time for considerable travel in order to support the expansion efforts. As CBDO, you'll play a key role both from an internal and external perspective. Internally, do the teams have what they need? Is the rest of the company responsive/supportive? What dynamics exist in the new market that don't exist at home? What roles do we need to fill and when? Externally, you as CBDO become a spokesperson for the business in that market and may have to play the role of Managing Director or country manager until someone can take your place. Again, the CBDO will be the ambassador of the business (both externally and internally) during the growth phase.

The final element to pay attention to if you expand geographically is culture. Specifically, what sort of corporate culture can you cultivate in a different culture? As important as the hard metrics are, ensuring that your company's values exist in the new region is critical. Early hires and leaders in a region set the stage for whether or not your company values are embraced, so it is imperative that the best possible candidates are chosen. While there will be a desire—and a need—to fill roles quickly, finding the right match is the bigger priority. By populating a new geographic expansion with people who have the skills, mindset, and values, you can ensure a version of your culture, identity and leadership framework will grow thousands of miles away. The easiest way to do that is to relocate an executive to make it all work. We couldn't do that and our first two hires were disasters—we had to let each of them go. But the advice you should give to anyone helping to plant a flag in a new geography is to tell them that their job is to blend company culture with local business culture.

Chapter 80

M&A: Buy Side

One of the best parts of the CBDO role is when you're considering mergers and acquisitions (M&A) as a growth strategy. Not every company will do that, but if you do, here are some pointers that I've found to make every deal work better. First, you'll need a robust understanding of the ecosystem and be able to provide hard data to others in the company. At the minimum, you'll have to keep track of all the players, and not just your competition but other solutions your primary personas are engaging with (share of wallet). As noted in the earlier chapters, once the entire management team is aligned on which solutions are needed to bring to market, the team must also exhaust all options (build vs. partner vs. buy) and align on an approach for each solution. Yes, you read that right. Even if you're not (at the moment) pursuing option #2 or option #3, you'll still need information, economics, pros and cons for those options, along with option #1.

All the events you've been to, all the people you've met, all the conversations that you've had in the ecosystem will come to bear on the M&A question, and you'll leverage the ongoing ecosystem findings to bring to your colleagues all potential targets for acquisition. This information should include things like relative size (revenue, # of clients, employees), solution strengths/weaknesses and solution overlap, etc. The CBDO should also bring any possible conflicts or concerns of each player along with market comparables with recent transactions in the related space. You'll need answers to the following likely questions:

- Are the targets using similar development methodologies?
- Similar technology stacks?
- Are there duplicative systems?
- Do they have similar or different buyers?
- Are they using similar sales philosophies, or similar sales systems?
- What does combining GTM efforts provide?
- Is it additive?
- Are there overlaps?
- Can the combined businesses open up new markets/buyers?

And that's not all. To provide a robust picture of the space you'll need to know, for each target, whether you have similar cultures and employee development, and whether or not there are cost synergies. How much redundancy can you expect, and will you be able to repurpose staff?

This is where your work on becoming a trusted and reliable resource to the ecosystem (especially counterparts and industry CEOs) pays dividends. To be effective, the CBDO and team ideally need to have access to counterparts and CEOs. Board members and VCs can help—maybe make some introductions—but the networking that you do, and the trust you develop in the ecosystem should be well underway before an acquisitive posture. The private equity world is a little bit different because companies get traded a lot more. But like I said earlier, if you're not trusted within the ecosystem, you won't be able to get anything done when you really need a deal.

In every M&A deal, you'll be counted on to provide an external view of the market to the rest of the company. CBDOs must bring an "outside in" view to the company. Client-facing teams can provide the micro (per client) view and while that view can be high quality, it's too narrow to base decisions on. That's where you come in as CBDO because you'll bring in the macro (ecosystem) view to the management team. You may be one of the only people who has access to the macro, ecosystem, perspectives.

You'll also be counted on to head up any due diligence. Even before interests are aligned, the CBDO is responsible for accumulating and organizing all details regarding possible targets. As target lists begin to narrow, the depth of information collected will increase dramatically. In addition, you'll be the primary communication gateway with all possible targets and you must ensure that all information requests and your responses to those requests are clear and flowing quickly. A key part of your role is keeping all diligence teams on time and on target throughout the process. It is imperative not to do this in a vacuum but to engage with the management team appropriately through any process.

When it comes to valuation, deal negotiations and execution, you'll most likely work with a small team that includes at least your CEO and CFO to work through all the data and decide when to bring in outside resources. So, while you might look forward to the deal-making aspect of the CBDO role, the actual work that goes into understanding whether a target makes sense for your company is significant. Being able to whittle down a large number of targets to a few good options will serve you well, especially in fast-moving markets. Working with a smaller list of potential targets will also help you be realistic in your estimates as well as timing.

If you do go through with an M&A deal, have an agreement and a signed contract, your work is not done and you'll have to turn your focus to integration. Often, the CBDOs will either lead integration efforts or work with an appointed integration lead. After making a considerable investment (cash and/or equity), it is imperative that the post-deal integration goes as smoothly as possible in order to maximize the deal value. Whether or not you're leading the integration, the CBDO must ensure all efforts and resources are being utilized for the best possible transition.

Chapter 81

M&A: Sell Side

You may never be acquired—you and your team might not want to do that, you might want to go public and maintain independence, or keep growing your business privately, but as CBDO you need to prepare for the possibility of sell-side M&A. You'll want to conduct an ecosystem analysis of both strategic and financial sponsors and keep track of all possible suitors (present and future), especially market leaders and acquisitive players. While it may seem like bad luck to start planning your exit, you can never plan for it early enough. It is important to know your options at any point in time should the opportunity arise, and being contacted as a target can happen at any time. Getting to know your competitors, market/category leaders, those showing acquisitive behavior, is never a bad thing. Your understanding of ecosystem dynamics also overlaps with the earlier sections on ecosystem and trust development. Depending on your size, you should consider both strategic players as well as financial sponsors. While the private equity landscape can look daunting at first glance, you can easily trim the list down based on the ones who have invested in businesses that are similar to you. It's also worth noting how life will differ after going public, or selling to a strategic buyer versus a financial sponsor.

To build your ecosystem impact, start by developing a networking strategy to put in front of key players. Similar to the buy side M&A discussed in Chapter 80, you cannot develop a network during a process. Ideally, your network is well established long before any possible sell-side activity. And the network you develop should be broad, and you ought to work on becoming a trusted/reliable resource to the ecosystem, especially with Corporate Development leaders and bankers. One avenue to help develop your network is to become a go-to resource for Corporate Development leaders in market leader organizations as well as the investment bank community. Providing them with your perspectives as they look at their ecosystem is a great way to get intertwined in their world. During my years running Corporate Development, I took countless calls to provide my viewpoint and opinions on certain market segments, players, trends, etc. While I ensured I answered all the questions being asked, I also made sure to use the introduction as an opportunity to call upon that person down the road.

Depending on the size and stage of your business, it may make sense to hire a banker to represent and advise you through the process. In some cases, it will be proactive—when you and the Board think it's the right time to test the waters. In other cases, it may be reactive after an unsolicited inbound inquiry. In any case, it's best to be familiar with a few bankers early.

Similar to the M&A buy side, you'll be counted on to provide an external view of the market to the rest of the company. As with partnership and buy side strategies, being able to convey who your potential suitors are to the rest of your management team is critical. I will say that not all CEOs will run their organizations that way, but you need the information on the ecosystem at your fingertips regardless of how broadly that is shared internally. People in sales and service leadership will only know specific divisions of companies, based on client interaction and feedback. While helpful, the information is fairly narrow and as CBDO your job is to take a more macro view and understand large multi-divisional/national platform companies and why your company may make sense as part of their world.

Another task you'll have to undertake is exit roadmapping: working with the CFO and other executives to prepare to sell the company, due-diligence readiness exercises, compiling the legal/corporate records, determining security/privacy readiness, and ensuring that operational data is available. Irrespective of your exit path, there is a series of standard work that needs to be prepared in order to sell the company. Like many other items, it is best to get a head start on them.

You'll likely be responsible for due diligence and response management. When a larger organization or PE firms look to acquire a target, it often comes with volumes and volumes of due diligence requests. These often come in waves and in non-structured formats. These requests could take the form of formal or informal data and/or a myriad of other requests. It is imperative that you organize the inbound requests, push back where appropriate and guide your colleagues and keep them on track/schedule, accordingly. Your banker can definitely assist in the process but much of the internal wrangling falls on the CBDO.

Finally, if you are acquired, you should expect that larger organizations will have integration teams at the ready. As CBDO you will likely be the initial gateway for that team as they look to develop an integration plan. In my experience, private equity firms may have a smaller onboarding team but the company operations will largely stay intact on day one.

Final Thoughts

I believe that the CBDO role is the most exciting role in a company. It's intellectually challenging, it requires nuance in people skills, it has elements of product and sales, data collection of hard metrics, information gathering of soft metrics, and analysis, and it is a role that allows you to develop lifelong trusted relationships with people inside and outside of your company. I said in the beginning that *how* you get things done is just as important, if not more so, than *what* you get done.

As a CBDO, you can do everything right, you can follow all of my advice about understanding the ecosystem and becoming a trusted, reliable person in the ecosystem, and a deal won't get done. On the other hand, you can ignore my advice, you can make every rookie mistake and get lucky and get a deal done. I don't mean to imply that what you do doesn't impact the outcome—it can and does—but sometimes your effort doesn't lead to ideal results. So much is unknown about how the ecosystem will respond to situations that it's difficult to foresee how things will play out. Also, there are a lot of things that are out of your control and you can do everything right to try to sell your company, for example, and something unknown to you can happen at the buyer side that you couldn't control or even see.

I think of the CBDO success rate as closer to baseball than to basketball: in the NBA the average field goal percentage is close to 50%; in the MLB the batting average is close to .250. The CBDO success rate is very much in the MLB ballpark.

That's why I believe that operating with trust, being a good partner, helping people inside and outside your company succeed, form the best way to approach the role. Ensuring that *how* you go about being a CBDO can never be a question.

The CBDO role isn't for everybody, but if you want to represent the company that everyone loves to work with, if you want calls from a bank, VC, or industry player about your views on the market, if you want to get out early on something and position your company for success, based on the information you're able to gather/compile and analyze, then the CBDO is a great role for you. There's nothing more gratifying than being trusted by your peers, CEO, and the board to take on tough challenges and to increase the valuation of your company.

Chapter 82

CEO-to-CEO Advice About the Business/Corporate Development Role

Matt Blumberg

What comes before a full-fledged CBDO? In most startups, not much of anything. Usually strategic partnerships and M&A are handled by the founder/CEO, or potentially by someone in sales. If a business is partner/channel heavy, that may be the focus of the sales team in general. Or for M&A, external advisors/bankers may be used.

Signs It's Time to Hire Your First CBDO

You know it's time to hire a CBDO when:

- You wake up in the middle of the night wondering about M&A buy or sell sides.
- You are spending too much of your own time discussing things with people outside your company.
- Your Board asks you for your M&A roadmap, and you don't have a great answer and aren't sure how to get to one.

When a Fractional CBDO Might Be Enough

A fractional CBDO may be the way to go if:

- You need help defining your partnership or M&A strategy, or creating a market map, and you don't want to rely on an external advisor or banker for it.

- You need help executing a couple of M&A transactions that are too small for a banker.
- You need help defining a major new strategic building block like "creating an indirect sales channel" or "international expansion," and you and your whole leadership team together have no experience in that area.

What Does Great Look Like in a CBDO?

Ideal startup CBDOs do three things particularly well:

1. They have a good balance of all three core components of the role that Ken outlined in this Part: partnerships, M&A, and strategy. Even if they started their careers out as an investment banker or a management consultant, they should be able to bring all three competencies to bear to help further the goals of their team—to optimize the company's place in the ecosystem.
2. They look at business strategy first before trying to solve a problem. Forget about being able to develop business strategy per my last point. CBDOs also have to return to that strategy constantly—especially the CBDOs who come out from banking. Otherwise, they risk becoming the proverbial hammer in search of a nail.
3. They see the whole system at a company: product (and all of its components) as well as go-to-market (and all of its components). Like the CEO, CFO, and Chief People Officer, the CBDO needs to have a holistic approach to everything and not only be closely aligned with the market-facing organizations.

Signs Your CBDO Isn't Scaling

CBDOs who aren't scaling well past the startup stage are the ones who typically:

- Throw everything over the wall internally. Some people in this role, especially ones who have been long-time bankers or consultants and who are used to having armies of junior resources at their disposal don't like or don't know how to roll up their sleeves and handle execution. The reality is that in-house BD teams are very small, frequently only one or two people, and the person leading the team needs to do a lot of the work, not just the planning and external meetings.

- Have an over-reliance on outside advisors like bankers and lawyers. The whole reason companies in-source this role is that they expect to have a fair amount of activity—developing partnerships, executing a roll-up strategy, building the channel. While external advisors are critical for a number of those activities, knowing when to (and when not to) hand things off, especially when the advisor bills by the hour, is critical.
- Focus on quantity over quality. There are times when it's important to be able to show a large number of partners, for example, if you're trying to run an industry-wide coalition or program. Or when it's important to show a lot of deals in the pipeline, for example, if you're pitching an M&A roll-up strategy to a potential financial sponsor. But you know your CBDO is in trouble when the focus becomes the number of deals in the pipeline as opposed to making sure there are a few larger ones with deeper, multi-faceted relationships that will move the needle on the business objectives.

How I Engage with the CBDO

A few ways I've typically spent the most time or gotten the most value out of CBDOs over the years are:

- Making ecosystem maps. Understanding exactly what ocean you're swimming in, which other fish are swimming nearby, and which ones are sharks you need to watch out for, is an important part of the CBDO role—and one that is important to engage with and help shape since you'll have specialized knowledge of some of the other players, their CEOs, and their strategies.

- Working sporadically. The deal world is intense. When one is going on, you may be talking to your CBDO 20 hours a day. When it's business as usual, you may go weeks without deep interaction.
- Working in-market and in-transit. As with the CRO, I spend time extensively with the CBDO since we are likely going to the same place at the same time a few times a year. Since the essence of the job is as a trusted ambassador on all fronts, as Ken identifies correctly in this Part, the CEO has to constantly be engaging the ambassador on the organization's most current thinking, positioning, forward-looking strategy. Over the life of Return Path, there's no question that I spent the majority of my "planes, trains, and automobiles work time" with Ken.

Part Seven

Customer Success/Account Management

George Bilbrey

Chief Customer Officer

Welcome to the New World

Nothing in my background would ever lead you to believe that I am a passionate advocate for customers—not even my part-time job working in a bagel shop during college. Most of my roles have been quantitative and analytical, from my first job as an economic analyst, to my role in consulting, to all of my roles within startups. Over the past 20 years I have been a General Manager, a Product Manager, and had a variety of roles on the service side of the business. It wasn't really until the last few years of Return Path (before our exit) that everything came together and I became the Chief Customer Officer (CCO) with the global service organization reporting to me.

I've learned that to be an effective CCO, you need to think like a generalist because you'll be collaborating across the organization, and also externally with channel partners, distributors, and sales reps, and constantly reminding employees that the company exists because of customers. The purpose of any business is to do the jobs that the customer is hiring them to do. It's easy to lose sight of that as you grow. A big part of what the leaders of a business need to do is stay laser-focused on what that job is, and make sure that you're doing it better than anyone else and making it as easy as possible for your customers to do that job. If you lose sight of that, you'll start to see your churn increase, you'll see customers defect to competitors and leave space in the marketplace for people to enter (and potentially beat you). It requires constant effort every week to keep the customer at the forefront of everyone's thinking. I saw my role as CCO as largely asking the company, every day, "Hey, but what about the customer?"

The role of the customer in business has always been primary, of course, but has changed over the last 30 years. In the early 1990s, if you wanted a salesforce automation platform, you would buy Siebel Systems at a very high cost. Siebel would be installed on servers that were in your data center. You might have bought some professional services along with your Siebel license and you'd probably be spending a little bit of money on a maintenance contract after you had that initial implementation. Over 90% of the revenue was up front and 10% might be in that maintenance contract.

Software migrated to the cloud. Very few software systems are "on premises." Now software is bought on a subscription. Software-as-a-Service (SaaS) has dramatically changed things. Most of the revenue comes from retaining and growing revenue. If you're selling Salesforce rather than Siebel, most of your revenue doesn't come up front, but comes every year with the renewal of the contract. In addition, there's an opportunity for you to upsell and cross-sell. An account executive at Salesforce might be looking at the opportunity to sell additional seats into the enterprise. If you are buying a CRM package from Salesforce, maybe they can upsell the Marketing Cloud to you as well.

So, what does that have to do with the CCO? Well, I believe a SaaS business, particularly businesses that focus on material enterprise customers, requires a C-suite executive who is in charge, making sure the clients reach their desired objectives through their interactions with the business. The CCO makes sure that we're doing the job the client paid us to do.

A CCO does that by leading the teams that allow the customer to get value from the relationship with the company. For example:

- The onboarding team might help configure and connect the products. The training team might build the curriculum that allows the client to learn about how to use the product in basic and more advanced ways.
- The customer success team might actually provide the training (or it might belong to a separate customer training team), while the support team might provide assistance in using the product.
- A knowledge team maintains the help center (knowledge base) so that the client can research more about how to use

the product and employ the product to help improve their business.
- The customer success team will perform the executive business review or the quarterly business review that helps your company see how customers are using the tool and see maybe new ways they can use it to drive more value in their business. These reviews may also highlight that there's additional revenue opportunities inside of that company.

A company needs an executive to lead these disparate organizations. It's a complex set of functions that are individually difficult to master.

Why is this a C-suite job? Why isn't this job just "head of service?" It's because the CCO also has to coordinate across multiple functions *outside* of service. For example:

- Sales should be selling to the optimal types of clients and set expectations appropriately for the jobs that the company will do for the customer.
- The product team must make it easy for the customer to onboard and achieve their objectives.
- The marketing teams need leadership to ensure that they help clients understand all the features or that they're aware of a webinar that is being used for training customers. The customer marketing function is something that must be coordinated across boundaries.

As these examples show, there must be a tight relationship with the sales team, the product management/engineering team and the marketing team (among others).

In summary, a business needs a CCO because they require someone who can coordinate across service functions and perhaps, more importantly, someone who can coordinate with the other senior executives to *make sure that the company as a whole is helping the customer achieve the objectives they set out when they signed a contract with a company.*

Chapter 83

Five Misperceptions

Customers

One of the things I wish I had learned earlier in my career as CCO is that there are common misconceptions about customers and the service organization and you can spend a lot of time dealing with issues that are not real, but perceived. I have identified five common ones, although I am sure there are more.

Misperception #1: Service organization fully controls churn (customer attrition). Frequently you'll see the service organization be measured *solely on customer churn.* If you really think about it, there are many elements that come into play that impact churn, including: how the customer is sold, the quality of the product, how easy it is to onboard the customer, how easy it is to use the product, or even how easy it is to understand what kind of value you're getting out of the product. The service functions do have a critical role, but they're not the only functions that impact churn. The responsibility for churn lies with all of these teams since they all have a critical role. One reason why you need a C-level senior person in charge of all service operations is because they have to be able to understand the customer experience broadly and they'll have to work cross-functionally to ensure customer retention.

Misperception #2: The service organization is just a cost center. In many businesses, if a function isn't generating new logo revenue, it's seen as "second class." I would argue that revenue retained is revenue gained. As noted above, the service organization doesn't control churn but they do have a big impact on it. It's also worth noting that in most organizations the account management portion of a service organization is in charge of up-sale and cross-sale opportunities. These two areas are an important source of growth. I believe CCOs should work within their company to alter that misperception of service as a cost center because the service organization can have a huge impact on revenues.

Misperception #3: Service teams should focus on preventing defections. I've recently found a situation where the customer success team is built to focus on churn. Specifically, the focus is on the clients who have raised their hand and said, "I want to leave." This reactive approach drives low job satisfaction and isn't the "best and highest use" of our service team's time. By the time a customer is frustrated enough, or isn't seeing the value enough that they want to leave—you've missed a window of opportunity. The right focus should be proactively helping customers reach their desired business objectives. If you can do that, most customers will stay. That's the theory behind the rise of the customer success team and that's what great companies are doing today.

Misperception #4: Service's job is to "paper over" gaps in the product. There is a widespread practice of covering for product issues by throwing service at the problem. That certainly can work, but it's not optimal. The superior approach is to focus the service team on becoming a trusted advisor for customers, helping those customers achieve their desired outcomes. To do that, the CCO will have to work cross-functionally with the product team, the marketing team, and the sales team to drive a more friction-free customer experience.

Misperception #5: Service is boring and tactical. There is a wide-spread misperception that working in the service organization is boring. It's mundane, it's tactical, it doesn't appeal to people who think strategy is grander than tactics. I don't agree with that at all. A great service organization starts with a strategy. It starts with an understanding of customer segmentation. It includes thinking about the different customer personas and how to define an appropriate and valuable customer experience. That core strategy actually takes a while to develop. Once the strategy takes hold, it is core to driving retention over time. And, while a lot of people perceive that the service organization jobs are boring, or just answering trouble tickets or reacting to client problems, that's not the whole role. It is a strategic role as well.

Chapter 84

Startup Customer Success Organization

If using the term Chief Customer Officer, you may think that only a large, established, organization would have a CCO on their team. If you're a startup, it would be easy for you to dismiss some of what I'm saying. You're likely to think, "We're not there yet. We haven't scaled to the point where we need a CCO—we're just four developers and a salesperson." The CCO title is more than semantics and as a startup, you should think about creating that role immediately, even if the title is "Head of Services." Customer Success is an extremely important part of the early life of a company. Getting client feedback and being with that client while they onboard, while they use your products, is critical. It's even more important to bring that feedback back into the company. The product organization needs to understand the "rough edges" of the customer journey that can be improved by better product design. The sales and marketing teams need to understand which customer segments are a fit and which are not. Ensuring the client journey is crisply defined and consistently drives customer outcomes *from the beginning* will ensure the scalability of your business.

When you have a senior person on your leadership team that is given purview over Customer Success, it's a clear signal to the market that the customer is at the center of everything you do.

Once you get beyond a single person responsible for the customer success organization, you'll probably add people in a logical way based on your needs. It's common to create a support team that's answering questions. There's definitely a need to build a knowledge base/help center. Support and "knowledge" can be one person initially—eventually, you'll need to specialize. You'll probably want a service operations function that runs the different software systems that you use to run your service operations (Customer Success platform, Support Systems, Chat, Call Analytics, and others). This team would handle tech touch communications out to clients with reminders like, "Hey, I've noticed that you haven't used this feature. Here's an email with a video explaining why this feature is so cool."

Even as a startup, you'll want to quickly add people who can help ensure that the customer experience is exceptional.

What I Look for in a Chief Customer Officer

Guy Turner, Managing Partner Hyde Park Venture Partners

I am an early stage investor, so what I look for in a Chief Customer Officer is for there not to be one. This may be contrarian to the point of the book, but companies less than $10 million (and maybe much higher) in revenue should have such a unified and pure focus on customer needs that a separate CCO isn't necessary. A CCO becomes necessary when some of the byproducts of success and scale—diversifying products, expanded geographies, maturing competitive space—put a company's customer mission and experience at risk. At that point, a C-suite hire with a singular focus on the customer may be necessary to harmonize the customer experience across business functions.

At the early stages, a company should have one product progressing systematically to greatness with each new customer as the customer's needs, concerns, and experiences are relayed from Sales and a basic Account Management function back to Product. At the same time, Product itself watches leading indicator metrics through a well-sensored product. Harmonizing this effort is a CEO who spends 65% of their time focused on customers—selling to them, talking to them, and prioritizing product needs for them.

If an early stage company has a CCO, it may be a sign that the CEO isn't as interested in the customers as they should be and has abdicated this singular focus in favor of other part-time CEO roles like fundraising, hiring, and culture. All three are important tasks but secondary to the singular task of understanding customers. No customer, no problem to solve, no company.

Chapter 85

Scaling the Service Organization

As you scale, the customer service organization can be very complex. I think a lot of people have the misperception that the customer success organization runs the support team and the account management team and nothing more. That's part of it, but in a mature company, especially a SaaS company, customer success is extremely complex. Here's a listing of some of the kinds of animals you'll find in the zoo.

- **Professional Services.** Professional Services are most common in enterprise organizations. In an enterprise organization, there's a fair amount of integration required with the client's other systems. Systems need to be configured to work with the complex workflows of larger organizations. Professional Services organizations tend to be mostly technical and consultative in nature, gluing together different sources of data to make sure that the systems work together properly. Frequently that work is offered for a fee. Like many consulting organizations, Professional Services is measured by utilization. How much of their time is billable? Another way to organize Professional Services is to have the team focused on best practices and strategic advice. This was the focus of our Professional Services organization at Return Path.
- **Onboarding team.** A second role within Customer Success, and the first person the client will probably see after they've signed a contract, is someone from the onboarding team. This team is focused on helping the client get their initial

value from the product. This team works with the client to configure the product so that it is functional for the client. Frequently the onboarding team will ensure that the client is trained on how to use the product to achieve their desired business objectives. The metric for the onboarding team is frequently the amount of time required to obtain "first value" for the customer. This metric is based on the primary value the client is trying to get out of our product and measures the first time that value happens. For example, at my current company, Signpost, our measure for clients is the first review that gets generated by our system. whether that's on Google, Yelp, or Facebook. For other companies it might be the first time that a customer actually logs into the tool and reviews a report. There are different types of value for different kinds of companies but the important metric is one of speed: How quickly does the client get value?

- **Support.** A third role as you scale is the support team, people who answer client questions and solve their problems—most frequently through an email ticketing system or through chat (and sometimes over the phone). Usually, the customer is encountering technical problems (the customer doesn't know how to login, can't figure out how to configure something, believes there's an error, or is actually getting an error message). Support teams are typically measured by the time to answer a question and by the percentage of questions that are answered on the first attempt.
- **Customer Success management.** This team is focused on making sure that clients achieve desired business objectives through their interactions with the company. The Customer Success team will take a look to make sure the client is using the tool and all the features that are available to them and getting all the value they possibly can. They also will actively show value to the client. For example, at Return Path, we'd generate reports that would say, "Look, your inbox placement has improved," or "You're making more money as an email marketer for your company." In most organizations, the Customer Service Manager (CSM) will be charged with

obtaining customer advocacy (referrals, case studies, presenting at company events, etc.).

In some cases, the CSM team will renew clients and other cases they won't. That's one of the eternal arguments that CS teams have: Should they be renewing or should they be a pure service function? In some cases, they also do upsell, which is the second major sort of religious debate in the customer success world. The third debate is around compensation and I'll return to these "eternal questions" in Chapter 93. The CS team is typically measured by renewal rates and by a net promoter score.

- **Account management (strategic account management).** Another role you *may* build as you scale in a Customer Success (CS) team is an account management team (sometimes called a strategic account management team). In some organizations, when the CS team isn't responsible for renewals and upsells, there is a separate group called account management. This organization is a sales team focused on the current client base. Account managers will be introduced to the client by the customer success manager and will try to understand the problem of the customer better, and they will then suggest additional products that can help them solve that problem. And you want to do that and hopefully close the sale.
- **Knowledge management and customer training.** In a more mature company, there is usually a training and knowledge management function within the CS organization. The customer training team creates the training material for clients and may run in-person training that could be virtual or at the client site to ensure that clients know how to use the product. Also, usually inside the same training part of the CS organization, there's a knowledge management team that is in charge of maintaining a knowledge base—the repository of all information about best practices and how to use the products and the technical details of how to use the product, as well as any kind of domain expertise that's required to use the product.
- **Service operations:** Depending on your business and scale, there's frequently a service operations team. In some

companies the service operations is a person on the sales operations team. Sometimes it's within the customer service organization. In either case, I highly recommend that service ops headcounts be fully allocated to CS (not shared with Sales) because you need someone to maintain the data and the "tooling" required to run Customer Success. It's a big job. If the resource is shared, I've found that Sales usually will command a disproportionate share of the resources because CS is "just a cost center." We'll cover service tools later, but every company has (or should have) a CS platform. At its heart, the CS platform is a workflow tool allowing the Customer Success manager and other people on the service teams to organize their day. The CS platform will provide automations that send out communications to clients based on the state of the client usage of your platform and their state in terms of contract status.

- **Service enablement.** Finally, you may want to create a service enablement team. This team is charged with making sure that all the disparate CS teams have the training to ably perform their jobs. The service enablement team should run processes to understand where the teams aren't following best practices (frequently by reviewing phone calls, chat transcripts, and emails with clients). Their major role is to assist the managers of each team to coach their employees to follow best practices and improve their performance.

As you can see, Customer Service is a complex organization in its mature state. As such, it requires a CCO who has enough domain expertise across all of those groups to effectively provide direction.

Chapter 86

Timing: When to Hire Your Team

Customers

Looking over this extensive list of roles you might be wondering, "Couldn't we just outsource a lot of these until we need them full-time?" Of course, that's possible and quite common. You could outsource just about anything on the list of service functions mentioned earlier, either on a project basis or on an ongoing retainer basis. However, most successful companies are very choosy about what they outsource. You wouldn't want to outsource anything that's core to your relationship with a client or core to making sure the client actually gets the job done they hired you to do. I would be *least* likely to outsource Customer Success since that team has the core relationship with a client, and they're in charge of making sure clients get value out of your products or services.

Most startups might have only a handful of clients. The typical path is to start with a more senior player/coach as a "lead" service person. This person will have to have a broad set of skills because they will be doing a large number of things: from onboarding and training clients to answering tickets. That person will then become a leader of a small(ish) service team. At first, hiring a few CSMs and then creating a support team that can answer tickets or chat. Typically, an onboarding team would be next because onboarding work is very routinized and focusing a team on onboarding leaves more time for the CSM to partner with the client to drive outcomes. It's easy to create a specialist in onboarding and it's very useful.

Typically, the other specialties emerge as you find the need to support the organization, such as service operations and service enablement. Typically, one would invest in these functions after they have a critical mass of people. The account management function you can create by assigning a salesperson to call into the base and you can create that pretty early in your scaling process.

Do not allow someone to sell into the base and also sell new logos. I've made that mistake and learned that they will default to 100% of selling into the base because it's so much easier than selling new logos.

Early on in a startup team, members will be forced to wear multiple hats: the CSMs may have to take care of onboarding. A support person may maintain the knowledge base. I make it a rule to hire more experienced people earlier in the life of a startup who can wear multiple hats. As the company matures, you can move to more of a specialist model.

No doubt, your mileage will vary. Each company is unique. In going from startup to scaleup, you'll have some choices on what functions to bring in and when to do that, and what functions you could outsource. The end result is that in its fully mature form, the customer service organization is large and complex with lots of moving parts.

Chapter 87

Customer Segmentation and Journey

I can't say this enough, so I'll repeat it here: the purpose of the service organization is to help the client achieve their desired objectives through working with your company. The customer hired your company to do something and your job as leader of the customer success organization is to help your company achieve that. But how do you consistently do that? How do you make sure that every client has an appropriate customer experience that helps them achieve their objectives?

The key word here is *appropriate* customer experience. A client paying you $100,000 a year has a different experience than a client paying you $1,000 a year. Their expectations differ, the degree of customization, of interaction, and the value they expect differ. As a startup and as you grow, if you're unable to differentiate customers and manage expectations and experiences, you'll either underinvest in some customers or overinvest in others. Either way, you've got problems.

There are a handful of things you can do as a Chief Customer Officer to create appropriate customer experiences.

Segment Your Customers

Define a segmentation scheme for your client base. There are a lot of different ways to segment customers. Frequently customers are segmented by revenue, by the product they purchased, or by industry. The proper segmentation will vary by company and there is no hard and fast rule. Generally, the criteria I would use to segment customers is based on simple questions: Are we providing more service, more hands-on service, to people that are paying us more? What is the expectation of the client? Is that expectation being met by the level of service that we're providing?

While I may have these criteria in mind, creating customer segments never happens in isolation. You'll certainly want to know how your competitors are segmenting their customers and what value they're providing. You'll also want to collaborate with other teams that touch the customer, like Sales, Marketing, and Product, because the segmentation scheme needs to be something that's shared across functional boundaries.

If the sales team has a different kind of segmentation from other teams, that will typically create problems. Having a shared segmentation makes it easier for the sales team to properly set expectations during the sales process.

Set an Appropriate Customer Experience for Each Segment

Once you have a segmentation scheme defined, you should define a customer journey for each segment. I think of a customer journey as the set of interactions that the customer has with the company that gets them closer to their desired outcomes (and for them to understand how you are helping them).

A typical customer journey contains the following stages:

- **Onboarding.** Getting the client set up for success. The product is configured for the customer's needs and the client is trained on how to use the product.
- **Activation.** An early use stage where a client becomes habituated to using your product on a regular basis (they become 'activated').
- **Use.** The client deepens their relationship with the company and the product—uses more features and makes the product an important part of their day-to-day.
- **Renewal.** As the client nears the end of their contract, you'll need to take a series of actions to remind them of the value you've created and get the renewal.

In each stage of the journey, the service teams will be undertaking a wide variety of proactive and reactive actions. You might be tempted to focus entirely on the renewal activities since retention is commonly used as the primary measure of effectiveness for service teams. Avoid that temptation. Experience shows that focusing on delivering the goods for your clients will ultimately drive better retention.

For segments that don't pay a lot (or prefer to self-serve), you'll want to build a "tech touch" experience. In a "tech touch" experience, most service is mediated through the CS Platform (frequently triggered by outbound messaging), the Help Center, the support team and directly in the product. The product management team and service operations teams become an important partner in tech touch environments. With the tech touch approach, the Chief Customer Officer becomes the Senior Product Manager of the customer journey. They know every message that goes to customers and every customer touchpoint in the product. Much of their work is focused on making sure that this end-to-end customer experience consistently drives the results that the client is paying for.

Other segments will warrant a higher touch customer experience. Here the focus is designing a customer journey with discrete "moments of truth" that are supported by members of the service team. Moments of truth are moments where clients are significantly moved forward in their ability to get value from their relationship with your company (or understand that value). In these moments, you hope to generate an "aha" moment in the client: the moment where they understand the value of your service. At Return Path, we spent a lot of time designing (and iterating) our kickoff meetings and Executive Business Reviews to drive more value for our clients and make sure that our clients understand the value. Making sure that the CSM and onboarding teams had the right materials, training, and coaching to deliver these moments of truth was a major focus of the management team.

Coordinate with Sales

One of the common conflicts in any organization is between the product, service, and marketing teams, on one side, and the sales team, on the other. The sales team wants to be able to sell to anyone but the marketing team finds it a lot easier if they can target specific kinds of customers. Similarly, the product organization finds it a lot easier if they can design a product for specific kinds of customers with specific problems and use cases. And the service organization can be much more effective for customers and drive higher retention if they are similar. One of the common friction points in organizations is this battle between a frontline sales manager ("why are we limiting ourselves?") and the rest of the organization saying, ("focus is really important here"). For the Chief Customer Officer, it's key to bridge the gap between Product, Marketing, and Service, on one hand, and Sales on the other.

At my current company, Signpost, we've recently made the decision to focus our efforts on the home services market. With this decision, we've been able to improve our marketing efficiency and build more differentiators into the product. The service, product, and marketing teams had to come together to make the decision. We're still in the middle of the transition and it hasn't been easy for everyone but we're all pulling in the same direction. The sales team is getting more skilled in talking to plumbers, roofers, and other businesses in the home services market.

The importance of customer segmentation and of creating a journey process map will pay dividends for the customer success organization and your whole company. The customer segmentation helps build a common viewpoint and language between all the functions that interact with customers, and the journey process goals allow you to create the conditions that keep customers renewing contracts. Customer segmentation should be "job one" for every CCO.

Chapter 88

Understanding Customers

If you want to create a great customer service organization, you'll have to do more than segment your customers and design a customer journey. That is a good start. However, if your organization doesn't understand who the customer is at a human level, then you'll always be a step behind the customer. You should understand what their problems are in the context of their business as well as what they have been doing before and after they're using your product. If you don't understand these things, you find yourself reacting to unexpected crises with clients rather than proactively delivering value.

One of the major challenges for many organizations is losing sight of the customer. Without the customer firmly fixed in the organization's collective mind, the organization makes bad decisions. As Chief Customer Officer, you need to make sure to drive an understanding of those customers as individuals deep into the organization. Your customers need to be understood as whole people who do a lot more than use your product.

In a customer-centric organization, employees at different levels and in different functions have regular interactions with customers. In these companies, client feedback is consumed across all functions. Common techniques for driving customer understanding throughout the organization include:

- **Executive sponsorships.** Executives across different functions "own" client relationships. Executives attend business reviews and develop relationships with the decision-maker customers.

- **Building and distributing customer personas.** The product (UX) and service teams jointly create customer personas. These personas are well distributed throughout the organization.
- **Having customers present at company events.** At Return Path and Signpost, I've made it standard to have clients present at all customer meetings. It takes a lot of effort to run this sort of program. That effort will pay off.
- **Publishing customer feedback/verbatims.** I've found a lot of value in publishing verbatims from online reviews, NPS surveys, and client meetings.
- **Publishing a summary of support tickets.** Support tickets have a wealth of information about customers and their problems. Make it someone's job to review tickets and summarize the kinds of problems that clients are coming to support to solve.
- **Temporary assignment to the service team.** I've never done this, but I've heard that some teams will assign engineers, product managers, and executives temporarily to the service team (frequently answering support tickets).
- **Ethnographies.** If you can, having someone from the company (often from Product Management) spend time in the field with the client provides extremely valuable information. How do they spend their day? What tools do they use? What do they do that seems inefficient or painful? If you can invest in building detailed ethnographic studies of your customers, it's well worth the effort. Make sure to publish your findings to the rest of the company.

Chapter 89

Understanding Customers Through Metrics

In Chapter 88, I wrote about qualitative feedback you should obtain from customers (and publish widely inside your organization) to develop an understanding of the client. You can also develop a deep insight about customers through looking at product and process metrics. I find the best way to develop and understand metrics is by aligning these metrics with the customer lifecycle stages. For each stage, you can choose to look at inputs, client actions, client outcomes, and client attitudes.

- **Input metrics.** Did we do the required work to drive desired client actions? For example, did we hold a training session for a client?
- **Client actions.** Did the client take the desired actions that set them up for success? For example, did the client attach our product to a particular data source?
- **Outcome metrics.** Are we driving the results we expect for our clients? For example, do email marketing systems drive opens, clicks, and conversion for our clients?
- **Attitude metrics.** How does the client feel about the results? Within each stage we will do surveys of clients either in the product or via email solicitation, to ask them how they feel about us in different stages. This is frequently a Net Promoter Score (NPS) survey.

Beware of the vanity metric. By vanity metric, I mean measuring things that make you look good instead of things that are important. What I find frequently is that people will report on things that are only tangential to what's important. They don't know why they're tracking the metric, it just seems in the neighborhood of what is correct and it typically makes you look good, or smart, but it doesn't help you create a better client experience, it doesn't help you develop a better client relationship. This is the reason that I *love* outcome metrics. If you choose these metrics well, you are completely focusing on the needs of the client.

Let's take a look at the type of metrics that are likely to be important by lifecycle phase.

Onboarding Phase Metrics

The onboarding phase typically focuses on an outcome metric: time to first value. There are a lot of other metrics available to understand where the client experience can be improved during the onboarding phase:

- **Input metrics.** Frequently the "work to be done" in the onboarding state is something simple: Did we have the kickoff call? Did we deliver the training session?
- **Client action metrics.** Are the clients taking the steps necessary to get to "first value" in the product? Are they attaching to required data sources? Did they configure the product?
- **Outcome metrics.** During the onboarding phase, I am partial to a laser-like focus on time to first value. For example, in my current company, we focus on how long it takes to drive a critical number of reviews at Google, Yelp, or Facebook for clients. There are other metrics that crop up during the onboarding process, like complaints, and you can use these to understand your customers and your service organization.
- **Attitude metrics.** In every organization I've worked in, we've sent out a NPS survey on the date by which "first value" should have been realized (for example, 30 days).

Activation Phase Metrics

After the client is trained, their tools are configured, and they've generated their first value; the next stage is what I call the "activation" stage. During the activation stage, we are most interested in understanding if clients have been "activated" (are habitual users). This is most frequently achieved by looking at some form of client action metric around habitual usage: average number of sessions per client, percent/count of clients that use the application more than a critical amount per week or month.

There are a wide variety of metrics to choose from in the activation stage:

- **Input metrics.** Did we send triggered messages to clients to drive them to use critical features? When clients do not respond to triggers, did the client success team reach out to intervene with the client?
- **Client action metrics.** As stated above, are clients regularly using the core features of the product?
- **Client outcome metrics.** Are we driving business results from core features for clients? For example, in my current company, we measure the percentage of time we are driving an increased number of Google, Yelp, and Facebook reviews for our clients.
- **Attitude metrics.** In addition to the NPS score, I also like to measure a "regret" score to see how hooked clients are on the product. I ask how upset the client would be if they no longer had access to the product.

Use Phase Metrics

At the start of use phase, the client is ideally a habitual user of the core features of the product. The focus then becomes to drive the client to use additional features that drive additional value for the client (and tie the client more deeply to your company). Ideally, this requires a little bit more investment by the client (time spent configuring, customizing, and integrating the product). In many cases, the additional features or services are provided for an additional fee—they are "upsells." To sell these new features, you will need to develop a better understanding of the client's situation and map that situation to your company's capabilities:

- **Input metrics.** Have you done the work to assist the client to make use of additional features? Have we performed the training sessions or assisted with configuring the client's product?
- **Client action metrics.** Did the client configure the product to be able to use additional features? Did the client buy additional features?
- **Client outcome metrics.** Did the client get the desired outcome from using additional features?
- **Attitude metrics.** How do clients feel about the new features?

Renewal

The renewal stage is about getting the client to recommit to a relationship with your company. In addition, there may be an opportunity to upsell additional features or get the client to advocate more for your company. Frequently, the renewal stage will start 90–120 days out from a renewal—assuming you're doing annual renewals. In many cases, you might hold an annual (or "executive") business review with the client. The annual business review is a way to show the client the great things you did for the client over the year. You should also identify additional opportunities to help the client. Rarely are decisions actually made in a meeting like that. It's the conversations prior to the decision to that meeting where the decision has already been made, which is why you've started these conversations 90–120 days prior to renewal.

At this stage it's important to make sure the client understands the value that you have generated and continue to generate. To be clear, you are not providing the client with information on the amount of work you've done for them, the number of meetings you've had, or the number of things you've attached or emailed. You're providing the actual outcomes that they hired you to do in the first place.

As a recovering product manager, I'm fond of reviewing the product roadmap with the client. In many cases the excitement of new features adds to the momentum to renew. When you review the roadmap, it's always useful to talk about the value in terms that are important to the client. How does it solve a problem for the client? It is a sales call after all. During a business review you may find it helpful to bring someone from the sales team with you or, if you have a separate account management team, the person leading that would come along. You want someone who's adept at sales. Frequently, if it's a big account, you will bring an executive like the VP of Customer Success or Head of Product.

What metrics should you look at?

- **Input metrics.** Have you done the work to review the value you've generated with clients? Have you done the work to identify opportunities to deepen your relationship by providing additional value to the clients?
- **Client action metrics.** Did the client renew? Did the client agree to customer advocacy actions (participate in a case study, act as a reference, present on a panel)? Did the client take the actions to take advantage of additional opportunities highlighted in this phase?
- **Client outcome metrics.** Are clients seeing better outcomes because of taking advantage of the opportunities that were highlighted?
- **Attitude metrics.** This is typically the point for measuring Net Promoter Score.

Chapter 90

Foundations of a Great Customer Service Organization

I've learned that there are three primary capabilities that need to be elevated to have a successful Customer Service (CS) operation. By "elevated," I mean that the three capabilities exist in nearly every organization but unless you invest in them they become a drag on performance, rather than an accelerator of performance. These capabilities are data, technology, and people. If you can manage those three elements well, you can build a foundation that will help catapult your company to high growth. And if you can't manage them well, you'll spend a lot of your time on internal issues, putting out fires, being a referee between CS and other functions, and losing sight of the customer.

Data

There are multiple data sources and types that a CCO should invest in to drive a good data foundation. You want to collect, aggregate, and analyze the data that is required to understand the health of the client, including the opportunities to upsell.

Product usage and feedback data are basic measures that the CS team will collect although other teams might be responsible for that, too. Feedback data commonly takes the form of a Net Promoter Score (NPS) but might be other kinds of customer satisfaction scores or verbatims from the client (what they said about you, your products, or service). You also need to know all account information like contract terms, when the contract is due, and what the usage limits are.

An important source of data is customer interactions summaries. Every time a client calls, emails, or chats with the company, the company should be logging that interaction. Ideally, those client interactions would be logged automatically. If the interaction is person-to-person, you might need to write some notes down and make those notes available to others and structured in a way that they can be searched and used.

You may collect customer data differently for different types of clients. In my current company, we're selling to small business owners like plumbers and roofers. We'll frequently talk to them on the phone and summarize notes in our CRM.

You might be wondering how you'll be able to convince clients to take the time to give you feedback. If you've done your job, have sold the client's expectations appropriately, and have shown value in the first 30 days, they should have time for you. In many cases, clients will find the time to tell you that you're doing a bad job if they think by doing so they can improve their situation. Gathering this qualitative client feedback data doesn't have to feel artificial. Many times it comes naturally from the interactions you're already having with the client. A "great conversation" is not you telling the client how great you are. A great conversation drives understanding of the client's business and how you can do better. For a great Customer Success Manager, asking questions about the customer and asking for client feedback come naturally.

Product usage typically is kept in some sort of database that's associated with the product itself while the feedback data might be kept in a separate "voice of the customer system." A lot of client interaction data is kept in the CRM system (Salesforce, Hubspot, etc.).

Technology

As the CCO you'll need to ensure the service team can answer important questions:

- Who needs intervention right now?
- What is the state of this client?

- What interactions have we had with this client?
- Where do I have an opportunity to expand my relationship with a customer?

Luckily, Customer Success platforms (Gainsight, Churn Zero, Totango, and many others) have been designed to make it easy to answer those questions by building a 360-degree view of the client data—the data that we discussed in Chapter 89.

Client cuccess platforms make this data actionable. You can create tasks for service personnel to perform in certain situations (for example, set up an annual business review at 120 days prior to renewal, perform an onboarding task). Alternatively, you can design automated actions. For example, an automated action can send an email or SMS message to remind a client to use a feature. At Signpost we have a built-in trigger to send an email out that urges the client to attach core data sources where the client has yet to attach. The links from these messages lead to a streamlined onboarding process with a great deal of contextual help to make it easy for the customer to attach this data.

In addition to customer success platforms, there are other tools that are required to do your job as a service team. Support platforms make it easier for support teams to answer email-based tickets or conduct chat-based interactions with the client. A support platform may include a knowledge base where clients can research and solve their own problems, usually with the help of troubleshooting and how-to guides.

I have found a product analytics package to be an important source of information. Product Managers will be the primary user of a product analytics package. However, as CCO, understanding the detailed paths that users follow when using your service is important. Where are clients getting stuck? Where could they use some extra help?

It is extremely important that you have sufficient budget for an operations team to help you wrangle all that technology. In very early-stage startups it's relatively easy to work out of your customer relationship management (CRM). In this early stage, you'll need someone to configure that CRM for you. That could be someone you bring in from the outside as a consultant, or there might be someone in the sales operations team that can set it up. That's only a short-term solution because when you get to more than a handful of customers, you'll want to invest in a customer success platform and the data that power a 360-degree view of the customer. At this stage, you'll want to invest in a support system that can provide more ticket-based support. You'll also want to get a platform that supports the knowledge base.

I've seen a lot of service teams struggle because they have underinvested in service operations. Don't make that mistake.

People

A common business platitude is that "people are our most important asset." For a service organization, this is absolutely true. Yes, clients will stay with you (maybe) because you have a technology that solves a problem for them. However, the main reason clients stay is because of the relationships they develop with their service team. The people who onboard them, train them, answer their questions, look for additional ways to help them, and understand them as a person and as a company. As CCO, that means you've got to recruit, train, and coach your team members so that they provide the appropriate experience for your clients.

If you want to build a great customer service organization, your first task is recruiting. My general rule is to look for a handful of personality traits—intelligence, curiosity, the desire to please. However, I have found that the most important characteristic is whether the person will go that extra mile for the client. You can assess candidates by asking how they have behaved in challenging client situations. How have they solved problems for clients in those situations? You're looking for someone who is looking to act as a trusted adviser to the client, someone who wants to solve problems instead of someone who's just there *only* to please. You want people who are willing to develop the expertise to tell the client what they need to do, even if it's unpopular or difficult for the client to hear.

Your second task is to train the people you hire in "process skills." It's not enough to just show them the customer journey you've created, you have to train people who are following this journey on what happens in each step. They have to practice and master each step in the journey. For example, if you have a standard quarterly business review (QBR), you could have your people mock present a QBR for a client and have other members of the team play the role of the client. The training for process skills also means training on how to use all the technology. What is the proper method to use the customer success platform after that quarterly business review, after onboarding, or after other key lifecycle events?

In addition to process skills, it is important to help your team develop their soft skills. How do they become that trusted advisor that earns the right to provide advice to the customer? People have to develop soft skills, whether it's listening skills, question-asking skills, writing skills, or speaking skills. There needs to be a lot of great soft skills development if you want to create trusted advisors. And that's something I think is extremely important.

The way to drive home training on "process skills" and "soft skills" is coaching. Training lasts a moment. If you don't require that your team iteratively practice a new skill (and give them feedback in the moment), the training will not stick. I'm a big fan of "mindful" practice. At the start of every week, first line managers should talk to each of their reports about what skills they are working on this week. Next they should map out what occasions they're going to practice that skill—whatever that skill happens to be. Is there a difficult client situation where the CSM can apply a soft skill? The manager should then provide the employee feedback on that skill as close to the moment they practiced it as possible.

So how can managers pull that off? Well, they might actually sit in on a call and listen to it and then provide feedback after the call, or they may record that call and provide comments and notes back to the CSM. To me, coaching is the thing that most managers don't have enough time for. However, coaching is the difference between a good team and a great team.

A key to developing your team is to define a career development ladder for your team. The defined progression might be CSM1, CSM2, or Support Rep 1, Support Rep 2. It doesn't matter what the labels are. Build defined criteria of what good performance looks like at each level. This allows service managers to provide solid development plans. A major influence that you can have as CCO is to require a regular set of performance reviews for your team. Comparing each employee versus the standards for their level will highlight where their development opportunities are.

The customer service organization attracts people from both technical and non-technical backgrounds. There's not an advanced degree that a person can earn in customer success (at least that I'm aware of). The people drawn to Customer Success share a passion for working with customers to help them achieve their objectives. In some cases, they are motivated by being the hero. In many other cases, the customer service team simply are motivated by helping clients be successful. Being excellent at service is a skill, like being a developer or an accountant or any other function. People can grow in their career inside a service organization. They can improve technical, interpersonal, and other soft skills. Service requires very real skill that demands respect.

Chapter 91

Building an Effective Team

Customers

The customer service organization is one of the most complex parts of any company with many different specialties—like an American football team. Like football, service is a team sport with the CCO as both coach and general manager of the team. It's important who's on the team; it's important that you have very clear roles on a team; it's important that you focus everyone on what they're really good at, which means that the roles are not cookie cutter roles defined in the factory somewhere but that you're defining the roles based on the people you have in play. Most importantly, it's critical that the team trusts one another. As CCO, a major area of focus is to work very hard at developing an effective team that has a high level of trust. Only good things happen when your team has a high level of trust. When you are a member of a high-trust team, it's a wonderful, wonderful thing. If you haven't read it yet, I'd suggest reading *The Five Dysfunctions of a Team* by Patrick Lencioni, to get a better understanding of what goes into building a high-trust team.

When I first started out, I didn't realize the complexity and specialized skills needed to create a great Customer Service organization. As a startup it's common to have roles coexist within a person and I didn't realize that you could break up these roles, or even that you'd want to do that. But as you start to scale, you realize that these are just different skills and different people doing the work.

As the CCO you have to understand the rudiments of all these disparate disciplines. You will find creative people, salespeople, writers, analytical people, and technical people on your team. I have found it impossible to be a subject matter expert on all of these areas. To handle this complexity, I have had to make sure that the heads of each of those functions within the service organization are strong—they know their function well and are solid managers. With the right folks in these roles, the CCO's job is to work with the team to set goals and to support these leaders. Create a small number of quantitative targets and support your team in reaching those numbers.

I would also recommend looking at how the employees in each function are being measured and what their incentives are. There is no one approach that works for all companies in all functions. However, there is a common pattern that I've found when incentives go wrong—the group of metrics used to measure the team do not include client outcome metrics. You should align your team's incentives (at least partially) with client outcomes.

Chapter 92

Partnering with the Organization

A healthy company behaves as if the most important person in that company is the customer. As the Chief Customer Officer, one of your most important jobs is to make sure that all the other groups in the company—Engineering, Product, Marketing, Sales—are thinking about the customer. I think that the Chief Customer Officer job boils down to three things:

- Running the service organization.
- Coordinating with other departments to ensure the end-to-end client experience is great.
- Giving the entire organization the "why" of why we exist, which is, of course, to help clients reach their desired outcomes.

When working closely with the other functions, you shouldn't expect that you'll be totally aligned all the time—but that's the goal. The more aligned you are at the executive level, the better it is for the people on your team and the better it is for the customer.

The following are a few of the common sources of friction that I've seen across functions:

- **Is the sales organization selling to the right customer?** If the company has done its job correctly, its products and services are tailored to serve a certain kind of customer, maybe in a particular vertical, sophistication, or size. The common tension between sales and the service organization is rooted in the fact that the sales organization is judged by the revenue they generate. Their incentive is to sell to as many different kinds of prospects as possible. An aligned sales organization will build an ideal customer profile (ICP) that they target. Ideally, the ICP is centered on the kinds of customers that your product and services are tuned for. A good Chief Customer Officer will work with Sales and Acquisition Marketing to make sure the ICP is well defined.

- **Is the sales organization setting appropriate value and onboarding expectations in new clients?** Another common issue is setting appropriate expectations for new clients. Since the sales team is being judged on the revenue they generate, the incentive can be to "write checks that the company cannot cash"—make promises about product functionality or service levels that are inappropriate.

  In addition, sales teams may have a tendency to minimize the expectations around the work required to configure and integrate the product. Ideally the sales team will make it clear to the client what steps will be required to onboard the product. In addition, the sales team should capture key bits of information required for onboarding. When the onboarding or CSM team hears "I've already told you this information"—that's a sign of a poor sales-to-service handoff.

  Ideally, your Chief Revenue Officer is a clear partner here. A strong CRO is also holding their team accountable for making the right sales to the right customers. I've worked successfully with CROs to change the incentive structure for the sales team, tying part of the sales team variable compensation to successful onboarding or retention for a certain number of months.
- **Is the product organization focused on a customer journey that smoothly drives client outcomes?** A good Chief Customer Officer will understand the end-to-end customer experience. They will show strong empathy for what the client is thinking and feeling at every step of the journey and work with the Product organization to prioritize fixing those parts of the journey that get in the way of the client achieving their desired business outcomes. When the client has to reach out to the Service organization for help with frequency—that is often a sign of a poorly designed

client journey (and a flaw in the product). It's also costly for the company.

I've found that there are several approaches that can be used to drive more alignment with product organizations:

- **Keep a record of the customer journey map.** I have a strong belief that the Chief Customer Officer should be the "keeper of the journey map." There are various forms of journey maps (you can research these yourself). I think the key aspects of a good journey map are that: (a) they show every interaction between the company and the client (product messaging, marketing messaging, meetings with the service organization, interactions with the product); and (b) they lay out what the client/user is likely thinking and feeling at each interaction. With a solid journey map, it's easier for the product team to understand *why* you are asking for an improvement and *what* changes are required.
- **Align on product team KPIs.** Good product teams have a series of KPIs that they are focused on improving. Work with the leadership of product teams to ensure that the metrics include those metrics that are closely aligned to clients achieving value from the product. Are clients configuring the product? Are they using the product on a daily basis? Is the product delivering the outcomes that the client engaged with your company for?
- **Agree on budgets for bug fixes and small customer experience improvements.** Product and engineering teams have a lot of different kinds of investments they could make: new features, paying back "technical debt," upgrading infrastructure and improving the client experience on current features. I've been most successful in working with product teams when we can agree on the percentage of hours that will be allocated to fixing client facing bugs and making "smaller" improvements to the customer experience (small = not new features but improvements on existing features).
- **Is the product team building for the right client? Does that client match who the sales and marketing teams are focused on acquiring?** Another coordination role of the Chief Customer Officer is to make sure that the "personas" that the product team is building for match what the service team

Customers

sees in our clients. These personas should also match the Ideal Customer Profile (ICP) that the marketing and sales teams are building for.

If your company has a somewhat mature product organization, it should have a set of user and buyer personas that it uses when building its product. Ask to see these. If there is a mismatch between who the product team is building for and who our clients actually are—it's incumbent on the Chief Customer Officer to resolve the issue with the product team. If there is a mismatch between the product personas and the ICP used by the sales team, that's another issue for the Chief Customer Officer to resolve.

From my experience, these are the *most* likely areas of friction between functions. There will be other sources of friction. How to have these conversations in ways that are productive? My answer has been grounding the conversation in the client experience and not making it about the other functions or about the CS organization. Don't make it about "my team has to work extra hard here" but instead make it about "the client is having a bad experience here." As long as you can ground that conversation in the client experience, good salespeople, good product people, and good marketing people will respond positively.

The Chief Customer Officer has to help other functions prioritize. Every other team is as busy as you are. I can empathize most with product teams (having run those teams for years). Product teams have an infinitely long backlog that they are trying to fit into a very finite roadmap. A key part of the Chief Customer Officer's role is to be very clear on what is important for you to be working on right now to improve the client experience.

Understanding the likely friction points, framing conversations in the client experience, and providing prioritization to other functions are the cornerstones to partnering well with others in your organization.

Chapter 93

Five Eternal Questions

Customers

There are a handful of questions that I see repeated in customer service newsletters and forums for CCOs and VPs of Customer Success. I've dealt with these questions my entire career. I call them the "eternal questions" and I can tell you right now that the answer to every question is, "it depends." If you are a Chief Customer Officer, you will need to be prepared to address these issues, because they will come up.

1 How Should I Structure Variable Compensation for the Customer Success Team?

It depends.

What percentage of an employee's total compensation should be fixed (base salary) and how much should be variable, and what is the variable based on?

My short answer is, all things being equal, I prefer more in the base salary and less variable. But everything is not always equal and in many cases—where upsell is a major part of the CSM role—having more variable compensation might make sense.

My general philosophy is to only use variable compensation if you are tying success directly to something that is in the employee's control, and where the additional money is a motivator to do something that otherwise they won't be motivated for. If there's any sense of unfairness in variable compensation—i.e., "the customer left because of the product, not because of anything I did"—then you'll spend a lot of your time navigating these conversations. Instead, I prefer to clearly focus on metrics, and hold people accountable to those metrics. It's a performance management issue if they can't achieve the metrics agreed upon.

There are a few ways that I translate this general philosophy to customer service organizations:

- Make sure to include inputs that are in the team's control in variable compensation consideration. Did the CSM have the quarterly business review? Did we reach out 120 days in advance of renewal to start the renewal process? These input metrics are completely under the employees' control and are, therefore, fair.
- Focus variable compensation on "new behaviors" that are uncomfortable. I've found that for some employees, compensating them for new behaviors that need to be learned will—on the margin—drive a more swift adoption.
- Focus on improvement, not average levels for client action and client results metric. If your CSMs have a book of business that they are measured on, you will inevitably hear that the clients they have been assigned have particularly poor health and that the assignment is not fair. I've found that focusing on whether the CSM has improved the health of their book is a better way to go. The book is not under their control, but whether the client takes desired actions (e.g., configures the product, attends training) and that those actions drive better client outcomes are *more* under the employees' control.

2 Who Does the Upsell?

It depends, of course.

It depends partially on the scale of business you're in. Most small companies can't afford a separate group of people that do upsell and cross-sell so the Customer Success Manager or Account Manager ends up owning that. It also depends on how common or important upsells and cross-sells are. At Return Path, we had a large number of products that we could cross-sell or upsell and we ended up having a group focused on that. But do you have something to upsell to? That depends on the structure of your packaging.

My experience is that people are good at sales *or* good at building trusted client relationships and solving client problems. If it makes economic sense to specialize, I'd do it.

3 Doesn't the Customer Service Organization Own Churn?

I have a very strong point of view on this and talked about it in the five misperceptions in Chapter 83. When I started at Signpost, I ended up running a company that believed that the customer success team was 100% responsible for managing churn. Of course, the CS team has only some of the control over retention and churn. By the time the client decides to leave, you don't know if they've been oversold, if the product is buggy, or if they weren't properly trained to use core features early enough to drive value. Some of those things are out of the control of the CSM. The organization that is looking to the CS team to control churn is one that doesn't understand what's truly driving it.

4 Shouldn't We Charge for Onboarding Services?

This is another conversation we hear all the time. I don't have a strong opinion on it because it depends on how complex and onerous the onboarding process is and how much customization is required. At Signpost, we are a startup that serves small businesses and we have a small handful of things to configure and a small handful of data integrations we need to make before the tool is useful. It's our choice not to charge for onboarding services because it's built into the tool and is largely automatic. In cases where you've got much more robust, complex, and intensive requirements, you should think about charging for it. In those cases with expensive onboarding, the length of the contract is another consideration. The sales team can (and should) trade off payment for onboarding for a longer contract if the financial math works.

So again, it depends.

5 Where Does Customer Advocacy Live?

Again, it depends.

I believe that it lives everywhere. When a salesperson closes a new client, they should request customer advocacy. I've seen plenty of sales reps be successful with "Hey, can you give us a testimonial? We'll give you a break in price." It also lives in the customer success team. The CS team should be asking for testimonials, case studies, referrals, and participation in conferences. Advocacy also lives in Marketing because that is where customer marketing lives and part of the marketing job is to get those case studies. Typically, you'll find there's someone on the marketing team (or the CS team) that is organizing that across functions. In the end, it belongs in many places.

Summary

I started this Part by saying that the role of Chief Customer Officer is a really great role for a generalist because you have to be able to manage a complex organization that extends into all functions within a company and extends externally to different customer segments, markets, and verticals. I've shared how to structure the customer organization as you scale, including the roles you'd want to hire, and I've suggested some of the metrics you'd want to put in place to measure both your team and their output. What I have left unstated, but which is actually a critical component of the role, is that the best CCOs are people-focused.

A lot of roles are "people-focused," of course, but as a CCO your ability to influence the company, the culture, and the individuals is far greater than you might think. Many people enter a company through a customer service role and you have the ability to impact the company through these first-time employees. In my last few companies, many of the leaders of the business started in the CS team.

The best CCOs are also people-focused when it comes to customers. Yes, you can think of a customer as a revenue figure, as a metric of some sort, but I've found that the joy in being a CCO comes from really understanding customers, connecting with them, and helping them solve a real problem that they're experiencing.

So, while I believe that being a generalist is essential to being a good CCO, people who thrive in this role are the people who remember that a company exists to provide something that the customer paid for, and it's the relationship with customers and others in the company that helps ensure that you can do this.

Chapter 94

CEO-to-CEO Advice About the Customer Success Role

Matt Blumberg

Customers

What comes before a full-fledged CCO? In most startups, you'll start with a "jack of all trades" account manager position where people handle all customer issues from basic support all the way through to true customer success ... and often some of these functions will be handled by the product team. Specialized roles and multiple teams (e.g., support vs. professional services) with their own managers will usually come before a full CCO, unless one of the company's founders happens to be playing that role.

Signs It's Time to Hire Your First CCO

You know it's time to hire a CCO when:

- You wake up in the middle of the night and realize you've never measured customer satisfaction—no Net Promoter Score, no basic customer satisfaction survey, no product engagement levels, nothing.
- You are spending too much of your own time putting out customer fires rather than thinking about how to make customers more successful by using your product.
- Your Board asks you which of your customer segments is the highest margin, or has the most opportunity, and you don't have a great answer and aren't sure how to get to one.

When a Fractional CCO Might Be Enough

A fractional CCO may be the way to go if you have a relatively contained/small customer success/account management organization, but it is already very diverse in its sub-functions (support, account management, success, professional services) and none of the team leaders of those teams have the range of experience to orchestrate the handoffs and synergies across the sub-functions.

What Does Great Look Like in a CCO?

Ideal startup Chief Customer Officers do three things particularly well:

1. They are the primary evangelist across your executive team—and entire organization—on, as George writes in this Part, doing the job the customer is hiring you to do. In non-professional service businesses where the bulk of the organization is not face-to-face with customers on a regular basis, it can be very easy for employees, teams, and projects to quickly become internally focused. The great CCO is the one who brings the Outside In, every day.
2. They are equal parts quantitative and qualitative. Almost all high-level work that the CCO and their team do includes math: Net Promoter Score analysis; customer segmentation; customer profitability. The CCO must nail these, and if partnering with the CFO or someone else, at least be fluent in them. And the greatest CCOs are also the ones with the most customer empathy, and the ability to relate that feeling to others in the company, and to other customers. They can recite customer experiences stories like a politician giving a stump speech.
3. They like designing processes. Account management, customer success, support, onboarding, professional services, knowledge management—all the different teams reporting

to a CCO—must work together in a seamless way to apply their specific areas of expertise to bring general solutions to the customer. The head of the team, the CCO, must be a rock star at process envisioning and design, and at engaging teams in the process. Otherwise, the teams will be inefficient, hand-offs will be missed, there will be no single source of truth, and customers will not be well served.

Signs Your CCO Isn't Scaling

CCOs who aren't scaling well past the startup stage are the ones who typically:

- Throw bodies at things like support instead of making processes more automated or efficient. This is true of other functions I've written about in other chapters of this book (Accounting, for example), but it's particularly important in Customer Success. As a company scales and takes on more customers, especially if the product team spends their building more new features and functions rather than automating internal tools or sunsetting old product modules, the support burden can get out of hand. And while sometimes, sure, it may make sense to open up a massive support location offshore, that may be just a less expensive way of avoiding a process redesign or system implementation.
- Fail to specialize the department as it grows. Just as a startup scales from its founding team as generalists, capable of pitching in on everything, to more specialized roles running different functional areas, CCOs have to grow their teams by increasing specialized roles. It's easy to get stuck in a pattern of hiring and training expensive generalists because they're really good, and they don't require a lot of training. It's much harder to break a role down into two or three smaller roles, figure out how to career path existing generalists into the more specialized roles, and redesign systems and processes to execute better and more efficiently.

How I Engage with the CCO

A few ways I've typically spent the most time with or gotten the most value out of CCOs over the years are:

- In person at "canary in the coal mine" customers. I always find that the largest clients, the most demanding ones, the ones who push you around, are the ones who make your company a better company. At Return Path, we had those types of clients over the years, from eBay, to DoubleClick, to Microsoft, to Groupon, to Facebook, to Bank of America. The demanding customer is the one who breaks things and forces you to own up to your lack of scalability. They also either take you to task or threaten to pull their business if you don't clean up your act. As painful as some of those meetings are, they are also ones I always wanted to attend in person with my CCO, both so I could eat whatever form of crow needed to be eaten as the Chief Crow Eater (which sends a very powerful message to the customer), and so the CCO and I could experience the chirping of the canary in the coal mine and learn from the experience together.
- Understanding the base. As the old saying from the hardware world goes, "God was able to create the world in only seven days because he didn't have an installed base." The new world of Internet technologies, SaaS, and agile development is one where your installed base of customers is your biggest asset, not a millstone around your neck. Some of the most meaningful experiences I had over the years with our CCOs was to be in market, spending time with all kinds of customers together in small groups and large ones, deeply understanding their needs and use of our product.

Part Eight

Product and Engineering

Shawn Nussbaum

Chief Product Officer and Chief Technology Officer

I never fully realized how much of an impact technical debt could have on a business until I found myself faced with an impossible challenge: I was responsible for a 100+ person product development team spread over three business units and two cities, supporting dozens of products on an aging platform, with a culture focused on maintenance and incremental change rather than on innovation. At the time, Return Path was a 16-year-old "startup," and I had been with them in leadership positions for seven years, so it wasn't like I was an expert brought in to change the product development organization—I had helped to create this problem. We were faced with slower growth and a changing market, and although we had a good business strategy to reorganize around our core competencies and we had overhauled our go-to-market machine with a compelling message, our Product Development organization was struggling under the load of a complex product set, an aging stack, and past decisions, and our product launches were often delayed, uncoordinated, and underwhelming.

Product and Engineering were in the hot seat, and we were being asked tough questions from others in the company like, "Why does it take so long to get product out the door here?" and "Have we missed something over the years that has caused us to fall behind in product innovation?" Those were the questions that I was pondering myself as I looked at this challenge of getting the Product Development organization aligned with the rest of the business and firing on all cylinders. How do you change the culture of a large team and improve the velocity, while reinventing the underlying technology without pulling over to do it? That's the proverbial "change the engine while the car is running" problem.

I never would have predicted that I would have a 27-year career managing technology businesses nor that I would be writing a chapter in a book on how to create a great product development team. My family or friends might have predicted it, though. I grew up with two entrepreneurial grandfathers and learned the value of taking risks, working hard, and treating people well, and those traits are common to leaders of any team. I also liked to play around with computers as a kid and I remember finally getting bored with *Oregon Trail* on my Apple IIc and printing out the code and trying to figure out how everything worked. Through trial and error and a well-worn Apple BASIC book from the library, I created my own adventure game. In high school I was the "tech guy" who helped friends and family with their computers and networks. In college, I actually studied communication and music but was always programming on the side.

My first tech job was managing a small computer department for an insurance company and writing my first large business software program, an auto claims and billing system. I've been a hands-on technical contributor about 80% of my career, whether that's as an individual programmer or technical co-founder, and I have over 20 years of experience managing teams. I provide this short background because my journey is not all that different from other product and tech leaders. Many of us come from non-traditional backgrounds—without formal training in computer science or engineering—and I've found that we share common traits of being determined, hard-working, curious, and ready to dive in on any task or problem. My background is one part entrepreneurial, one part hands-on problem solver, and one part leader—and I'm most comfortable at the intersection of tech, business, and people.

The most formative parts of my career have all happened in startups, which I found to be places where you can work really hard with other people who want to create something. I've had three startup experiences: I worked in a startup that failed, a startup that scaled to $100m and a successful exit, and a startup that went into zombie mode. As a startup, your favorite outcome is a successful exit, but the second favorite is an outright failure—so that you are free and clear to learn from the mistakes and go on to the next thing. The worst outcome is the zombie startup, the one where the company doesn't grow but it looks interesting enough to keep sucking you in for one more year to turn the crank and try to get it to be successful. In the zombie startup, I either felt like I had too much invested to leave or that success was right around the corner. So I stayed when I should have cut my losses.

Fortunately, you can learn lessons in all three cases and I want to share some stories, examples, and patterns I've used to run effective teams and drive technical and product strategy in this Part.

Chapter 95

The Product Development Leaders

Products are the lifeblood of a company; it's how customers experience and think about a company. A company's product is often synonymous with the brand—from Coca-Cola to Salesforce to Nike. And while a company can survive if some functions are ineffective, a company will never survive if their products are poorly designed, suffer from quality issues, or simply aren't useful. Getting your products right means getting the Product Development organization right, and that falls squarely on the shoulders of the Chief Technology Officer and the Chief Product Officer.

I'm referring to Product Development quite broadly as the section responsible for designing, developing, and operating software products as a whole team. This means Engineering and Product Management as well as other functions in the product development process like data science, user experience, quality, infrastructure, and others. It is important to see this as one team since all these functions have a shared goal of building products that customers will use.

One of the challenges in Product Development is getting these separate functions to be one team, to communicate, collaborate, and to understand their roles/responsibilities so that they can create an effective product and customer experience. Part of your role as leader is to help make that partnership between the groups effective and, ultimately, make the partnership with the company effective.

The role of the product development leader changes depending on the stage of your growth. You might start with one generalist who manages everything "product" and add more functional leaders as you grow. Table 95.1 outlines the core functions within product development and what they could look like at various stages of growth.

While a company might start out with one leader responsible for Product Development as a whole, it is important to understand that there is value to having dedicated owners of the Product and Engineering functions. Building a great product requires managing tradeoffs between the functionality and value of something versus what is feasible. Often what the business needs and what Engineering needs can be opposite from each other, and this creates a natural tension that is good and forces tradeoffs and collaboration that can result in better products. World-class product companies have become good at managing this push/pull between Product Management and Engineering. It is important to cultivate managing this tension somewhere in the organization even at the early stage. It doesn't have to be between two C-level leaders over Product and Engineering initially; it could be just between a strong Product Manager and a Technical Lead.

One other point I want to add about technology leaders. There seems to be a lot of opinions about the differences between Chief Technology Officer (CTO) and VP of Engineering and I've seen organizations debate and cargo cult this for my entire career. I've heard, for example, "the CTO is always the founding technical engineer and should always be the CTO (no matter if they can't manage their way out of a paper bag)," or "the CTO has to be the most technical person in the company," or "the VP of Engineering doesn't have to be deep technically because they are just a people manager and makes sure the trains run on time, and they should always report to the CTO," and on and on. The reality is that the most senior person who is responsible for the technical strategy for a company should be a leader. They should be the person who can handle the responsibilities of this role and at the size/stage of the company.

Let's break that down. They need to be a leader, not the strongest programmer at the company. They need to be able to communicate deep technical details to internal and external business stakeholders. They need to be able to grow a leadership team and effectively manage people. They need to be a delivery manager and make sure that the work of designing, building,

Stage	Product	Engineering	Design/UX	Data/Analytics
Startup	Having a product leader is the most critical role for a company at the startup stage. The key responsibilities of the product leader are to determine what is going to get built, develop early adopters into paying customers, and get to product-market fit. In a raw startup the initial product leader could be a technical co-founder with deep product experience and knowledge of the business/industry. Or it could be the CEO paired with a strong engineer who is building the first version of the product.	In the early stage, the technical co-founder/CTO is a hands-on engineer responsible for building the minimum viable product (MVP) and the initial engineering team. They need to define the initial technical choices for the business, balancing pragmatism and the need to be lean and get to early product-market fit, with enough of a long-term technical strategy not to be buried in technical debt as the business scales. This individual could be a more experienced leader who is also responsible for product management in the early days or a strong technical contributor paired with a product leader.	The role of a product designer is also critical at the startup stage and they work closely with the product manager to ensure that the product is successful—specifically the interaction between the customer and the product and the usability of the product. They are responsible for product discovery, UX, building UI mockups and prototypes, and user testing. An early-stage company will likely have Product Design report into the head of Product instead of a dedicated leader of UX.	Data includes engineers responsible for collecting and processing data and data analysts/scientists responsible for extracting meaning from data. Data is increasingly becoming core to businesses—not just for decision making, but as part of the product, or it is the actual product. Depending on the business, the early Data team could be a core strategic function, or it could be one or more analysts who report to the Head of Engineering initially.
Growth	As the business grows, Product Development and Technology Development start to become more independent from each other (but still highly collaborative). Product leadership is now splitting features across multiple teams and needs to balance maintenance, growth, and innovation across the product. While there are individual product managers at this stage, they still might report up to one business leader that is responsible for Product Development as a whole.	In the growth stage, the focus is on scaling the initial team, enhancing the early product, and becoming more efficient. Hiring and team composition start to become more specialized instead of the focus on generalists that were part of the initial company. Multiple teams are built to take on parts of the product and focus on key technical areas, so it is important to build front-line leaders and an effective project/process framework to manage and measure work. While there are distinct front-line leaders in Engineering at this stage, they still might report up to one business leader that is responsible for Product Development as a whole. If your initial engineering leader was a strong individual contributor, you should be hiring a technical leader to drive strategy across the organization.	Depending on the complexity of the product and number of product lines, you will be scaling the product design team and likely looking for a dedicated leader for this function, or at least a team lead. One area that becomes critical at this stage is having consistency or a style guide across multiple product lines. Structuring this team to be able to pair with a Product Manager for day-to-day work, but also have a functional design review process can help balance this.	At the growth stage, the data leader is responsible for scaling the team and collaborating with the rest of the product development functions. A key decision that often needs to be made at this point is whether to structure this team as more of a service bureau or embedded in the product teams and navigating the tradeoffs of each approach.

(*Continued*)

Table 95.1 (*Continued*)

Stage	Product	Engineering	Design/UX	Data/Analytics
Maturity	It is important at this scale to have a dedicated product leader who owns the overall product strategy and roadmap across the business and is effective at delegating strategy ownership to the individual product managers and collating and communicating those strategies across the business and externally. A core responsibility as the Head of Product is to build a great team of product managers and set the product culture for the organization.	It is important at this scale to have a dedicated engineering leader who owns technical strategy and has dedicated leaders owning key functions within engineering (infrastructure, quality, etc.). The engineering leader is responsible for communicating and getting buy-in to the strategy/philosophy for technical direction and development practices and should be able to articulate and defend a business plan on how technology can be used as a force-multiplier for the business to drive innovation or efficiency as the business grows.	At this scale, the product design team will generally have a dedicated leader and report to the C-level product leader.	At this scale, the data team will generally have a dedicated leader and report to the C-level product or technical leader, or in some cases be a C-level role directly (i.e., if data is the product).

delivering, and managing the product gets done efficiently and effectively. And, maybe the key part: they need to understand how to use technology as a strategic asset to help the business succeed, and, by that I mean, they are in the room when the big business decisions get made. The technology leader can't let the team just be the technical arm of a company that is handed specifications to go build. They have to ensure that Product Development is integral to the business. That is, they need to help set the product vision and strategy and execute Product Development across the organization in collaboration with internal and external business stakeholders.

Not every technical leader fits the bill, and not every technical leader is competent across all the stages of growth. The upstart hacker/CTO may be fine for building the minimum viable product (MVP) and getting the company to Series A, but if that's what you have, you should look for developing leadership around this person if you plan to scale anywhere important. The IT leader who spent their career at large companies might be good in a middle management job with a lot of structure and great at managing products in maintenance mode, but if your scaleup business has a deep technical product and a strategic roadmap that requires technical integration with partners, they will soon be out of their depth around driving technical vision and strategy.

What I Look for in a Chief Product Officer

Scott Petry, co-founder and CEO, Authentic8

When I started my career as a product manager, the role of Chief Product Officer didn't even exist. My PM group oscillated between reporting up to the VP of Engineering or the VP of Marketing, depending on the latest powerplay or recommendation from "the consultants." But since those early days the Product function has become more influential and today Product is the nexus between all parts of the business—operational, technical, and go-to-market.

With that increased influence and dependency between product and other critical functions, the organizational structure of most companies, especially tech companies, has changed. Today it's common to have the product management function report up to

(*continued*)

a Chief Product Officer. That CPO typically reports to (or will report to) the CEO, and sits at the executive table next to the CTO and the CMO.

CPOs are invariably senior technical leaders but having the title "CPO" doesn't necessarily reduce conflict or confusion with other technical leadership positions. Who is responsible for the product, the CPO or the CTO? Who do employees need to consult or inform on decisions related to product, the CPO or the CTO? Confusion on the role of the CPO is rampant in many companies and there have been a number of wide-ranging and informative blog posts trying to differentiate the role of the CPO from the CTO. One common theme on the difference between the CPO and the CTO is that the CPO should be responsible for the "why" a company invests in a particular product, leaving the "how" to the technical teams.

While this is a good model for differentiating the roles, I've seen too often that CPOs are assessed and performance-reviewed based on their raw technical chops rather than their ability to assess a market opportunity and rally the organization behind the cause.

CPOs are typically classically trained engineers who have expanded their skills beyond technology and product development into business-related issues. But even with the addition of business acumen, I have seen many CPOs revert to their comfort zone. For example, where the ambiguity of the market obscures clear answers, CPOs may hesitate, delay decisions, or ask for more data; but where issues are more concrete, say, in selecting a new tech stack, CPOs will become highly engaged, debate issues, and influence the choice.

It's not just CPOs themselves who struggle with the "why" of product; non-technical people who evaluate CPO candidates or peer performance reviewers also place a greater, asymmetric weight on the technical assertiveness of the CPO. In my opinion, that's the wrong measure.

Clearly, the CPO needs to understand the technical aspects of the product and its place in the market—that's the baseline, the bare minimum that a CPO ought to know. Another big "why" the CPO needs to understand is "why are customers buying (or not buying) our solution?" The answer to this question is more difficult to figure out. The CPO needs to have enough business sense to understand market white spaces, enough technical expertise to understand technical trends and direction shifts, and enough social media knowledge to understand the role of analysts, influencers, and societal trends impacting why customers buy or don't buy a solution. Mapping what Engineering can develop to what the customer will buy requires soup-to-nuts understanding and expertise, not just familiarity with the latest SDLC model or development stack. The intersection of social, technical, and market forces on consumer behavior is where the most fertile answers are found, but this is the area fraught with ambiguity and uncertainty.

It is the place where CPOs need to spend their time, but the place many deeply technical people try to avoid.

I look at the CPO as being an emotional leader in the company more than a technical leader. Of course, the CPO needs significant technical chops to be credible and to focus the product development efforts. But the full measure of their emotional leadership capabilities needs to be applied to rallying the organization behind the initiative. The "why" needs to resonate with each internal team. Why should Finance care? Why will HR benefit? Why will differentiation from competitors help Sales and Marketing improve their performance?

Technical chops are table stakes for a CPO. But good CPOs are evangelists, and with great CPOs, you might never even know they come from a technical background.

Chapter 96

Product Development Culture

Good culture is a strategic advantage for a business and this is the key in the product development organization too, because everything else hangs off of your culture. Your ability to create remarkable products is a lot more likely if you have a culture that embraces innovation and creativity. Your ability to balance the team between pragmatism proportional to the business while hitting real outcomes (not just output) is accelerated with a good culture. Your ability to adapt and change as the company grows is a direct result of the culture you create.

While the product development team benefits from the overall company culture, it is a distinct subculture within a business and it's important to understand those differences and cultivate the culture you need if you want an effective product development organization. The first things to get right about your culture is exactly how you'll fit within the larger company.

The product development team is not just the execution-arm for the company. They own the roadmap and they're the ones shaping the product to have the biggest impact on customers. You can't do this alone and you'll need to cultivate a collaborative relationship with other functions like Sales and Marketing to understand their strategies. It's not uncommon to get requests from the business to experiment on things, but I'd suggest that you keep this in perspective because you'll have your own needs and requirements to manage innovation, technical debt, and product lifecycle.

Another thing to do to ensure that the culture of the product development team is focused on the right activities is to create space so that the team is a product group, not a project group. Some companies still see and treat product development as if it were similar to manufacturing where processes are repetitive and outcomes are reasonably predictable. If you try to utilize a product development team like you would a team in a manufacturing facility, you'll set the team up for failure. A product development team runs into many problems that are unpredictable and your teams will need time to experiment and learn through discovery and feedback from users.

Types of Development Cultures

I've experienced three different types of development cultures in my career and only one of them is what you'd want to strive for. One development culture is what I call an "over-engineering" style. In this culture, the leadership and engineers want to solve everything with engineering and, although it can be good, the cumulative engineering investment over time is far greater than what the business needs at that stage of growth. The result is that you have disproportionately invested too much in engineering relative to the business need. You built a Mercedes when what the customer really needs is a bicycle. One exception to this is if you are building safety-critical systems (avionics, pacemakers, self-driving cars, etc.) and need to over-engineer—especially on quality.

Companies that over-invest in engineering are often trying to create a culture that attracts great engineers who want to solve difficult, complex, or interesting engineering problems. What you really need to create a great development culture are engineers who want to solve business problems with engineering. I've seen companies create the over-engineering approach, thinking that they could out-guess where the business is going and they end up over-optimizing before the business gets there. Most of the time this guesswork is just completely divorced from any business reality and what you have is a group of engineers, isolated from the market and customers, thinking about their own problems, or their pride in work or modeling things after what Google or other successful high-profile companies do.

Another development culture is what I call an "under-engineering" style. In this approach, companies disproportionately invest too little in engineering relative to the business need. This can start out noble and appropriately pragmatic which is what most raw startups should do. But, if left unchecked, the product development team stays overly pragmatic and scrappy too long and they become reactive and in service to the business instead of in partnership with the business. Chapter 97 covers this problem in detail.

Good-enough solutions sometimes lag behind the business and then you'll need more of a step-function to catch up. The problem is, if you operate with an under-engineering, scrappy style, you'll always be in crisis mode, your team will be overworked, and morale will plummet. I've seen the under-engineering model play out in a couple of ways. One way is that the business and the product development organization are misaligned and not

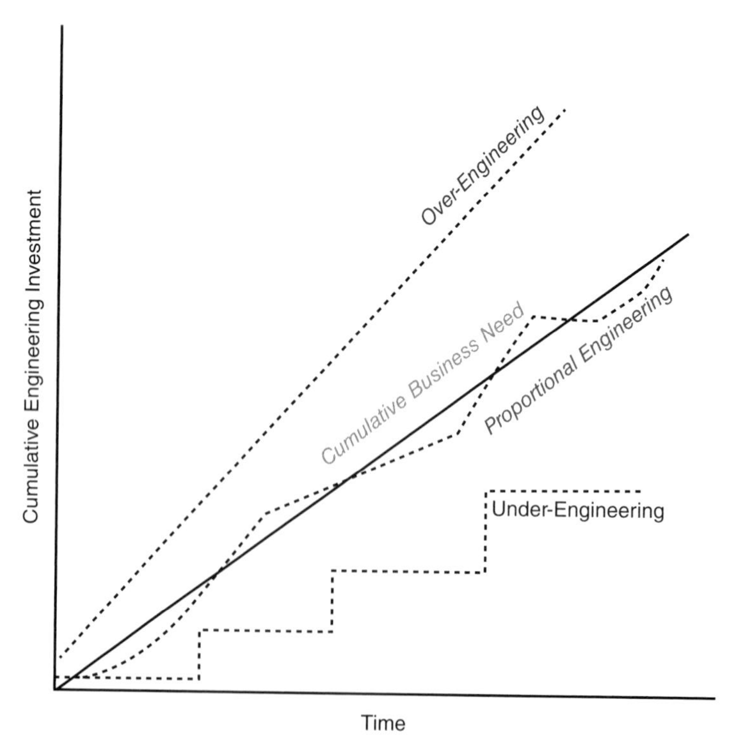

Figure 96.1 Type of development culture.

communicating on where the business is going and what role product and technology play in that overall strategy. This can happen because there's a weak product development leader, or it could happen because there's a really strong business leader who doesn't understand the cost of cutting corners. Or this can happen because a product development leader or engineering team is disconnected from the business and values an overly scrappy approach technically without understanding the real cost of unchecked technical debt.

The third development culture that I've seen, and the one I'd recommend, is what I call a "proportional approach." In this model you invest proportionally in Engineering relative to the business need. You'll need a strong vision on technology being a strategic advantage to make it work. The result will be that your product development team will be highly aligned with the business and in lock-step with the product organization. Sometimes you may even be able to get ahead of the business and experiment to drive innovation, but in this model your experimenting will be focused on things that the market needs, not just innovation to be innovating. Figure 96.1 presents the types of development culture.

Chapter 97

Technical Strategy: Proportional Engineering Investment and Managing Technical Debt

Product

As a product or technology leader at a startup, you have an opportunity to build the first version of the product, grow the team, and set the culture and strategy for your organization. That's a great and humbling responsibility. I started this Part on Product Development by introducing a challenge that Return Path was facing around getting the product development organization aligned with the rest of the business and firing on all cylinders again. Startups don't begin with those types of problems but I think it's important to look at the impact that decisions have on a fast-growing company and learn from those lessons. No one will get their product, infrastructure, or culture right—even when they have the opportunity to start it from scratch. However, being intentional about where you want to go and diligent about changing things that break—or end up being wrong—is the key to long-term success. You can't avoid technical debt as you grow, but you can be intentional about amortizing it as you go so that it doesn't compound and negatively impact innovation and velocity.

So, how did Return Path get there, with a big pile of technical debt? We held on to an approach that worked really well as a startup and then we didn't change it as the company grew, as markets evolved, as technology choices broadened, or as new competitors emerged. We didn't invest proportionally.

When you're working in maintenance and incremental improvement modes, you can easily lose sight of the fact that healthy companies grow and change—quickly. The product development organization that you start with has to also continue to adapt and support (and sometimes get ahead of) the business as it grows. As a Product or Engineering leader, you'll need to get comfortable with the idea that your current philosophies and approach will be wrong at some point and you will be continually challenged to be the leader that the organization needs at that stage of its growth.

For our product development team we had two things working against us that slowed us down and made it difficult for us to generate the velocity and innovation that we needed: unaddressed technical debt and complexity.

Like many startups, our founders were technical and pragmatic and they had a scrappy approach to building products and a business. It's actually what attracted me to Return Path in the first place, since the engineers weren't just focused on solving technical problems but were interested in solving bigger picture business issues. But while the scrappy, pragmatic approach that we had as a startup helped us survive, we didn't have a strategy or the time to manage around it.

There's a saying that, "In engineering, you pay for it now or later, but you always pay for it." As a startup, we were scrappy and, on the positive side, that meant that we were fast, efficient, practical, did prototypes, and hit the outcome without caring as much for the methods used to do it. But being scrappy also meant that we took shortcuts, made temporary fixes, were short-term thinkers, and had scalability issues. While the leaders at Return Path knew how to balance scrappy with areas that needed more rigor, that same thinking didn't always translate to the teams doing the work. Teams often cited "scrappy" as an excuse to cut corners and borrow too much technical debt to hit a goal or just get something out the door. Those decisions created a culture and a foundation that later became an issue when we put stress on the system to move faster and innovate.

Another aspect of unaddressed technical debt is the real operational cost of maintaining active products. Business and product development leaders are generally good at understanding and measuring the time and cost of new product development, like having a team of six work for six months to accomplish X. What is less understood—or is taken for granted—is what it takes to operate a product that is in market. I find this to be especially true at companies that are early stage, high growth, or have limited resources. I'm not talking about core products that are undergoing active development because those are typically staffed with full teams that are balancing new development and ongoing maintenance. I'm referring to products that are still supported in the market but don't have large ongoing development and are more in maintenance mode—on the shelf.

There are at least three components of operational costs to support products that are actively in the market but are not core:

- **Ongoing maintenance and bug fixes.** This is the main area that leaders know about and consider when keeping a product active but on the shelf. Operational costs include monitoring, production support, and bug fixes.
- **Incremental development.** This is also understood well by most product development leaders, but it's unlikely to be accounted for from a time and budget standpoint. Most products that are on the shelf might not have a team assigned to operate them; instead, those products are just being monitored by another team or a dedicated engineering support team. But even a product on the shelf will at some point need an incremental update to support a marketing launch or a message or visual change. Any change usually requires people who can dive back into a product and learn about the product and make whatever changes are necessary. In theory, it sounds easy to just throw more people at it, but it can be more challenging than it sounds, depending on the change and the challenges with older technology and finding people to go deep on it. Incremental changes also can introduce additional bugs or support requirements that make it harder for the product to just be put back on the shelf when the incremental change is made.
- **Technical updates.** This third area is often overlooked by business leaders and can sometimes be overlooked even by technical leaders if they aren't paying attention to their technical debt or products on the shelf. If big changes happen

in technology, security, and privacy that affect the platform, then all of the products—not just the core products—are affected. For instance, if the business moves from a data center to the cloud, then every product has to be updated to support containerization; if a data feed or data processing platform changes, then every product has to be updated to consume from this new service; if the business has to adopt a new privacy law or audit protocol, then every product has to be updated. When you have multiple products in maintenance mode on the shelf, this last type of ongoing maintenance is a killer and introduces a lot of thrash and distraction into the development process—pulling teams apart to work on old products that need to be reworked to support these deep platform changes. And when old products haven't been refactored or updated recently, it takes time and expertise to change them. In addition, often there is no time to refactor or rewrite them, so ad-hoc changes are made and bugs or scaling issues are introduced.

How do you manage the challenges of shelf products? Ideally, you would have dedicated teams for each product that is active and in market, but smaller, scrappier companies that are moving fast and trying out different solutions to get to product-market-fit are often not staffed to dedicate people this way. You could have a dedicated maintenance team that manages all of the on-the-shelf products, but that solution is challenging because it is seen as maintenance work and doesn't attract senior engineers and what you really need, especially with larger platform changes I mentioned above, are senior engineers with deep familiarity with those products to make changes. You could have each active product development team take on a set of on-the-shelf products, but be prepared for that team's velocity to slow down as they have to move six things forward an inch at a time instead of moving one thing forward six inches at a time. It also eats into team morale and creates a large set of technology requirements for that team to be experts in.

There are no really good answers when it comes to managing a large set of on-the-shelf-products, but there are a couple of best practices. The best strategy is to prevent a company from getting in too much of a hole by having clear ownership and strong product lifecycle management. Be diligent about measuring and articulating the real cost of operating things and aggressive about killing things even if it impacts ongoing revenue. Limit the number of technologies (as much as you can) and be vigilant about keeping products up to date technically so you don't have big catch-up moments.

While unaddressed technical debt was a contributing factor to our product development challenges at Return Path, the other factor was complexity. We had multiple businesses with a vast number of technologies in use. Some of the technology issues came from a dozen acquisitions over the years, and we didn't properly address how those would be managed in the long term. And some technology issues came from our approach where each dedicated product team made its own technical decisions as long as they owned them. That independence afforded to the technical team worked really well in the early stages of our growth because we had several large teams responsible for a few key products. As the company grew and developed many products and as people moved between teams, our approach wasn't that great. Sure, from a talent standpoint, we had our experts distributed to the individual product teams and we had a few central teams that were responsible for operations, support, and platforms. But the central teams were staffed by mid- to junior engineers, and they weren't seen as important as the product teams. We couldn't easily standardize or create common tools because of the complexity of each individual team and because of the inexperience of the central teams.

After looking deeply at our current reality this is what I found: Not every team had legacy systems they were dealing with, some were already in the cloud, up to date, had high business alignment, no distractions and great velocity. But the teams differed enough that it slowed down the company. A bigger problem was that with many products (and a large number of them on the shelf), coupled with years of putting off technical upgrades, and lower morale with people having too many technologies to be experts in, we had a culture where we dragged everything forward and continued to take shortcuts on top of shortcuts. And without us realizing, "scrappy" had become "crappy." One team had 30 products that were important to the business with a team of 30 engineers responsible for them, and this was the team we expected to innovate and drive high velocity changes to the business.

Sometimes when you are faced with a velocity challenge, you can just add more engineers and power through—and that works if your systems and processes can support moving faster with quality and efficiency. At Return Path, we didn't have that and we were dealing with a perfect storm of people, process, and technology issues that were gumming up the works.

I started by setting three big priorities that needed to change to get the product leadership team and development team aligned:

- **Team/Talent.** First, we needed to step up and lead. We needed to be driving the technical and product strategy for the business and not just reacting to things or building whatever was asked of us without considering the costs. We got traction on this by having technical conversations and partnering with the business on decisions.

  Within the product development team we started driving standards/practices across product development so that everyone was aligned and rowing in the same direction. We started a mentorship and onboarding process for new engineers so that they could become productive sooner. And we created an Architecture Council to get the senior-level engineers involved and working on core issues and not just sequestered on their individual product teams.
- **Tooling.** We needed to be more efficient and stop using different technologies on each team so we created an engineering efficiency/tooling team to drive DevOps patterns, automation, and central tools and services that every team could benefit from. If done well, using a few really good engineers in a centralized team like this can be a force-multiplier across the organization.
- **Technology.** The third priority was around modernizing our aging systems. By taking an iterative approach, we slowly moved data centers to the cloud, servers to containers, scheduled jobs to services and workflows, data processing from large batch processes to streaming analytics.

It wasn't a fast or easy change to implement in the product development organization, but a nice side effect of prioritizing teams, tooling, and technology was that we were able to hire more early-career engineers on product teams because they could learn the stack quickly and not have the complexity that existed in each team before. Another benefit was that people could move between teams and still be able to use the same technology, processes, standards, and practices that they learned on the first team.

Getting the product development organization aligned behind all these changes wasn't difficult philosophically because engineers are fairly motivated around good practices and newer technologies. The main pain points were around organizing the people and the work to balance the technical debt work with ongoing maintenance and new feature development. We also had a small group of good engineers who unfortunately decided to leave the organization during this change because they either saw this change as an indictment on them being too scrappy or they just had a different philosophy around engineering. So, that was a bummer.

The biggest challenge in any large technical refactor will always be how to communicate it to the Board, the CEO, and your peers on the executive team. I found that I needed to get their support to make these changes in the first place, and after getting the support, I had to manage the timeline and expectations during the journey.

You'd think, given how the product development organization was widely perceived to be the "problem," that the Board, the CEO, and executive team would be super-supportive of our strategic plan to make the product organization fast and innovative, but a perceived need is not enough to move the needle. After all, we were the ones responsible for the situation we were in. So I created a story and used data to drive home the point that without a dramatic change in the product development organization, we would be unable to overcome our situation and get back on track for high growth.

The first part of the story presented our current reality of unaddressed technical debt and overly complex systems (lots of them!) for the type and stage of our business. It's important, especially if you're making a big change, to make sure that everyone else sees what you see and agrees that it's actually a problem. The second part of the story captured and presented our challenges with data, quantified in a way that would make it impossible for a reasonable person to disagree with our plan.

I found two artifacts that could really help drive our point home. The first artifact was a table I created listing the full product development team members with the category of work that they spent time on. I cross-referenced that time spent working with the terms "tomorrow, today, and yesterday." Every healthy product has Today and Tomorrow work. Today is the cost of maintaining a product, making incremental changes, and refactoring the technology stack to stay up to date, balancing incurring technical debt with paying it off. Tomorrow work includes the active development to add functionality and continue to evolve the product. Older systems in maintenance-only mode might just have Today work. At Return Path we had Yesterday work as well, and that's not good. We had team members spending time dealing with old systems that were important to the business but not cared for actively by a team. Even worse, these systems generated all types of effort that the business wasn't aware of. When I first generated the table, we had 30% of our team working on Yesterday, 60% on Today, and 10% on Tomorrow. That was an eye opener! We had a visual story of why we couldn't innovate and move faster.

The second artifact was a stacked area chart that showed the allocation of engineers across product lines looking back four years along with release milestones. It showed where the business put its focus and where some of the teams were running slower or missing outcomes because they had so much drag. I added 18 months of forward-looking details that included an investment in Yesterday work to illustrate the impact and timeline of digging out of the hole.

These two artifacts, along with the narrative around unaddressed technical debt and complexity, were used to tell the story to the Board, the CEO, executive team, and everybody in the company. We also used the data on time spent and the categories of Today, Tomorrow, and Yesterday to update everyone on our progress as we went along. The metrics (% of engineers on Today/Tomorrow vs. Yesterday) became part of my KPIs as we embarked on the journey to "Slow Down to Speed Up." Ultimately we were able to finish that work and move the product team back to a high velocity, efficient, and effective team that was focused again on Today and Tomorrow work.

Chapter 98

Shifting to a New Development Culture

You don't want to wait to shift your development culture until your technical debt is so high that you're in a big hole, like we were at Return Path. I can tell you from first-hand experience that being the part of the business holding everyone back and having all the fingers pointed at you is not a good place to be. But what can you do that will actually change your development culture? Fortunately, there are steps you can take to alter your development culture and they don't have to be dramatic—you just have to be consistent in your approach. One of the best examples of changing a culture is from Ed Catmull's (2014) book, *Creativity, Inc*, where he tells the story of Pixar. Catmull highlights that the right leadership, transparency, and candor can foster a great culture and lead to innovative results. In 2006, Disney acquired Pixar not only for the technology and talent, but to take the Pixar culture to Disney. At the time of the merger, Disney was in a 16-year trough since their last number 1 hit. They had a lousy work environment, a leadership imbalance that diminished creativity, filmmakers who had lost their voice and were afraid of pouring their hearts into something that wouldn't succeed. They saw themselves as hired guns, not owners of high-quality films, and they weren't empowered to fix what was broken. In short, they were playing it safe and being a utility. The Pixar culture of innovation, candor, freedom, and change was brought into Disney and four years later Disney released *Tangled*. It became the second highest grossing film from Disney Animation ever, after the *Lion King* 16 years earlier. The key was that the studio was still populated by most of the same people from before. What changed was that the leaders applied Pixar's principles to a dysfunctional culture and team and changed them—unleashing their creative potential.

Developing products, like making a film, is an inherently creative process and creating requires making thousands of small decisions that result in a unique outcome. Some companies approach product development as a utility: functional, practical, pragmatic, do what's expected, no frills, no drama, don't piss people off—just get it done. And for certain products and companies that works and is good enough. But if the key goals of a product are to inspire, solve a problem, be fun to use, be efficient, and amaze the users of it, then the people creating that product have to be innovators, creators, and be deeply empathetic about their users. Empathy and usefulness are the keys to remarkable products. Like Pixar, if you want to create an environment that is able to harness innovation and creativity, you have to do more than have a culture that makes people feel valued—you have to create an environment that makes them valuable to the business and to make that happen, you'll have to connect the dots from the work your team does to how it impacts a customer or drives a sale. You have to give people freedom to fail and make mistakes. You have to be candid/transparent with them and give them the data to make decisions. And, you have to let them have real ownership over something.

Setting the right set of values is important in changing a culture. With the under-engineering culture we had a value that said, "Focus on achieving the outcome, not the methods used to achieve the outcome" and that is what we delivered—the outcome, and nothing more. We didn't think much about the technical choices we were making, and we weren't intentional about the product lifecycle. You can't just keep making products and putting them in market without understanding the ongoing cost of operating them (or helping the business understand that). Building a partnership between Product Development and the business and investing proportionally in engineering was the key to maximizing our effectiveness.

Here are the Product Development team values we ended up with at Return Path to drive the type of culture we wanted:

- **Be effective.** The business outcome matters. Our number one job is to create value for the business—have a bias for action.

- **Be intentional.** Think about the methods behind a problem; think end-to-end; build something that works for today but scales for tomorrow; use data to learn and inform decisions.
- **Be efficient.** Creativity is important, but don't waste time; try something and learn quickly; don't reinvent the wheel; use patterns and tools.
- **Be awesome.** Build stuff that is awesome and works—we care about quality and driving technical innovation in our products and platforms.
- **Be iterative.** Take an incremental and proportional step to achieve the business outcome based on learnings.
- **Be collaborative.** Seek counsel from peers and contribute your knowledge.

Things to Consider

- Rewriting and migrating big systems are risky and time-consuming. The better alternative to a complete rebuild is to slowly and systematically take over the old application. You gain immediate benefits and reduce risk. Martin Fowler calls this the "Strangler Pattern," and a blog post of his contains a great metaphor of strangler vines around a tree that is a perfect visual for this work.
- Creating an Architecture Council with senior-level engineers across the organization has multiple benefits for engineers and the company. It provides a career path for engineers; it gives teams a framework for making decisions instead of relying on the CTO to make all decisions; and it allows the Product organization to coordinate and work on overarching initiatives across the company instead of just on individual product teams.
- Hackathons can be a good way to foster collaboration and innovation across the company. We ran hackathons a couple times a year and encouraged employees outside of Product Development to join teams, providing a broad range of ideas and experiences. Some great product ideas came out of this and it set a culture of creativity and experimentation.

Chapter 99

Starting Things

As I mentioned before, if you have the opportunity to start an idea and a company from scratch, it's a great and humbling experience. You get to solve a problem for a customer in a specific way and get them to pay you for that solution. You get to build a team and set a culture, and you get to pick the technology and build the product that will change the world—or at least you hope for all that!

In this chapter I want to outline a few of the practices that I use when starting something from scratch.

Startups face two initial challenges before they can shift to growth mode. The first challenge is to define their idea (that will become a product) and determine if it addresses a problem that's worth solving. The second challenge is to build the first version of that product and get to product/market fit, which is the moment where customers are paying for your product and you are retaining them.

If this is a raw startup, then you are creating a product and a company at the same time. I like to use Ash Maurya's Lean Canvas, from his (2012) startup book *Running Lean*, to document the initial business plan. The Lean Canvas is a one-page business model template that is simple and quick to use to capture the building blocks of your business model and then systematically test each element until the idea is sufficiently validated or de-risked to continue. For defining and validating the idea, you have to get out and talk to potential customers to understand if they have this problem, how they are solving it today, and if they care enough about you solving it for them. Ash Maurya has a great framework in *Running Lean* that I like to use to do structured problem interviews with customers.

Here are a few things to keep in mind when validating an early idea and talking to potential customers:

- Problem interviews are mostly about setting the context and then listening to customers. Do they have this problem? How do they solve this today? Is there a possible competing product that they are using to solve this problem?
- Is this problem a must have (pain killer) or a nice to have (vitamin)?
- The bigger the pain point for a customer, the more likely you have found an early adopter, especially if they have hacked their own solution to solving this problem.
- Pay attention to anything that holds a customer back from moving to a new solution or anything that prevents them from using a solution.
- Listen to what customers say, but more importantly, watch what they do. When you build a product, you have to balance listening *to* customers with innovating *for* them. Pay attention to the problem they have, not necessarily their specific idea about how you should solve it.
- Make sure you are aligned with the benefit or outcome the customer is looking for and not some intermediate solution or experience. The old business school adage is that "people don't want quarter-inch drill bits, they want quarter-inch holes."

Once you have determined that you have a problem worth solving, you can then move on to the second challenge which is to build the first version of your product. This phase isn't a single process to build and launch the product, but rather an iterative process of continuously refining your product until you achieve product/market fit and are ready to optimize and scale your business.

In my startup experience, getting to product/market fit is the hardest part. There is something special about an idea in your head that always initially gets worse when you start to build a product. Early versions of a product are intentionally minimal, with rough user interfaces (UI) and limited functionality, but early versions create a gap between the end state you envision in your head and what you see being developed. You'll eventually (hopefully) close this gap as you continue to iterate and complete the product, but just starting can be a big challenge. I find that the best remedy for this stage is to just keep moving (even if the direction doesn't feel right) because you learn by moving and you incorporate that learning to make the product better.

Having a framework or methodology is useful for getting into this validated learning loop. I like to use the Build-Measure-Learn loop that is described in *The Lean Startup* by Eric Ries (2011). It relies on validated learning as the key unit of progress with the goal to eliminate uncertainty (fail fast) and narrow down the core problem and solution and then iteratively build out the minimum viable product (MVP).

So, now we have a high-level plan in our Lean Canvas and a feedback loop framework to guide our experiments to iteratively build something, test it, learn, and build something else. The missing piece is a project management tool for managing the details. I typically use something like Trello (at the early stage) and then something robust like Jira (as the company gets bigger). The problem I found is that the project tool is flat and is just a prioritized list of cards to work on. I was missing something in between the Lean Canvas and Trello. That's where Story Maps fit in. Story Maps were created by Jeff Patton (2014) and are two-dimensional maps that contain "big stories" or user activities across the top and then have progressive levels of detail on the vertical axis to define specific cards that relate to that user action. I find them incredibly useful to define the outcomes that users want to accomplish with your product—kind of like a living requirements document about the product that you are finalizing. The big stories at the top (which are basically "epics") are not prioritized and don't change as much, but the stories under them are prioritized and are part of various release swim lanes to organize and plan the work.

Here are a few things to keep in mind when iterating on your MVP and getting to product/market fit:

- MVP is really about the minimum functionality that you need to show to validate learning. Some people refer to the

"P" in MVP as "prototype" instead of "product" to make this point. Sometimes you don't need much more than a user interface mock-up to validate an idea and then you can start building the product.
- When you start building early versions of the product, I find it is useful to think about each release as a complete working product—albeit limited. For example, if the goal is to end up with a car, it is more useful releasing a skateboard first, followed by a bicycle, and finally a car. If you are iterating on a car and your first version is a car with one wheel followed by a car with two wheels, it isn't useful to anyone.
- Be prepared to change direction or pivot many times in your validated learning loop before settling on a version that you want to build a product around. Pivot before getting to product/market fit and then optimize after that.

As a technology leader, one of your key responsibilities in the early stage is choosing the technology, building the MVP, and delivering the first version of the product. Here are some thoughts about technical choices:

- At the early stage it is all about supporting validated learning. This usually means that where possible you should use UI mocks, simple prototypes, or simple web landing pages in your Build-Measure-Learn loop.
- There are a number of services that you can use to build fully functional applications quickly. For instance, a combination of Webflow, Zapier, and Airtable can be used to quickly spin up a templated web application with integrations to most SaaS services and an experimental data model.
- As you move from prototype to building the application, you can get many benefits from using cloud providers. I typically use managed services on a cloud provider for the database, queues, containers, logging, and deployment so that I can focus on delivering application functionality without the distraction of managing infrastructure. Another benefit with cloud providers is that they have startup programs where you can apply and get credits worth tens of thousands of dollars to offset your initial hosting costs.

- The specific technologies that you choose should make sense for the type of business/product you are building, the level and makeup of your team, and the support for the technical strategy you are driving. Keep in mind the principles I've described around proportional investment in engineering. With these initial choices you want to balance pragmatism and the need to be lean with enough of a long-term strategy not to be buried in technical debt as the business scales.
- Part of the validated learning loop is measuring experiments. Being able to define, collect, and communicate key performance metrics is important from the start. At this stage, the important metrics are around measuring the effectiveness of the MVP, product/market adoption, and customer success (things that determine if we are doing the job the customer is paying us for). Another key area (if it is important to the business) is to be able to track the ROI of the product investment and understand when things aren't working.

Things to Consider

- The definitive book on creating technical products is Marty Cagan's *INSPIRED* (2018). It is also one of the best books for understanding what a Product Manager does (or should do). It is a book that should be on every product leader's shelf.
- A great resource for the raw startup is *Running Lean* by Ash Maurya (2012). This contains the Lean Canvas and useful guides for running customer and solution discovery.
- *The Lean Startup* by Eric Ries (2011) defines the Lean Startup method and is a must read for all product people.
- Jeff Patton's (2014) book *User Story Mapping* is the definitive source on story maps.
- *The Four Steps to the Epiphany* by Steve Blank (2013) is the book to read to understand your customer.
- As you scale your company, I recommend the following books to think about how to keep your culture innovative,

to make sure you don't get stuck just protecting the core, and how to learn to start new things in an existing company: *Escape Velocity* by Geoffrey Moore (2011), *The Innovator's Dilemma* by Clayton Christensen (2016), and *Creativity, Inc.* by Ed Catmull (2014).

Building an Initial Data Team for Your Startup

David Wilby, Former (at various times) Chief Product Officer, Chief Operating Officer, Chief Data Officer, Return Path

Even though data is critical for technology companies, I've seen startups delay key activity involving data until "we reach a certain size." When they do address it, they go all-in in a big bang which can cause more problems than it fixes. A phased approach to data management is what I advocate because it will yield higher results.

Step one is to map out your data flows, where it's collected, at what frequency, in what format, and where it's consumed and surfaced. It's important to have one team, like a Data Pipeline team, be responsible for all data rather than to have each function collect their own data, otherwise you can have a mess on your hands. For example, take something as simple as counting the number of emails that have been rejected by a mailbox provider for any given customer. Suppose you want a weekly total—just sum seven days, right? But what happens if two teams choose different days of the week as the starting point? You'll end up with two parts of your product team telling your customer two different values from what should be the same data. That could easily lead to a lack of trust in the system overall and increase the churn risk.

The Data Pipeline team can create standard roll-up tables to transform the data with tags as it transitions through the pipeline. That way the Product teams can focus on delivering customer value rather than data manipulation.

Step two is to hire a data analyst, a generalist who can pull data from anything, like Google Sheets, Salesforce, Hubspot, or anywhere else. You don't need to invest in heavy BI tools at this time; instead, spend your energy (and money) on establishing more important things like a unique key that spans your data sources—CRM, Platform, and Support. That will unlock more value than fancy graphs and charts and will give you insight into the health of the business and guide investments.

Step three: consider hiring a data scientist. This person needs to be carefully managed and will typically sit on a project within engineering because they'll need access to data and a way to implement their solutions. But I've seen a tendency for data scientists to quickly build an initial model and tune it to produce "good enough" results and then watched them spend weeks and months trying to perfect it. A trap you can all too easily fall into is getting swept up in the excitement of what the data can tell you and to keep pushing for more without ever releasing the value along the way.

Finally, once you've got the output from your data scientist, you'll need to integrate it back into your mainstream engineering teams. There might be initial resistance from the engineering teams since data scientists rarely align with them, but a data engineer with roles in both data science and engineering can smooth the implementation. A good data engineer should have enough experience to highlight work with the Production Engineering teams to resolve data issues, but you need to ensure that somebody in this chain of data custody is thinking holistically.

Chapter 100

Hiring Product Development Team Members

Product

Hiring and building teams are the most important things you will do as a leader. People make an immediate impact (positively or negatively) and they make a long-term impact; again, positively or negatively. There's no better way to create the culture you want than to hire people who will help you accelerate it. On the other hand, there's no better way to ruin the culture than to hire people who are not going to help you create or promote the culture you want. I certainly don't have all of the answers and I'm still learning and experimenting with what works, but after interviewing hundreds of candidates over my career and hiring 100+ people into product development roles, I've learned a few things that I'd like to share with you.

First, you need to be clear about what you're looking for in a new hire and this involves close collaboration with your team and with other functional areas. Obviously, you won't consider hiring someone to just plug a current hole or gap; you'll want to think about how the role can grow with the company as it scales. I'll talk a bit more about this in career pathing, but you need to have a roadmap in mind for the product development organization, the individual, and the company before you even start vetting candidates.

I've broken this topic up into some chapters: interviewing (Chapter 100), onboarding (Chapter 100), increasing the funnel and building diverse teams (Chapter 101), retaining and career pathing people (Chapter 102), and growing your bench (Chapter 103).

Interviewing

Interviewing is more than asking a few standard questions or putting a candidate through a stress test to see how they respond. It's an opportunity to understand a person better and an opportunity for you to provide a clear picture for them on what your culture is, what you value, what exciting projects you have coming up, and what their future might look like if they join your team.

Interviewing involves the following, and you ought to be very prepared and organized because you're driving the process. That means asking good questions and keeping things on track even if it means stopping someone from talking with the statement, "Let's move on to X."

The Audition

With product development roles, having candidates demonstrate their skills is a core part of an interview—especially with engineering hires. How you learn about a person's skills is important though, because it defines your hiring culture and drives the type of results you get from candidates and from your hiring process. For instance, a number of technology companies have an interview process that values deep algorithmic expertise and puts candidates on the spot on a whiteboard, asking candidates to regurgitate their computer science knowledge. While that approach might be required to find the absolute smartest engineers you can find, you will also overlook other key attributes that you want in a team member. The "let's figure out how smart you are" approach attracts certain types of engineers, but excludes a bunch of others who are great problem solvers. Maybe they are more intimidated or don't think as well in high pressure environments.

At Return Path, a problem-solving audition was important but we went about trying to make it as close to a normal experience as we could so that we could assess the candidate in their natural environment. We wanted to learn how they work but more importantly, how they work with others. Are they collaborative? Do they seek advice and counsel from others? Before we brought a candidate onsite, we would give them a small project to complete to whatever level of detail they wanted. The project was intentionally vague so that we could see how creative they would be at interpreting the requirements and getting a start on the solution. And, isn't that the reality of most startups, where everyone is involved in figuring out what they are building with loose requirements? What better way to see if candidates struggle with not having detailed requirements? For us, if they had a difficult time in this situation, that's probably not a good fit anyway.

When candidates come onsite, we ask them to bring their computer and development environment with their code example. Having their own computer is a key item for creating a natural environment for an engineer. I always hated doing technical interviews on someone else's workstation with a development environment that I had never used; I didn't want candidates to have to do that. I want to remove the variables that are introduced from the stress of the interview and the lack of comfortable technology as much as possible. If the candidate is struggling, I want that to be a strong data point and not me guessing that maybe they just can't type because of the split keyboard we put in front of them.

Next, we have them walk through the code they wrote. This is a great icebreaker for them because they are in the driver's seat and are leading the discussion. They aren't being asked hard-ball questions right off the top or staring at a problem for the first time in front of a room full of strangers. And, there are so many data points that you get from their pre-work. How much time did they spend on the problem? Did they introduce any novel approaches to solving by writing their own algorithms or did they rely on libraries? Did they write tests? Did they comment their code? Then we give them a new dataset and ask them to run their code against our dataset instead of the 10 toy records that they got with the original assignment. Of course, our data typically breaks their code and we spend the rest of the time in a debug loop with them and a couple of engineers who are part of the team interview process. This brings in a totally different part of the process because we then get to understand how they debug and collaborate with the team around them to solve things that they haven't seen before.

The audition provides us with a really good picture of a candidate's ability to design systems, solve problems, fix bugs, talk about quality and scaling, and collaborate with others. It is a well-rounded development experience that is more like a day-in-the-life of an engineer than a few "gotcha" questions proving a person remembers their computer science theory from college. I credit this practical audition process as one of the key reasons we hired well in engineering at Return Path. We also used a form of an audition for Product Management, UX, Data Science, and leadership roles. They were all designed around a practical problem with some work at home ahead of time and a presentation or collaborative whiteboard design session.

As I'm writing this, the world is dealing with the COVID-19 pandemic, and I'm thinking about how this changes things regarding interviewing and remote collaboration. The good news is that the process I described for interviewing can be done remotely—as a matter of fact, we hired a number of engineers that we interviewed remotely using this same audition process. The difference is that now these engineers will be working remotely as well, so you may want to incorporate a few other practices in your interviewing process. The interview process will be fully remote and you might want to spread out the schedule to accommodate the different interviewers that will be involved—besides, everything doesn't have to be scheduled back to back since the candidate isn't coming onsite. That schedule could be better for the candidate too since it might be hard to schedule a full-day for interviews. There will also be more of a focus on how the candidate communicates and manages their schedule since they will be working more independently and will need to overcome the communication challenges of being remote.

The Practical Interview Process

Another part of our interview process, after the audition, was a series of team-based interviews that covered experience, culture, work style, and fit. We always approached this as a team, because we were such a collaborative group that we wanted to stress this during interviews and make sure that candidates were getting a full representation of what it would be like to work with our team.

It also gave us the ability to mentor employees who were new to interviewing since they were encouraged to attend and listen and maybe just ask a couple of questions that they had prepared ahead of time with their manager. Having new and experienced team members participate in the interview kept the quality of the interview high and provided a better experience for the candidate. Another benefit of getting the team involved is to showcase your culture. A candidate who only interviews with a handful of employees doesn't have a full picture of what it's like to work at a company, what the company values, and how diverse and creative the teams are. It was typical for a candidate who interviewed at Return Path to talk to 15–20 people (in small groups of no more than four people) and to meet many more during their onsite visit.

Finally, the other thing I will stress with interviewing is the importance of respecting people's time and being the type of company that the candidate wants to work for even if they don't get the job. I always reminded the recruiting and interviewing teams that I wanted a declined candidate's reaction to be, "That's too bad I didn't get the job—I hope I get another chance sometime and definitely want to work here in the future." That means that every candidate who applies should receive a nice personal note that the timing wasn't right and that we will consider them in the future for other roles. If a candidate doesn't make it past a phone interview or in-person interview, they should still receive a personal note or a call saying that it didn't work out but that they should try again later. Respecting a candidate's time and effort also meant that we never interrupted an interview in the middle and walked someone out (unless there was an obvious behavioral issue). We interviewed and were fully engaged with a candidate until the end because they had invested a bunch of time in preparing for the interview and it was important for us to spend time with them and provide thoughtful feedback about why it didn't work out. If we had bad candidates after the initial phone interview, it was because something broke down in an early stage and that's on us. That's something we need to fix, not something that we take out on the candidate.

Onboarding

Your job isn't done after the candidate accepts the offer, it actually continues through the first 90 days. It is critical to have a plan beyond providing them a desk, laptop, monitor, and a ton of paperwork. Creating a strong product development team culture starts with a robust onboarding process. In Product Development we didn't just rely on the HR team to do the onboarding, we had our own process that kicked off on Day 1. Some of the things that we did included a 2-hour meeting with the hiring manager going through a deck about the business and our products, introducing the new hire to the product development team and our practices, providing information on how to find things out on their own, introduction to their mentor, and taking them out for a casual team lunch. At the end of the day the new hire would write up their 30-60-90-day goals based on conversations with their manager.

As a manager, one of the best things you can do to get a new hire off on the right foot is to set expectations and the framework for communication. In my experience, most conflict is created by mismatched expectations and miscommunication, so getting this right at the start is the key to having a great team member. In communicating with a new hire, I find it helpful to lead by sharing my values and leadership style—and include within that discussion my "operating manual" for how to work best with me. I will also cover the interaction points between me and them, and between them and the team they will be working with. These interaction points include one-to-ones, stand ups, time with their mentor, peer reviews, and collaboration across teams.

One of the best frameworks I found to set clear expectations is visual, where I draw a horizontal line on a page or whiteboard. I tell the new hire that the line represents the expectations for their role, something defined that they are responsible for. I place a dot just under the line that represents their current position (or performance) with an arrow leading up and to the right from the dot that represents their trajectory. I make it very clear that the new hire owns their current position and they "own" their effort and trajectory for their growth. My role, my responsibilities, are to help, coach, find resources, and clear any obstacles to make them successful, but they are responsible for their success. I own the expectation line and ultimately some point in the future where we admit that it's not working. In cases of performance issues, that timeline for a decision on staying with the company or leaving is made very clear to the new hire so that they understand it. For new hires that initial timeline is 90 days and it isn't about scaring them, but providing a really clear framework with a bunch of support to help them ramp up, contribute, and be successful. Finally, I always use a detailed plan with the framework where we can collaborate on goals and outcomes that drive the specifics and create accountability for the new hire. The goals include specific tasks and learning items for each 30-day period with a key outcome or demonstrated deliverable.

Here is an example of the types of things that we included in our engineering 30-60-90-day onboarding goals:

First 30 days. Complete all onboarding documents, build relationships within and outside the team, and learn and understand all systems. Work on story cards around distinct pieces of work to support learning the system and to contribute to quarterly goals. At the end of the first 30-day period, the new hire will teach back to the team the key systems and conduct a reverse interview with leadership.

Second 30 days. Demonstrate that you can work independently in small distinct areas of the system. At the end of the second 30-day period the new hire will run a demonstration to the larger team of a feature/change they implemented.

Third 30 days. On their own, the new hire must demonstrate that they can drive larger projects/epics independently and they need to participate in the on-call rotation. At the end of the third 30-day period the new hire will provide a tech talk or product update to the business or their peers.

Bootcamps

Another useful process to introduce to new hires are bootcamps. Bootcamps are particularly useful in engineering where you have certain central processes and practices that you want everyone to be aware of and trained on, but bootcamps are also useful across other teams too. A bootcamp is a more informal setting where you can have courses on leadership values, innovation, and agile processes that are key for product development teams to know and key for others across the company to know, too. Courses on the business domain are also really valuable to new hires (and provide updates and refreshers to others) that are taught by others with expertise. While bootcamps typically require a larger employee base or consistent hiring flow, they can be done informally (even one-to-one) and have the side-effect of making sure things are documented and consistently taught. A couple other key side effects of bootcamps are that they are really good for recurrent training across the company or for employees who want to learn about other areas of the business. Because bootcamps provide a broad view of the company, they provide a window to the new hire and others about career pathing and they expose new hires to other employees across the company who lead the training.

That expectation line that I draw for a new hire in their first few days is more than a line and their starting point. It involves lots of deliverables by the new hire and by the product development team to ensure that the person not only has the opportunity to learn about the company, but the opportunity to interact with others so that they don't live in a bubble. One of our core values is collaboration and our onboarding process will highlight any red flags that a person is not going to fit with the team.

Chapter 101

Increasing the Funnel and Building Diverse Teams

A lot of people talk about building diverse teams, a lot of people wish for that, but if you're not intentional about it, it will never happen. We were intentional at Return Path about building diverse teams. I believe that Product Development is a profoundly creative profession and that the quality of the work is lower if you lack diversity—diversity in people, in experiences, in ideas, and in approaches. Without diversity you're worse off as a product development team. Creativity depends on life experiences that we get when we broaden our stereotypes of who an engineer, product manager, or data scientist is. While building diverse teams takes longer on the whole, in my experience, it returns better results in the long term. Non-diverse, homogeneous teams of like-minded people who have deep experience working together are naturally faster at collaborating and getting something going quickly, but their solutions can often be one-dimensional, and non-diverse teams lack the creativity and tenacity to think differently about problems and approach problems from a different angle. When we broadened our candidate pool, removed biases, and searched for people who didn't fit the traditional stereotype of an engineer, we generated a number of key (positive) side effects:

- We had really interesting candidates to choose from to continue to help us build even more diverse teams.
- Diversity has a positive impact. When you have diverse teams and you interact with people at career fairs and interviews, they want to learn more, especially when they

see an engaged, diverse team of people who are proud to work for a company. The best way to attract diverse talent is to be diverse yourself.

- It allowed us to be competitive in areas (New York, Denver, Austin) that were difficult to hire in because we weren't just trying to go up against Google, Amazon, Facebook, and others all the time—we were looking for non-traditional candidates and had more to offer for some candidates.

To find diverse talent, you don't compromise on your hiring criteria, but open your funnel and find people who have been marginalized by the traditional tech screening process. I talked in Chapter 100 about how we designed our audition process to be more friendly to candidates who were burnt out on the traditional interview screening process that puts them in high pressure situations where they have to regurgitate all of their computer science knowledge. That helped attract different candidates and helped them relax so that we could get the best performance out of them during the short time that we had to gauge their skills. It also helped to include a diverse mix of employees in the hiring process, including at the top of the funnel at career fairs and phone screens, as well as at in-person interviews. You gather new data points when building diverse teams. One day I was in an interview with a male candidate and the technical audition was being led by a strong female engineer. Every response or follow-up that the candidate gave was directed at me or the other male engineer in the room instead of to the female engineer who asked the question. I would redirect him by saying, "Don't look at me, she asked you the question," but I had seen enough to realize that this candidate would not be a strong fit with the type of culture and team we were building. I shudder to realize the hiring mistakes we would have made if we didn't have the right interview team in place to pick up on stuff like this.

Other ways to increase the top of the funnel and find overlooked candidates include building internship/genius-academy programs, using returnships, and incorporating blind auditions. Here are a few examples.

- **Paid internship.** We ran a paid internship called the Email Genius Academy at Return Path every summer and accepted 5–15 interns. The whole company participated, but the main concentration was around product development roles. We attended key career fairs every spring with a diverse group of employees who represented our culture and values and were warm and inviting—attracting similar candidates. By creating a welcoming and inclusive screening process, we were able to bring in interns each year who weren't just traditional engineering students but were problem solvers, curious about the business and eager to learn. We didn't give them "gopher" internship work either, they got a meaty business-focused project that would take up at least two months of their internship and a mid- to senior engineering partner to mentor them through it and help report back to the team and business at the end of their internship. We also accepted junior and some sophomore interns who would return to school the following year and tell their friends about the great experience they had working at Return Path, and word would spread, reinforcing our values and ensuring that we had interesting candidates seeking us out each spring.

- **Returnship.** Returnships or re-entry programs for women who have been away from the workplace is another way to find really strong candidates who have been overlooked simply because they don't have recent experience but were really strong employees earlier in their career. Return Path started a re-entry program in 2014 (that was eventually spun out to a non-profit organization called Path Forward), which was a four-month paid internship program designed for women who had been out of the workforce for more than two years to re-enter and build credible and relevant experience, and to expand the talent pool for our organization. Lisa Stephens is a great example of a talented software engineer who left IBM in 1992 to stay home and raise her children and was looking to go back to work over 20 years later. She had successfully brought her skills up to speed through coding schools, and online programs, but wasn't seeing any interest from employers. Lisa was part of the Path Forward inaugural class, and we hired her as a full-time software engineer at the end of the internship. Having a candidate who comes in with this much experience and only needs to update their skills to the latest technology is a win-win. Lisa was a strong engineer at Return Path and has gone on to be an engineering manager, expanding her impact and influence.

- **Removing bias.** Another program we used at Return Path was our partnership with GapJumpers. GapJumpers is a company working to remove workplace bias through diagnosing and improving your hiring practices. We used GapJumpers to host our audition process and do the initial screening of some of our candidates to experiment with blind auditions and to generally expand our funnel to find non-traditional candidates. This didn't work for more senior roles, since those candidates wanted to talk to a hiring manager earlier in the process instead of doing a bunch of work first without a guaranteed interview, but it worked for some of our associate and mid-level roles. Having a hiring manager review working code or a project plan before meeting a candidate or reviewing their resume flipped the whole process around and brought candidates to the forefront who would easily have been screened out by just looking at their resume. It tended to find people who had the tenacity, drive, and work ethic to stand out—people who were taking charge of their career and not just coasting on entitlement. It challenged our thinking about where problem solvers and programmers come from, and through it we hired engineers, product managers, and data scientists who were previously retail store managers, salespeople, and account managers.

Chapter 102

Retaining and Career Pathing People

What about after a new hire has been with the company for 90 days? How do you retain a great hire? Well, assuming you have the right foundation in place, you're most of the way there. That means that the culture is right, the employee is on a great team, they are solving interesting problems, and they can connect the dots between the work that they do and business outcomes. But, that isn't quite enough. Having a career trajectory is important and people want to know what comes after the role that they are doing now. Early stage companies often fall into the trap of thinking that people will just be satisfied to contribute in their current role for a long time if the projects are interesting and the company is growing. We had that issue at Return Path too, initially giving everyone the title of software engineer even though it was clear that there were big differences between people's skills. Now, of course, we paid people differently based on their experience and responsibilities, but there was no published career path in Engineering that people could track against, and it started to cost us as more senior people realized that in order to move to the next level, they had to move out of the company. We learned an important lesson that even if titles weren't important to us, they were important to people who wanted to know where they were at in their career progression and the next level they could aim for as they improved their skills.

We tried over a few years to put a career development framework in place and learned as we went along. The first version only introduced two roles (software engineer and senior software engineer), had 60+ attributes, took 90 minutes for engineers to assess themselves against, and unintentionally created a game where engineers would try to improve a few attributes to get to the next level instead of looking at it holistically. A second version simplified it, tied it to additional roles, and involved the manager more. Finally, we created a third version that really simplified it to five core attributes consistent across levels, tied it to our core values, and incorporated team feedback and development planning. We also rolled out versions of this across Product Management, UX, and Data Science.

What we ended up with in Engineering answered the question, "Where do I stand with regard to my professional growth as an engineer, and what do I need to do to develop my career?" We framed it as only one part of your career progression to another role. It reflected your skill proficiency in your current role and your development in the next role. It was to be used as a holistic set of expectations to be successful in that role and give you aims on your development areas, not as a game to be "won." The percentages that were tracked in the framework were there to be a rough measure of proficiency in those expectations and should be set and aligned with conversations between employees and their manager. We wanted to see above 90% proficiency in their current role and at least 50% proficiency in the next role before considering employees for promotion. This ensured that people had enough skill in the next role to be successful and supported the "working out of title" to prove interest and initial skill at the next level. But that wasn't all. Just "hitting the numbers" on the career development framework didn't mean that you were automatically going to move to the next level. It was fine for engineers to remain in their current level at full proficiency and as a high performer. A promotion should happen when we have a good overlap of interest, opportunity, and qualification. Other factors that were required for a promotion were:

- Engineers should be high performers to be considered for promotion to the next level.
- Promotions were reviewed every 6–12 months and reflected sustained performance over that period.
- There needed to be a business need for additional senior level roles.

- Years of professional experience don't have to line up exactly, but provide some pacing for career progression, and experience is a factor.
- Engineers eligible for promotion were reviewed by a group of their peers and engineering leadership before receiving a promotion.

The last point was key to getting everyone involved in making sure that there were going to be no surprises on promotions and was important as we got larger as an engineering organization. The biggest challenge was in having a manager who "graded easier" than another manager in another part of the company, and we made some missteps early on where we promoted someone and had a big backlash across engineering because that individual was good at managing up to their manager but didn't have credibility across their peers. Having a group of peers and alignment across engineering leadership were key in making sure that promotions made sense and had alignment.

We ended up with the following roles and responsibilities in Engineering:

- Associate Software Engineer: entry-level role, curious, learning quickly, solves scoped problems, and contributes to the team.
- Software Engineer: significant contributor, owns projects from beginning to end, collaborates, gives feedback, incremental, and proportional.
- Senior Software Engineer: recognized leader and key decision-maker on the team, solid stakeholder manager, collaborates between teams, and takes a rigorous engineering approach.
- Principal Software Engineer: recognized leader in engineering, responsible for core systems and projects, collaborates across engineering, and carries the engineering brand within the company.

Chapter 103

Hiring and Growing Leaders

Next to hiring well, the next important thing is to make sure you have the right leaders in place and that you are continuing to develop the bench of future leaders. I'll start off by saying that hiring leaders externally is difficult to get right. You are looking for people who are highly aligned with the culture and practices that you are building and who have the ability to build trust and rapport on the teams they will be responsible for. They also need to be a good leader, with both the skills to work effectively with people and the ability to get the team to execute and hit business outcomes. Finally, especially with technical roles, the leader needs to have deep experience as a contributor in the area that they will be managing. When you take all of this together, you can see how challenging it can be to find the right person who covers all of these areas. I find that it is easier to hire a leader into an organization that already has a strong leadership team—that peer team provides a center of gravity for the new leader and a chance for the new leader to pair with an established leader on a team for a period before taking over. Even then I've had mixed results with hiring leaders externally. The biggest misses have been getting people who weren't fully aligned with the culture we were building and creating pockets in engineering with different values or approaches to engineering. Or, we'd get people who were strong technically but weren't strong people managers and lacked the empathy to build a deep connection with the people they were leading. Or, we would get people who were only strong people managers and lacked the technical experience to build respect with the team. You get the point. It's not easy to fill senior technical leadership roles!

One option that I've had good success with is to find a strong individual contributor who is trying to take a leadership track in their career and is blocked at their current company either because leadership there doesn't feel they are ready yet or because there just isn't an open opportunity right now. I'll hire them and have them join a team as an individual contributor for a few months to learn the systems and build trust with the team. When (and if) the timing is right, I move them into a leadership role on that team—it could be a tech lead initially and then to a full front-line manager role.

By far, the best success I've had, though, is developing leaders internally. Again, it starts with hiring—looking for those extra leadership qualities beyond someone's experience and skill to just do the individual contributor job. It is about looking for the engineer who loves to solve business problems, not just technical ones, or the product manager who wants to start their own business someday. It's about finding a UX engineer who is completely passionate about product design and can't stop asking questions about your customers or the data scientist who is all about data-driven decision making. It's about looking for candidates who can do the job and are especially empathetic, connect well with people, and who are wired to lead. This doesn't have to be true for every hire, but what I'm saying is that if you start to look for these additional qualities, you will begin to build a bench of talented leaders with high potential to be future leaders as you help them develop.

And speaking of helping them develop, you need to provide multiple opportunities for individual contributors to take on ownership and leadership roles to give them experience and opportunities to work "out of title." At Return Path, we developed leaders through leadership development bootcamps, through mentoring relationships, through technical leadership opportunities on teams and certain technologies, and through opportunities to lead and present during hackathons, engineering summits, tech talks, and lunch and learns.

When we had an opportunity to put a front-line manager in place and we had a good internal candidate, we followed the 30-60-90-day plan that was part onboarding and part testing that this was a good fit. Our approach ultimately became a framework to turn individual contributors into managers. This is what that plan looked like—with credit to David Loftesness, Twitter's former Director of Engineering and the original inspiration for this:

Day 0. Discussion to make sure you know what you are getting into as a first-time manager. These are the truths about the job that you have to accept and the adjustments that you have to make—coding less, having a calendar full of meetings, leading by enabling others to do the work, setting the ends and letting the team figure out the means. We also want to make sure motivations are right: Don't become a manager in order to just please the boss; don't become a manager only to advance your career; do become a manager if growth for you involves others; do become a manager if you channel empathy; do become a manager if you can give the trust you ask of others.

First 30 days. Block off time to learn. This includes a reading list and our Engineering Leadership Playbook; it includes joining the bi-weekly engineering leadership peer support meeting; and working with your manager and mentor to shadow them as they manage the team—including joining them on one-to-ones. At this point you are still taking 60% of individual contributor work.

Second 30 days. Balancing between individual contributor and manager. Take on one-to-ones by yourself; lead more of the team ceremonies (planning, standups, etc.); and balance your current development work, including ramping down the number of cards you were picking up as an individual contributor to about 30% or so.

Third 30 days. Ramping up to be the full-time manager. In the final 30 days, you have ramped down your individual contributor work and are taking on all of the management responsibilities for the team.

During these 90 days we let new managers slowly take on more responsibility for the team, but as a practice didn't make them have performance conversations, write reviews, or handle compensation matters. Along the way I tell the new manager that three things have to line up at the end before we finalize the role. That is: (1) I have to like how it's working; (2) the team has to like it; and (3) most importantly, you have to like how it's working. As the leader who is developing this new manager, I spend my time mentoring and meeting with the team to gauge the impact that the new manager is having and getting to this decision point on day 90. Over my career, I've done this process with over a dozen new leaders and only had one person elect to go back to being an individual contributor at the end—and in that situation the process worked and saved the team from having a leader who wasn't passionate about being in that position. Actually, the feedback I got from that individual was that he was overwhelmed with how much the team vented to him and wasn't sure he wanted to deal with all of the needs they had. Without that person doing actual one-to-ones by himself, he would have never figured this out along the way before he became the full-time manager. As you scale your organization, the temptation will be to quickly promote people into front-line managers to handle the growth. I would just caution you to take your time and pay attention to the process and wait until the manager candidate is really ready. It's far better to overload a good manager temporarily with too many direct reports than putting a mediocre manager in place or prematurely promoting a new manager. You have to get this right as you grow—the number one reason people leave companies is still because of their manager.

Developing leaders internally this way has been the most successful model I've used. Leaders I've promoted into first-line managers have gone on to become Senior Managers, Directors, and VPs.

Other Things to Consider

While "smart" and "get things done" are two key characteristics I look for while hiring, there are a couple other things I always look for, and here are a few questions I would use to determine if those attributes are lacking:

- You can typically figure out how senior or experienced someone is by having them articulate what they hate about a process or company practice or programming language. If they are just getting started (or started a long time ago but haven't really grown up), little things will bother them, things that they could fix if they were more experienced or mature. The more experienced a person is, the more likely that what they "hated" was a big problem, something that shows that they really dug deep and addressed some hard problems.

- Ask individual contributors about someone's career they improved to test how empathetic and collaborative they are. This is also a good proxy to determine their potential to move into a leadership role.
- I find that the more details someone remembers about past projects they worked on, the better they are at caring about details and the better they are at getting things done. It always surprises me when I ask someone to talk about the hardest problem they solved or tell me the details of a recent project and how they went about it, that they can't quite remember or articulate the key parts of the project.
- I often ask engineers to tell me about the last piece of code they wrote to see how they think about decomposing problems and how they structure their code. It is also one of those questions that either gets immediate engagement or a quizzical look that reveals their passion (or lack of it).
- Look for someone (or encourage them) to teach you something. I always have found that the best hires are ones I learned something from during the interview process that makes me want to work with them. It doesn't just have to be in their area of technical expertise, but in any area.

People are the core of your business and as I said at the beginning of this section, hiring is the most important decision you will make. I found a post on Quora years ago by Michael O. Church where he answered a question about firing an employee with this simple and brilliant summary about the types of people in an organization:

The best employees are multipliers who make others more productive, and next are the adders (workhorses). Subtracters are the

> *good-faith incompetents who cost more than they bring. Dividers are the worst kind of problem employee: they bring the whole team (or company) down.*

I've used this quote in my career and refer to it as the "Math Model to People Management." The goal is to hire, retain, and develop people to be multipliers and adders. Subtractors are developing contributors who cost more than they produce—and all new hires start as subtractors. As I outlined in the Onboarding section in Chapter 100, your goal as a leader is to turn them into adders and multipliers or be specific about how long they have until it isn't working. Dividers damage morale and bring the whole team down—fire them immediately.

Chapter 104

Organizing, Collaborating with, and Motivating Effective Teams

If you are trying to build a development organization where people are engaged and solving business problems along with engineering problems, you need to get a few things right.

First, you'll need to create a culture of ownership that is accountability-driven. This is deeper than just defining responsibilities for people. While responsibility and accountability seem similar (and are often used interchangeably), there is a fundamental difference between the two. While people can have responsibilities, accountability is something that you take—something you do to yourself rather than something done to you. Organizations sometimes try to fix things by redefining responsibilities, but that doesn't change the way people think like personal accountability or ownership does. Ask yourself, "Is more responsibility going to lead to success here?" Accountability is empowering and whenever people get personally involved, it produces better results. This is why Agile ceremonies like Fist-to-Five (if taken seriously) are key because when team members are personally committing to the team to deliver the goals, you get better results. For leaders, this requires transparency, clear communication, and direction, and getting out of the way. It means that you centralize vision and decentralize control.

Second, to build a great team, you'll need people who are curious, self-starters, and intrinsically motivated. The good news is that if you create the culture described above, you have the right foundation to attract and retain people with these qualities. I've found that the people who are accountability-driven like to work at companies that give them ownership and freedom. In my experience, the key differentiator between the good and bad hires I've made has centered on the extent to which people want to hold themselves accountable.

Third, are people valued or valuable? It's pretty much table stakes now that companies "value" their employees and have cultures and perks that make people feel valued working there. I think the higher bar is whether or not a person understands their value to the company. In other words, can a person draw a direct line between the work they do and a company outcome? This is harder for a company to define for every person, but in my experience a deeper way to get people engaged and taking ownership is when they can connect the dots between their work and the company's success.

So, assuming that you have the foundation right, you have a team with intrinsically motivated people who take ownership, where leaders are setting outcomes and giving people the freedom to control the methods to accomplish those outcomes, then what? Your job as the product development leader is to understand how teams of people interact. The same principles apply for teams as for individuals. Teams need a vision and an outcome they can hit; they need to be motivated and have control over the methods they use to hit that outcome; they need ownership and accountability for the successes and failures of a product; and they need to have some roles and responsibilities that govern interactions with the business and other teams that they are collaborating with.

It's easier to keep teams all rowing in the same direction when the company is small with a single product that all the teams are collaborating on. It becomes increasingly difficult as the company grows and gets into additional product lines and business units. The effectiveness of each individual team taking ownership and having all of this control is now starting to backfire because the individual team outcomes or motivation are in conflict with those of another team. This is where understanding and designing the operating system at the team, department, or business unit are key to keeping a culture of ownership and accountability and pushing control to the edges while still collaborating effectively.

I'll now provide three different examples that illustrate how to align teams so that they focus on outcomes, not output. In one example we drove higher software quality by disbanding the quality assurance team; in another example we created a unified product development organization rather than separate product and engineering organizations; and in the third example we had to manage the impedance mismatch between product and go-to-market.

Example 1: Quality

Ensuring quality is a key requirement of an effective product development team and there are many ways to do that. One common approach is to have a dedicated quality assurance (QA) team (depending on the type and stage of a company), but it comes with a cost. A QA team needs another team to coordinate with to make sure that the QA team's motivations, values, and operating system are aligned with the company's and also make sure it is working effectively with the rest of the product development organization. When it's not working well, it can slow things down, create issues between teams, and actually reduce the overall system quality that the team was created to ensure in the first place.

That was my experience while working with a dedicated QA team in a rapidly scaling, early-stage startup. We were just getting to that point of scaling out teams to handle specific products and had a shared service team that was responsible for manual and automated testing before code was released. The team was working well when the company was smaller, had less product contexts, and had a predictable release cadence. There were two key KPIs that we noticed changing that pointed to the ineffectiveness of the QA team. One was that they were becoming the bottleneck in the development sprint cycle. The second issue was a spike in production support work, which was the work that teams did to fix bugs or other issues after code was released. When we dug into the data, we noticed three core issues on the QA team that were causing this slip in performance:

1. **Context switches.** Our high growth coupled with a trend to push more decision making onto the teams created deeper technical and product contexts that the central service wasn't equipped to keep up with. The side effect

was that the development engineers were having to come over and help QA engineers test the code they had just written.

2. **Lack of staffing and capability revealed we were doing one-off testing processes rather than building a suite of automation that could be generalized for all teams.** The QA team didn't have the expertise and influence to drive a standard here since most of the deep senior engineers were on the product teams and just went around them. The side-effect was that we had more conflict between teams. Teams would either go around QA, or QA would push to centralize a way of doing things to remain relevant even if it wasn't in the best interest of the business.
3. **Disconnected team members and a focus on output instead of outcomes.** I didn't like the culture we created where development engineers didn't have to care about being thorough and caring about quality because it was the dedicated QA team's job. I didn't like the fact that QA felt disconnected from the business problem and was just focused on the output of their work. I didn't like that QA was more concerned about looking relevant than being able to move the business KPIs and outcome forward.

While there are ways to fix this to keep a central team functioning well and aligned with the other teams, we decided to dismantle the dedicated QA team. This wasn't easy and it required a lot of effort framing a new philosophy around quality, how we would approach quality as a company, and then executing the change well. Not getting the framing right will work against any change you make and although we tried our best to frame our new quality approach, we still got accused years later of "not caring about quality ever since we shut down QA." We cared a lot about quality! We fixed an issue that was causing a spike in production support, but we were still misunderstood.

We ended up moving all of the QA engineers and embedding them in the product teams, and we pushed the product teams to incorporate quality into their team's process. The message was that every engineer was a QA engineer. They could build unit tests or full integration tests or incorporate more rigorous peer code reviews, but they had to come up with a process that fit their size and scale and was proportional to the business. Some teams were working on early-stage prototypes and could cut corners on quality and come back later to fill that in. Other teams were managing mission-critical systems and needed a higher level of quality. One side-effect of dismantling the QA team and embedding quality into the product teams was that teams got creative and incorporated various testing into their development flow (peer code reviews). I figured out that if you give an engineer a solution, they will complain about it, but if you give them a problem, they will solve it. The other side-effect is that we had more teams building testing automation so that they didn't have to test things manually like they were before. So, we improved our quality and we got our teams to be more innovative—we improved effectiveness with a proportional investment in quality that fit what the business needed at that stage.

Example 2: Unified Product Development Organization

I think one of the keys to a growing, dynamic, creative business is harnessing and managing the natural conflict and tension that exist between the functional areas inside the business. Sales is pushing for something more to sell; Marketing has an event schedule planned out months in advance and wants features launched at specific dates; Customer Service wants more functionality in the product and higher quality; Product wants time to deeply solve a customer problem; Engineering wants time to make a system more efficient; and Finance wants it done on budget. Does this sound like every company you've ever worked with? This natural push and pull between functional areas in an organization creates resistance and friction that, if harnessed, makes an organization stronger and causes the leaders to learn how to challenge and compromise to get the best out of each other and get the best for the business. On the negative side, if you don't harness the friction, it can cause a business to take the path of least resistance and not challenge each other to become great. Obviously, at its worst, friction and discord can devolve into unhealthy conflict and turf wars between functional areas. If you are unsure of which type of friction you have in your organization, look for finger pointing and the equivalent of "your side of the boat is sinking" comments.

Nowhere is your approach to friction more important to get right than in the product development organization. The product development team has to navigate all of the conflicting goals between the various functions, collaborate well, make tradeoffs, and bring a great product to market. They also have to manage the natural internal conflicts and tradeoffs between engineering and product management, like roadmap prioritization, technical debt, team organization, and the balance between innovation and maintenance. Some amount of friction and healthy conflict within the product development organization can be good and can drive more engagement and better results, but if this devolves into stalemates where leaders only care about their functional areas, it can really slow things down and become toxic.

At Return Path, scaling rapidly created some dysfunction across Product Development, but it was more about teams optimizing around the path of least resistance than outright turf wars between functional areas. Fighting and disengaging—though they are opposite reactions—can have the same effect on a product development organization, making it less effective at collaborating, and at navigating the hard tradeoffs that need to happen to create great products.

There were two symptoms that stood out and pointed to a larger issue here:

1. We had strong product managers (some of whom were fairly technical) paired with engineering managers who were oriented toward being in service to the business. The engineering managers didn't always consider, or speak up about, the long-term technical implications of decisions. As we scaled, we didn't push hard enough on the full cost of technology on the business and what we needed to manage was both the current cost of products as well as the cost of building new ones. That was a bad combination and led project managers to really drive the product and technical roadmaps and optimize around what they needed to get product out the door without

a strong engineering partner to navigate the technical implications of these decisions.

2. Another symptom that emerged was that teams became more output-driven than outcome-driven. If you looked at the goals for various teams, they were capturing the work that they were going to do for the quarter—and often hitting it—but, they weren't always able to tie their work to a business outcome and drive their goals around that.

Symptoms like these can sometimes reveal an inexperienced or immature leadership issue. But sometimes it is just that roles aren't defined well and there is confusion and overlap that create tension that can either be fought and defended or leaders will pull back and let another leader or team take the lead. For Return Path, it was the latter. There wasn't crisp role definition and it caused Product Management to take too much authority in some areas (especially around technical choices) and let Engineering off the hook for areas they were responsible for but not being accountable for.

A RACI model (responsible-accountable-consulted-informed) is a good tool to use to identify roles and responsibilities and to help you manage teams with high collaboration requirements. We've used that at Return Path before with good success. Another model we introduced that was simpler to communicate across teams was the "What and the Why" and the "How and the Who." Product Management owned the "What" and the "Why" and Engineering owned the "How" and the "Who."

How does this work in practice? Well, for Product Management, the "What" was obvious (our products) and they were already owning that. They were responsible for questions like: What are we going to build next? What is the priority of this? What story cards are in this next release? And what the heck do we do now? The "Why" was also obvious—to the Product Management team—but it wasn't expressed or articulated concretely so that it was difficult for the team to set goals, drive motivation, and just connect the dots between the work the team was responsible for and the business outcomes. A simple explanation to the team of "why" something is important and "why" the business cares about it goes really far in unifying a team and getting them to own that we are all in this together. The product management group knew why they were doing something, but ensuring they were accountable to articulate the "why" to the rest of the product development organization was really useful.

On the engineering side, the "Who" was obvious, too. Engineering leaders owned hiring, onboarding, and creating teams of engineers to work on products. Engineering was definitely on top of building teams and tuning them to make sure they were high functioning teams. What was more surprising to engineering leadership—and especially surprising to Product Management—was that Engineering also owned the "How." Now obviously, Engineering always decided how they were going to architect and code something, but the "how" Engineering was dealing with was bigger. Engineering was accountable for how technology was going to be used to solve business problems; they were accountable for how technical debt was going to be managed; accountable for how tradeoffs were going to be made between the technical and product roadmaps; and accountable for how (and when) time was going to be spent on technical innovation to take on a new market.

Product managers were owning the tactical "how" and feathering work into the roadmap that engineers said they needed to deliver a feature, but there was a gap in the larger technical strategy that required time and a business case to be made. The technical strategy needed to understand and account for the collaboration and tradeoffs between product and engineering leadership that were beyond just sprint or quarterly planning. Acknowledging this larger "HOW" and getting Engineering to own it were the start to help us drive a more dynamic and creative product development organization. And, not having this role definition and accountability in place earlier was part of the reason for getting into the hole that Return Path was in. It caused us to have high technical debt.

By having ownership of the "how" and "what" together, Product and Engineering can manage and agree on "when" things get done. This is generally dictated by priority and deadlines on the Product side and scope and headcount on the Engineering side. But this is where healthy conflict and discussion between Product and Engineering really drove the most efficient outcomes for us.

Example 3: Impedance Mismatch

The third example of how to build greater collaboration stems from understanding an "impedance mismatch." An impedance mismatch is an electrical engineering term to describe when inputs and outputs of an electrical load don't match and cause signal reflection or an inefficient power transfer. That phrase is often used in technology to describe problems that occur due to differences between systems, the most popular example being the object-relational impedance mismatch. That mismatch describes issues between the database model and the programming language model. In short, it's a mismatch between what you have and what you want. An impedance mismatch is how I would describe the interface between Product Development and Go-to-Market (GTM) organizations at Return Path.

We had just started to get the product development organization firing on all cylinders again. There was good alignment and collaboration between teams, our leaders had good role definition, and were accountable for the what and the why and the how and the who. We were starting to innovate again, and we were making big changes to products that had been in market for a dozen years. But then the problems just started to move downstream. In some ways, you could say that because the business had been making small incremental changes in product over the last few years. The adjustment to innovating again and delivering large cross-company launches was surprising and stressful to the system. Sales and Service needed training, Support needed administration tools, Marketing needed predictable launch dates to schedule events around, and Finance still wanted to know how long it was going to take and how much it was going to cost. The main challenge was around coordinating the project work across all of the functional areas in the company, and what I noticed was that Go-to-Market operated more around a waterfall project planning model while our product development model was Agile. This was our impedance mismatch.

Although the mismatch was between waterfall and Agile methodologies, both models have a place in most businesses. First of all, waterfall isn't always bad even though it is often seen as a negative from the product development side of a business. It is a valid project planning framework, and you almost have to use something like this when you are doing any type of event or conference planning and have hard dates. It isn't a negative thing that sales planning, training, and marketing events were planned out a year in advance with a tight schedule. That's just the way an enterprise business needed to operate. Second, while Agile processes are optimized for adaptability and delivering small chunks of functionality continuously, Agile is also good for maintaining a constant pace and being predictable. I've had to remind many development teams that you can commit to a launch date and still be Agile. Again, it's all about tradeoffs.

So how do you reconcile waterfall and Agile? The framework that was most useful to our GTM and product development organizations was "fixed scope" vs. "fixed date." If we are going to commit to a hard launch date, then the scope of work might change to hit the date. That flexibility allowed the GTM teams to set a sales plan and a theme and date for their events. The development teams could then commit to the roadmap and get as many of the key features in as possible while still hitting the date. So the product development teams had flexibility to incorporate discovery and manage unknowns along the way. On the other hand, if it was important to launch with specific functionality, the GTM teams had to relax the date and work around a rough month or quarter timeframe and the development teams would agree to a set scope of functionality to deliver. Working through this framework had big impacts on the expectations of both groups and helped manage the impedance mismatch.

Another challenge with complex product launches across a large company is that you don't just have to make sure that you have horizontal alignment across functional areas. You also have to have vertical alignment between the individuals doing the work and their senior leaders. Coming from the product development function, we had processes to make sure that individual teams rolled up to senior leaders who were involved in setting goals, tracking progress, and giving feedback. Expectations were set and aligned, and communication was frequent and bi-directional. But what I figured out was that the product development organization already worked in verticals; it was our normal practice. Pushing code, maintaining products, delivering functionality were what we did, just like breathing. Other organizations, not so much.

Our initial solution was to have a couple of salespeople attend a weekly launch meeting to help plan and coordinate their needs for launch because that would be an effective way to get stakeholders involved to make sure we weren't missing anything. And that's what we did—individual contributors from every function that was impacted by a company-wide product launch were involved. The thing we missed is that those departments aren't optimized around managing launches (like Product Development is) and they don't have a process for managing requirements and making tradeoffs and communicating with senior leaders about this work. What happened is that while product development contributors and leadership were aligned on the launch, other functional areas were not and the senior leaders in those areas were surprised and out of the loop. This caused a number of launches where things didn't go well: Support and Service weren't trained to handle inbound tickets, Sales wasn't fully trained on the functionality in the product, and Marketing didn't have collateral on the website. Obviously, this was not ideal and we needed to address the root cause and resolve it.

Our second attempt at solving the problem led us to create an executive level Product Council that met every week. It was driven by the product development executives and attended by other functional executives who were involved in company-wide launches. We also had key individual contributors who were involved in driving that specific launch also attend. The goals of the Product Council meeting were to get alignment and to over-communicate details around the launch. Part of the meeting was strategic and included talking through the roadmap, addressing the next scheduled launch, discussing tradeoffs and generating alignment across the functional areas. But part of the meeting was digging into the details for communication plans and training materials, reviewing support tickets, and doing product demos. We wanted to make sure that this group lived and breathed our roadmap, products, and launch process. In turn, we wanted them to communicate with their peers on the executive team and through their teams. Getting alignment at the top and getting vertical involvement with the individuals and teams responsible for execution were key. This Product Council didn't replace our annual or quarterly planning meetings, but it augmented them because the need for over-communication and high coordination and change happens in between the planning sessions. And launches across enterprise products in a large/complex business are hard and require a lot of planning and communication.

Things to Consider

- When you have a healthy team, it makes your job easier as a manager. It is much easier to identify performance and fit issues with an individual in the context of a team. It is much easier to challenge someone to step up because the team is counting on them. It is much easier for individuals to understand their roles and responsibilities in relation to their peers than trying to understand their manager's development plan. Individuals tend to work harder in order not to let their peers down rather than letting down their boss. Integrating a team and collaborating fit my management style well since if I feel like if I have to manage a person extrinsically, then I've failed. I'm not going to tell a person what time to come to work or tell them they can't take vacation time, but I can tell them to coordinate with the team and don't let the team down.
- In 2015, a group in Google's People Operations released some research answering the question "What makes a Google team effective?" They found a few key dynamics like dependability, clear goals, and impact, but far and away the most important factor was psychological safety. It makes sense—if team members don't feel safe around each other to admit mistakes, ask the dumb question, or try something crazy and fail, then it limits the potential of the team. On the flip side, a team with high psychological safety is more likely to engage in taking risks, experimenting, harnessing creative ideas, and being effective collaborators. Getting the culture right and building effective teams (across the whole company) is foundational to success.

- Large cross-functional planning sessions can be really good to get stakeholders from across the company to collaborate together, build empathy for each other's departments, and make important tradeoffs that build ownership. We did this quarterly Agile Roadmapping exercise, and used a massive whiteboard where each team laid out their own roadmaps in horizontal swim lanes. Once all of the teams were done, they linked cards together with string to represent interdependencies. This model quickly identified bottlenecks and helped prioritize work items that multiple teams were dependent on early in the planning cycle.

Chapter 105

Due Diligence and Lessons Learned from a Sale Process

As the company grows, you will have opportunities to accelerate growth or fill a hole by acquiring another company. At Return Path, I ran technical diligence on over 10 companies—acquiring five of them. Here are some of the things that a Product or Engineering leader needs to know about doing buy-side diligence.

First, you have to understand the purpose of the acquisition. Are we doing this to fill a hole in our product offering, do we need this technology, or are we really interested in the team as more of an acqui-hire? Next, you should evaluate if buying this company is the only or best way to get this product, technology, or team. Can we grow the team a bit and build it ourselves? Can we partner with a company or license the technology? Really think through the long-term implications of acquiring a company. Sometimes the opportunity and acquisition come together so easily that it distracts you from the deeper issues of integration—and that's where the real challenges are.

The product could look great on the surface but be hard to use or hard to sell; the business could have legal, financial, or data privacy risk deep in the past that you aren't aware of; the technology could be a pile of shortcuts held together by a hero engineer who isn't planning to stay after the acquisition; the team may look like a culture fit on the surface, but are used to operating differently and aren't going to make it through the integration. Make sure to walk through these scenarios and talk to peers who have experience with successful and failed acquisitions before you proceed. Finally, if you choose to acquire, make sure that you have a rigorous diligence process that uncovers as many of the risks as possible and provides the data you need to evaluate the tradeoffs between building versus buying.

When running a diligence process on a potential acquisition target, you typically have a small team covering the key areas of Finance, Operations, Legal, and Product/Technology. For larger acquisitions with many employees, we would include the People team. I would evaluate privacy and security during the technology diligence, but often bring dedicated team members in for larger or more complicated acquisitions. Here are the key areas that I covered in product and technology diligence and some of the details that I looked for:

People

I always start with people in a diligence process. As I've mentioned before, people are the key to any company and by talking with them first, you remove the potential awkwardness of a diligence process and you learn so much about the business. I start with one-to-ones with each team member in Product Development and ask about their background, strengths and development areas, experience with the company, and concerns that they have about being acquired. I also spend time selling them on my company, how past acquisitions have gone, and what the integration process looks like. I find that these initial conversations help us learn about each other and build trust so they know what to expect and that helps them feel comfortable opening up and sharing more about the business. The main things I'm looking for in these conversations is to understand the strength of each team member, if any of them are critical path, a sense of the quality of the team and culture, and—most importantly—any challenges, concerns, and operational or technical deficiencies that provide a roadmap for how to navigate the rest of the diligence.

Team

It's important to understand the culture and health of the team—their team operating systems, communication styles, decision-making process, any conflict, the way they manage ownership and accountability, and their relationship with failure. Most of the qualitative details are discovered in my one-to-ones with individuals, and the rest through conversations with product managers and the leadership team. I also look for quantitative details by having someone walk through recent project and Agile team stats and look for healthy velocity and a breakdown of the type of work the team is spending time on (feature, production support, technical debt, etc.). Another key area to dig in is around overall resource allocation and how teams are staffed and how planning is done. I always ask the leadership team what is missing and how they would use more money or people if they had them. There are often a number of key things that emerge when you understand the corners that are being cut by an understaffed team.

Product

The key thing I'm looking for on the product side is efficacy—how well does the product work? How well does it solve the problem? And do customers love it? I start with having someone do a product demo and walk through the core functionality. I'm interested in the origin story of the product—how was it started? When did it get to product/market fit? How has it evolved? Who are the key customers? How do they use it? Is the product a "vitamin" or a "pain killer"? What does the near and long-term roadmap look like? If it is an early-stage company, pay attention to the discovery process, how well they have achieved product/market fit, and talk to their customers. Make sure you understand if the company has traction or just looks like it does. For later-stage companies, you should be able to look at product and financial metrics to see evidence of a strong product/market fit and a healthy growth trajectory.

System Overview

The key thing I'm looking for on the technical side is efficiency—how well is it designed? Where is the technical debt? How does it scale from a cost and capacity standpoint? I have the technical team walk through the technical systems and outline the core functionality and how each service works in the whole. For each service I like to understand the key metrics (throughput and latency), any recent production support issues (and any recurring issues), the capacity plan for scaling, the overall complexity of that service and the effort to operate it, and what the technical roadmap looks like. Holistically, I'm looking for how well everything works together, what shortcuts were taken and how much technical debt exists, how complex the code base is and if the team generally is focused on outcomes or only on solving technical problems.

Privacy/Security

A lot of this is uncovered during the system overview, but I'm calling it out separately because it is really critical and areas of potential risk. You also learn a lot by a team's attitude and transparency around talking about these issues. On the security side, it is important to uncover any past security breaches, security processes in place, penetration test results, any audit controls and results, and the general process that is used to ensure security is built into the design and development process. On the privacy side, look at what data is stored and transmitted, the systems involved, and encryption controls that are in place. Work with your legal team to review the privacy policy and alignment of the company's data behaviors to that policy. With larger or international companies, I recommend working with outside counsel on complex deals to ensure that there are no hidden data privacy risks related to mistakes the company may have made around consent and use of data in the product.

IP and Legal Risks

Similar to security and privacy, a number of risks exist around intellectual property and software licenses. Again, you will be partnering with your internal legal team on this, but you will want to verify some key items. Make sure that you get a list of all of the employees and contractors that have been involved in building the product and make sure that contracts cover a release for work product and, if not, make sure you can secure a release from those individuals before you inherit this liability. Ensure you look at all commercial and open source software licenses in use by the product to make sure that everything is in good standing. There could be hidden costs to operating systems when you find out that they are using a community edition service from when they were pre-revenue that now requires a large licensing fee. For large and complex acquisitions, you can pay external companies to scan code bases to find license violations or whole blocks of code copied from a licensed product and being used unlicensed. This area is critical for really understanding what you have and making sure that you don't have a ticking time bomb legal issue that an external party will bring up after the acquisition, creating unexpected costs and legal issues.

Costs

Most of this comes up in the system overview as well, but it is important to get a clear picture of the ongoing fixed and variable costs to operating the systems. The headcount costs will most likely be modeled by your team members responsible for the financial and operational parts of diligence. You should still collect and evaluate employee comp and titles and how it fits within your organization for integration down the road. On the system side, try to get a detailed breakdown of the costs per service. Pay attention to the variable and unit costs incurred for scaling and make sure that you are comfortable with the cost curve and any one-time technical costs required to scale.

Risk Assessment

The last step is to do the analysis, risk assessment, and recommendation. This includes the financial model—assuming additional investment to fill in the gaps on the team and scaling systems—to justify the acquisition and show the payback and return on the investment. You'll likely be doing this in collaboration with the financial diligence lead. You need to compare and contrast this acquisition against other options like building it yourself, licensing a solution, or partnering with another company. When doing the buy/build analysis, make sure to consider key trade-offs like opportunity cost, technical skill required, operational requirements, and whether technology for this area needs to be a competitive strategy or more of a utility. Make sure you detail the risks in rank order including integration risks around culture and people. You will be collaborating with the rest of your diligence team and the business on this analysis, making sure to understand the impact this decision has on various stakeholders, and being able to present a business case with the tradeoffs. Finally, make a recommendation backed up by your evaluation and results of your diligence.

Having done a few acquisitions in my career, I can say that it doesn't always go well and I regret at least two of the five companies that I've been part of acquiring over the last 10 years. Here are some lessons learned and things to watch out for:

- It's not over until everyone has signed. A smart acquisition target is pursuing multiple offers and negotiating to get the best value. Be detailed around Letters of Intent and no-shop agreements and make sure you are paying attention to the timing and boundaries of these agreements during diligence. I've seen a company walk away at the last minute—legally—to pursue a better offer because we weren't on top of things.
- Make sure you know where the technical bodies are buried. Reviewing and understanding critical services and their scaling characteristics are important. Don't just take the technical co-founder's word for it. We experienced an order of magnitude volume of growth on a service after acquiring it that had an exponential cost curve to operate. It took us months to find the time to properly refactor it to get to a cost curve that we were happy with and had to eat the cost in the meantime. A deeper understanding of the system and capacity plan would have identified this during diligence.
- Don't buy the technology without the team. Unless it is just a competitive takeout acquisition or they use the exact same

stack as your engineering team, you are going to have a challenge integrating and operating a company's systems without any of their technical team. Don't ask me how I know this.

- Really understand if they actually have product/market fit or it just looks like it. With early-stage acquisitions, it is important to really get into the product discovery details and talk to customers. Some founders are really good at selling up (kind of like managing up) and have a really good story about their solution and how much customers love it. I've seen startups with special pricing for beta/charter customers or continuing to keep customers connected after they've churned. Make sure you deeply understand how valuable the solution is to customers and how much they are really paying for the solution.

Chapter 106

Selling Your Company: Preparation

At some point in a company's journey there will be an opportunity to sell the company or raise another round and you will be pulled in to sell-side diligence. The time to prepare for that is months or years before, not when it happens.

Now, of course, companies get opportunistic offers before they are ready to sell and get the process done without preparing well in advance, but from my experience, there are certain areas that you want to track from the beginning that make a sale process go easier. And, since the goal of most businesses is to grow and eventually sell, why not invest in this work upfront? So, one key for a successful sale is preparation. The other key is being able to tell a good story about the stage the business is at and the processes that are in place that make sense for that stage. Let's dig into both of these. Chapter 107 covers telling the story.

Preparation

The product leader needs to start at an early stage in the company's existence with preparation. The things to prepare are a mirror of the items I listed in Chapter 105 that are typically included in a buy-side diligence engagement:

Licenses and Intellectual Property

Track any commercial and open source licenses in use. It's also a good practice to have a policy across Engineering to enforce the types of licenses that are allowed and which ones require leadership sign-off. If you publish open source as a company, you should also have a practice or policy to follow a set of standards and make sure libraries and license types are approved by engineering leadership and Legal—mainly to make sure that you aren't exposing any business or technical IP and that proper licenses are in place. You also want to track all contributions to your internal IP and make sure that employees and contracts all have signed releases in place.

Privacy and Data Governance

Companies are increasingly using data throughout their business, and with the rapidly changing privacy laws, companies need to be aware of how data is acquired, used, and stored. As someone who has been involved in a complex data business, you definitely want to create a detailed data map and keep it updated as you scale the business. It needs to include the following attributes for each data source the company is responsible for: acquisition process (mapped to user consent policies); whether the data is public, private, or confidential; where and how it is stored and transmitted (type of encryption and physical security); the lifecycle/retention policies; and governance—or what products it is used in and how it can be used. This last point is really critical and important to enforce and document. Just because data is sitting around in the business doesn't mean it can be used in a new product idea. You have to tie the data consent by the user to the way it is being used in the product (can this data be used only in aggregate or anonymously, can it be exported to a third party, etc.?). A business that is lax about how they track and manage data can quickly get a number of products in market that are valuable and lead to an acquisition opportunity only to have it crushed when they realize they are using data inappropriately or illegally.

Security

Document your security policy (in coordination with your SecOps team) and show how secure practices are used in system design and development. Have a security diagram for your network and systems and track the entry points and the individuals with access. You'll also want a document that details how training is done and what controls are in place for determining and limiting employee access to systems or tools that provide access to customer data. For example: Do customers have to allow an account manager to access their data through internal tools, or can any employee access it whenever? Do engineers have access to production data or just operation team members? And is that full-time access or leased team when there are system issues? You also want to log access to production systems and be able to report and track this. Finally, make sure to document any security breaches and any penetration testing and results.

Product

For early products, document the discovery process and path to product/market fit. For mature products, track the customer usage data. You'll also want to show recent product enhancements, what drove those decisions, and the current breakdown of time spent on new innovation, feature enhancements, and maintenance. It's also important to keep an updated long-term roadmap and resource allocation document that shows the individuals and teams working on various products across the organization.

Documenting Systems and Processes

I like to keep an up-to-date system diagram that serves as an internal design document for engineering, but also can be used for security reviews, audits, and diligence. It's also important to track project work and be able to report on the flow of a reported bug on a support ticket, through a product management Agile tool, to deployment in production. This process is important to show during diligence, but also as the company gets larger for audits.

Metrics

Tracking systems and product usage metrics are important for any business. Make sure to focus on uptime metrics and long-term tracking of customer impacting events—especially if the company has any service-level agreements in place.

Costs

This includes detailed costs around running the product development organization. Breakdowns on people/system costs per product are really interesting to show investment and return. On the system side, it is important to show fixed and variable costs around licenses and SaaS tools, as well as hosting/data-center costs and the historic and projected cost curve.

Chapter 107

Selling Your Company: Telling the Story

Next to being prepared, you need to be able to tell a great story about where the business is at—and specifically, how the product development organization is doing. We just defined the building blocks of a great story above in the preparation part in Chapter 106, so your job is to weave this data together in a way that tells the story of the product evolution up to the current success of the product in market. The story will of course detail the processes that you have implemented that make sense for the stage of growth for the company. More importantly, you should be able to describe the processes that you are planning to implement that aren't in place yet. This presents the business in the best light and builds confidence that the leadership team knows exactly where to go, but is being practical and iterative in getting there. It also presents risk up front instead of the buyer digging in and finding things that are a surprise to the process.

Other than being able to walk through the data, processes that are in place, and the risk, the product and engineering leaders' key job is to sell the value of the product and the team. Assuming you have a great product that has found product-market fit and is scaling well, you should be able to talk about the current value, ROI, future roadmap, projected growth, and know enough about the buyer to articulate how the product fits into their overall strategic plan. Similarly, for the team. If you've spent time creating a great culture and efficient team operating system that is innovative, executing well, and hitting outcomes, then showcase that and show how this team is key for the deal.

In the second edition of *Startup CEO*, Matt tells the story of a failed sale process at Return Path. We received an unsolicited inbound offer from a large company and while we had some preparation in place from my list above, we hadn't done enough. The weeks-long diligence process was exhaustive and wore me and a couple of my leaders out. The process was highly confidential and we were trying to keep the internal team small so it created a lot of work for us. The day before the scheduled signing of the definitive agreement, the buyer got cold feet. The deal was off. I felt crushed, not just because the deal was dead, but because I was a core member of the team responsible for telling the story of the business and we had failed. Ultimately the reason the deal failed was because of a number of risk elements that the acquiring company felt were too much for them to absorb and integrate.

We had complex international data privacy agreements and a large technical organization that was in the middle of a big reinvention, migrating to the cloud from data centers, and tackling GDPR. Sometimes you can get prepared for a sale and tell a perfectly good story about where the business is at, but the timing is wrong. However, this was also a big lesson for me on how to communicate risks to an acquiring company. While we were transparent and clear about the stage the business was at and all of the work we were doing, it came across looking like more than we could handle on our timetable and contributed to the risk. The key lesson is that you have to tell both sides of the risk story—be transparent about the work and the risk, but also sell your team and their creativity, resilience, and historical performance at being able to solve big problems. It turns out that our team was successful in reinventing the business, migrating to the cloud, and handling GDPR on the schedule we proposed. We were also very prepared for the next opportunity to handle an incoming offer, and eventually a successful exit to Validity in 2019.

Conclusion

Leading a product development organization is a challenging and rewarding job and really is the intersection of technology, business, and people. I want to conclude this Part by reiterating the core foundational items to get right to build an effective product development organization and scale it as the company grows.

The role of product development leadership is not just to be the execution arm of the business but to be an integral part of the business, setting and communicating product strategy, connecting other business functions together around a shared vision, and using technology as a force-multiplier and strategic asset for the business. Success is measured not by the output from Product Development, but by hitting outcomes.

The right culture is key to cultivating the type of environment that allows people to have ownership, a deep understanding of the business, influence on outcomes, freedom to experiment, and a place to innovate and create remarkable products. Effective teams are diverse teams that are psychologically safe and empower people to trust, make mistakes, and do their best work.

Build a product development organization that is aligned with the business and has a culture of solving business problems with Engineering, not just solving engineering problems. A proportional product development approach maximizes effectiveness and scales with the business as it grows.

Chapter 108

CEO-to-CEO Advice About the Product/Engineering Role

Matt Blumberg

What comes before a full-fledged Chief Product or Technology Officer? In most startups, this role is played by one or more founders. Sometimes it is engineering-focused, sometimes it is a product visionary, sometimes it is both in one role, and sometimes it is two roles. Because of that, I'll address each role separately here.

Signs It's Time to Hire Your First Chief Product Officer

You know it's time to hire a Chief Product Officer when:

- You wake up in the middle of the night and realize that your product vision has gotten stale because your leadership of it has had to take a back seat to other issues around running the company or fundraising.
- You are spending too much of your own time managing the connectivity between Engineering and Product Management relative to the connectivity between product and go-to-market.
- Your Board asks you what the long-term roadmap is, and you don't have a great answer and aren't sure how to get to one.

When a Fractional Chief Product Officer Might Be Enough

A fractional Chief Product Officer may be the way to go if you have solid mid-level product management leadership but need extra experience to dig into product-market fit issues and create a new high level roadmap, or even to lay out a new process for creating that roadmap.

Signs It's Time to Hire Your First Chief Technology Officer

You know it's time to hire a Chief Technology Officer when:

- You wake up in the middle of the night worrying about a potential system outage.
- You are spending too much of your own time thinking about technical debt, engineering throughput, and security issues or audits.
- Your Board asks you what your Business Continuity Plan is, and you don't have a great answer and aren't sure how to get to one.

When a Fractional Chief Technology Officer Might Be Enough

A fractional Chief Technology Officer may be the way to go if your Engineering leader is a very strong coder and troubleshooter who needs some mentorship to become a strong architect and cultural leader.

What Does Great Look Like in a Chief Product or Technology Officer?

Ideal startup Chief Product or Technology Officers do three things particularly well:

1. They create product by starting with the business objective, not the technology. Even the most hardcore technology enthusiast in the world can't produce the best product by starting with technology. The selection of platforms, languages, databases, and architecture must fit the needs of the business and customers and available workforce, not the other way around.

2. They know the balance of taking shortcuts to get things to market quickly AND going back and cleaning up the shortcuts methodically after the fact. Shawn's whole point in this Part about "scrappy, not crappy" was a phrase that drove us over the last few years at Return Path. No one wants to get things into the hands of customers faster than a typical startup CEO. No CEO will ever get equally fired up about paying down technical debt. Great heads of Product, Technology, Engineering must be the standard bearers for this.
3. They are deeply concerned about the culture and career paths within the product or engineering organizations. In most technology companies, the two largest groups of employees are Sales and Product. In consumer-facing technology companies, the single largest is usually Product and, specifically, Engineering. The leader who knows how to optimize for engagement and retention, who spends real time thinking about compensation bands and career paths, who mentors engineers and trains them to be principal engineers or managers—that's the leader who knows how to get the most out of the team. There is also a role for the strong senior engineer who focuses only on the code or only on the development process—but that is probably not the person you want to rely on as your main executive in charge of the whole of the department unless they have a superb #2 who is almost a co-leader, or who knows how to partner very effectively with your Head of People.

Signs Your Chief Product or Technology Officer Isn't Scaling

Chief Product or Technology Officers who aren't scaling well past the startup stage are the ones who typically:

- Focus and thrive on individual contributor work like bug/code troubleshooting or pulling all-nighters on coding or releases. Similar to Sales, where you have a problem when your CRO gets more excited about closing a deal than with a rep closing a deal, Product leaders need to be focused on building the overall product machine, not on writing code. If what Bill Gates says is true—that the best engineer is 10x more productive than the average engineer—and I believe that to be true, then companies need to reward that engineer with a generous compensation package, not with a senior executive role, unless the engineer promises to make everyone else 10x as productive!

- Are not interested in engaging with sales reps and customers. Even the most introverted engineer can figure out how to do this in ways that aren't incredibly uncomfortable, like sitting in the back of the room at every user group or listening to sales calls on Gong. Even better to be an executive sponsor on a key account or speak at a customer conference.
- Downplay advice from nontechnical CEO or leaders because "they don't understand the technology." This is a bright red flag, and one that you only need to see once as a CEO. It's bad enough when advice on product from a nontechnical CEO is rejected, but it's even worse when the same happens with advice on other things like employee matters or interpersonal issues between executives. If you have a nontechnical CEO who respects and gives you a lot of leeway on technical matters, pay attention to the things they question or offer up. They may have some good insights because they're further away from the technology. Heads of Product or Engineering are clearly in trouble when they use this argument as a way of not doing something that someone outside their organization notices and suggests.

How I Engage with the Chief Product or Technology Officer

A few ways I've typically spent the most time with or gotten the most value out of Chief Product or Technology Officers over the years are:

- Challenging a paradigm shift. Years ago, when our business at Return Path started to take off and our CTO was thinking about how to scale 1.5x in the coming year, I challenged him to think about what he would do if we had 10x increase in volume. After his head stopped exploding, he began to think very differently about how to scale some of the trickier aspects of our system architecture. That led to a really different plan, and an outcome of a more efficient system, in the short term, that we might not otherwise have experienced.

- Bringing business strategy and acumen into the product team. Our various product leaders always asked me to do dedicated roundtable sessions with their leadership teams to riff on the business. Although these meetings were usually unstructured, product leaders always came with massive amounts of questions and opinions about the company. At Return Path anyway, they were more inquisitive about some of the ins and outs of strategy and finance than other teams, and these conversations almost always resulted in some new ideas for our product roadmap that made them worthwhile.
- Applying learnings from Product to the rest of the organization. Engineers are trained to think differently than most other people in the company, so they frequently think up new ways to do things that others might never come up with. At Return Path, our CTO was the one who came up with the idea to create a structured return-to-work program for moms who had taken a career break to raise kids. He was dogged in his pursuit of the concept. We funded and nurtured it, and it attracted national attention and grew to the point where we spun it out into a new nonprofit company called Path Forward. I'm not sure if we would have dreamed that up without a systems thinker pushing it. The out-of-the-box thinking by the CTO is a great example of "Agile Everywhere" (See also Chapter 27).

Part Nine

Privacy

Dennis Dayman

Chief Privacy Officer

The Rise of the Privacy Officer

A Chief Privacy Officer (CPO) is a relatively new role in business, even though ideas about privacy have been germinating since the 1970s. Back then, before the explosion of digital information, most of what individuals left as a footprint was based on things that could be attributed to us personally. So, buying a house or a car, or being arrested—anything that you did personally that left a paper trail could be considered a privacy issue. For the US, the U.S. Privacy Act was innovative legislation, incorporating ideas like data minimization, right to access, and right to correct—it is limited to data collected by the US government from its citizens. The Privacy Act had no impact on private industry or what data could or could not be collected on the Internet by companies, but it was a start in addressing personal privacy. Where things started to change was with health care information.

For the European Union (EU) in the early 1970s, countries like the United Kingdom began adopting broad laws intended to protect individual privacy. Throughout the world, there is a general movement toward the adoption of comprehensive privacy laws that set a framework for protection. Most of these laws are based on the models introduced by the Organization for Economic Cooperation and Development and the Council of Europe.

In 1995, conscious of both the shortcomings of the law, and the many differences in the level of protection in each of its Member States, the EU passed a Europe-wide directive which provided citizens with a wider range of protections over abuses of their data. The directive on the "Protection of Individuals with regard to the processing of personal data and on the free movement of such data" sets a benchmark for national law. Each EU Member State had to pass complementary legislation by October 1998.

The EU Directive also imposed an obligation on Member States to ensure that the personal information relating to European citizens is covered by law when it is exported to, and processed in, countries outside Europe. This requirement resulted in growing pressure outside Europe for the passage of privacy laws and over 80 countries and independent territories, including nearly every country in Europe and many in Latin America and the Caribbean, Asia, and Africa, have now adopted comprehensive data protection laws. More laws are being enacted by states in the US and are similar to the General Data Protection Regulation (GDPR), which is now one of the toughest privacy and security laws in the world. Though it was drafted and passed by the EU, it imposes obligations on organizations anywhere, so long as they target or collect data related to people in the EU.

There's nothing more personal about an individual than the status of their health since that information could be used by others to harm that person. For example, AIDS is a very personal thing to people and, especially in the 1980s as it became more rampant, it became something actually used to keep people out of jobs and out of communities because it was a deadly disease. The same could be true today with COVID-19. Informed consent was originally developed for rare, high-risk and potentially life-threatening situations, like surgery and medical research. In health care, a person cannot get assistance without disclosing intimate physical and behavioral facts to others. If you don't disclose this otherwise private information, death or serious permanent harm might result. Rooted in the Hippocratic Oath, this ethical rule frames a fiduciary role that exists between the physician and their patients. The physician receives health information in trust to be used only for legitimate health care purposes. Using that health information for other purposes would be a breach of trust, contrary to the Hippocratic Oath's first principle, "do no harm." An excerpt from the Oath administered says, "I will respect the privacy of my patients, for their problems are not disclosed to me that the world may know."

So, the idea of personal privacy rights was most understood by US citizens when it started with health care information and culminated in the Health Insurance Portability and Accountability Act of 1996 (HIPAA), which protects personal health care information. With HIPAA you're basically giving the health care professional or hospital the permission to share information about you and your diagnosis with other entities who may treat you. So, privacy laws really became entrenched in protecting people, giving them a good sense of feeling safe when they give sensitive information. People began to understand what their rights were as individuals.

When you leave health care and get into the digital realm, what is the footprint that you leave today as an individual? You might not use Facebook or Instagram so you'd think you're safe from tracking, that your personal information is safe, but you'd be wrong. Today, we leave a digital trail of every interaction—from visiting a website, sending a text, making a purchase, sending an email, or even participating on a video call. Privacy has become more important today because of digital technologies. We used to talk about six degrees of separation. Six degrees of separation is not just a party game, it's a reality. Everyone in the world is interconnected, and thanks to social media, that connectedness gets tighter and richer each day. It's now one or two degrees of separation. Anybody can find anything about you as an individual because we leave a digital trail from everything that we do.

That's a short, truncated history of a longer discussion of privacy laws, but today it's imperative to have a CPO in a company, from the first day of your founding until the end (whatever that may be for you). The privacy officer is responsible for any organizational privacy programs. That can be anything from keeping track of daily operations, to developing and implementing programs for maintenance, to creating policies, controls, and procedures. As a CPO, you'll also have to create monitoring programs because you just don't set privacy and then forget about it, or hope that it will be followed. You have to establish privacy metrics and use them to measure and monitor the compliance of the organization that you're working for. Of course, issues do pop up, like an inappropriate use of, or breach of, data. Your primary purpose is to mitigate any privacy issues for customers. You need to ensure that you're in compliance with all relevant privacy laws that apply to you, whether it's in Europe, the US, or elsewhere, and whether it's at the level of federal, state, or some other jurisdiction.

Chapter 109

The Role of Privacy Officer

Today the role of the privacy officer covers everything within a company that creates data, everything that touches data, everything that analyzes data. You'll look at it from multiple perspectives. From a consumer perspective, if I were the CPO of a social media platform, my job would be to make sure that the data that's being entrusted to us by that individual is protected, or that we follow the promises we gave the consumer when they entrusted us with that data, such as "we're using your data to market to you." Hopefully, you have won their trust, they're going to interact with your platform, and you'll do good things with that data. If you don't, then you'll be like Facebook and Cambridge Analytica who definitely broke the trust of consumers. Cambridge Analytica should never have received personal information from Facebook but they did and they exploited it. So that's two problems right there—breaking the trust of the consumer by not keeping their information private, and not monitoring companies you partner with.

Being an advocate, a protector, and consumer are the chief roles of the privacy officer, but there are internal data you'll have to manage as well, like HR information. People in HR have access to highly personal information about, for example, me as an employee. They know how much I get paid, what my benefits are, what family members have access to my health care plan. They know my employee evaluations and if I wasn't a very good employee, they would have access to my performance improvement plans and access to any complaints filed against me. HR knows how much I have in my retirement fund, they know financial information, and they know my history through a background check. Do I pay my bills on time? Do I have any bankruptcies? Do I have a criminal record?

A third role for the privacy officer, beyond consumers and employees, involves other relationships like partners, Sales, and Marketing. Anything that creates data that can be traded, moved, transferred, or used, that can easily identify me at any one time is considered personal identifiable information or PII, and absolutely requires a privacy program or a privacy officer to oversee adherence to the policies.

A first step for a CPO is to create strict guidelines on what data each person can access and use in their role. For example, for HR data, you might allow only the payroll and benefits person to have access to the employees' benefits selections, and another HR employee would not be allowed to access that data.

As the privacy officer you won't necessarily work in a bubble and you'll have to collaborate with others. For example, I am not an attorney. I started off in the law enforcement arena with a degree in that field, but as a hobby became a self-taught computer and internet geek. What really got me interested in privacy and security was spam—I hated it. The idea of someone bothering me or using my email address without permission really upset me. One of my first jobs out of college was to create a compliance role at AT&T and basically you needed to understand law, but also technologies and how they intersect at times. Over the years I took all I learned in anti-spam and compliance and began to work in the marketing vertical, in privacy roles, and helped brands and marketers. So, while I don't have a law degree I've been doing this for 25 years and I have enough training and enough experience to understand what the laws are, what the regulations are, and what the contracts or legal documents should have in them. But I don't work in a silo. I will interact with general counsel or corporate counsel at times because they are trained in a certain way to make sure that the right language and the right protections are in these policies and procedures. In some cases, a privacy officer is the general counsel, and they have formal legal training with a law degree. They specialize specifically in privacy law, or they specialize in corporate law, and they also have the same privacy certifications as a non-attorney CPO.

It's possible for early-stage startups to outsource some of this work to outside vendors. So when you're a smaller company with five to 50 people you probably can't afford a full-time general counsel or privacy officer. In that case you might outsource the privacy program to a law firm that specializes in privacy or have a fractional general counsel or CPO.

Chapter 110

Privacy Advice for Startups

Data is the new gold standard. Let that sink in a minute. For some people data is irrelevant, it is a non-issue for them. Yes, they know they're being tracked but they don't care; yes, they know companies know how they got to the company website, how long they stayed, what they did there, and where they went when they left, but it doesn't bother them; yes, they know that their mobile phone knows where they've been, and it's always been that way, so it seems normal. Because of its durability, gold has been used to store value in the form of money. In many countries, printed and minted money used to be linked to stored gold, which was reserved in the central bank. This way, central banks used to print as much money as they had gold reserves. However, the idea of gold as the most important economic asset has passed. Instead of entrepreneurs going to California to look for gold, as they did in the past, they now go to Silicon Valley to create companies that exploit the acquisition and sharing of data. The most valuable asset of Uber, the ride-hailing service that does not own a single taxi, is the data of the people who use it. The prime asset of Google, the largest library, which owns no physical library, is the index that takes customers to their desired websites. Therefore, the new valuable asset is not gold but it is data. Data has become the new gold!

As a startup, you absolutely cannot have a whimsical or lackadaisical approach to data because, as I just said, it is the new gold standard. And if you don't protect your data, the value of your company can drop significantly. That's right, if you don't protect your data, then you could face fines and other penalties.

For example, under the data protection regulation, GDPR, the less severe infringements could result in a fine of up to €10 million or 2% of the firm's worldwide annual revenue from the preceding financial year, whichever amount is higher! GDPR defines a "personal data breach" as "a breach of security leading to the accidental or unlawful destruction, loss, alteration, unauthorized disclosure of, or access to, personal data." Organizations that collect and control data (controllers) and those that are contracted to process data (processors) must adhere to rules governing data protection, lawful basis for processing, and more.

The more serious infringements go against the very principles of the right to privacy and the right to be forgotten that are at the heart of the GDPR. These types of infringements could result in a fine of up to €20 million, or 4% of the firm's worldwide annual revenue from the preceding financial year, whichever amount is higher. For a company like Google, with $160 billion in revenues, that fine would be $6.4 billion—not a trivial amount! So, in other words, data processing must be done in a lawful, fair, and transparent manner. It has to be collected and processed for a specific purpose, be kept accurate and up to date, and processed in a manner that ensures its security. Organizations are only allowed to process data if they meet one of the six lawful bases listed in Article 6. In addition, certain types of personal data, including racial origin, political opinions, religious beliefs, trade union membership, sexual orientation, and health or biometric data are prohibited except under specific circumstances. When an organization's data processing is justified based on the person's consent, that organization needs to have the documentation to prove it. Individuals have a right to know what data an organization is collecting and what they are doing with it. They also have a right to obtain a copy of the data collected, to have this data corrected, and in certain cases, the right to have this data erased. People also have a right to transfer their data to another organization. Before an organization transfers any personal data to a third country or international organization, the European Commission must decide if that country or organization ensures an adequate level of protection. The transfers themselves must be safeguarded.

The time to start thinking, and acting, on data protection is Day 1, when you're working on a minimum viable product, or when you're first meeting with customers trying to develop your value proposition. Privacy and protection should be baked into your systems from the start—into your engineering, your corporate engine and your go-to-market strategies. If you do that you'll be following best practices on data protection called "privacy by design." Incorporating privacy-enhancing technologies as a team right in the very beginning will help you be successful in the long run with your company and help you avoid huge problems later.

It used to be possible to start a company with an idea for a product or service and just start building it and getting it done. Entrepreneurs wanted to be first in the market, and they just developed their app and put it out there. They weren't really thinking about privacy or security. The opposite approach is to think about privacy and security very early and to operate like the best companies, in any industry. For example, General Motors doesn't theorize about a car and then build it piece by piece. They have a blueprint, they run tests, they go through thousands of processes so that when they're ready to build and sell the car, they can do it quickly because they have a tested, repeatable process. They've gotten all the kinks out from creating an early design model.

The same thinking ought to be applied to privacy and security, and it should be at the forefront of any development process. Whenever you get an idea and you start to doodle on a napkin, you'll get to the point where you can say, here's the idea, here's the widget, and it's going to create all this data and make me a billion dollars. But in that drawing, what you never see is any thought about how to keep that information private and secure. Privacy by design is the idea that, instead of building an application and then bolting privacy at the end of it, you build privacy inherently and by default into the function of the app at an earlier stage. Privacy by design is based on seven foundational principles:

- Proactive not reactive; preventive not remedial.
- Privacy as the default setting.
- Privacy embedded into design.
- Full functionality—positive-sum, not zero-sum.
- End-to-end security—full lifecycle protection.
- Visibility and transparency—keep it open.
- Respect for user privacy—keep it user-centric.

When I join most companies I begin to look out for what can be improved on product and process but also, who can I rely on to help in that endeavor? This is never a one-person job, but it takes a village sometimes to address the ever-changing landscape of Internet products and services. Products are increasingly being judged based on the privacy and security they provide, while organizations are judged on the knowledge they share and the commitment they have to ensuring personal data is protected. This presents immense opportunities in the market for organizations or products ahead of the game. Every team or organization will integrate privacy and security by design into their products in their own way. All will face different challenges and accrue lessons learned, which will grow best practices for the benefit of all.

With all the hype around Big Data, personalization, better ad targeting and customized user experience, data has tremendous value in today's economy. Every product manager wants to understand their customers better. Amidst all this, privacy is often an afterthought. Product managers focus on how best to collect and exploit user data before creating a system to protect user privacy or write a privacy policy. Privacy protections are a late addition, added on to an app, program, or website only to comply with legal requirements or to satisfy user concerns. This is changing now, thanks to laws such as GDPR.

At Return Path, we were committed to improving our products by incorporating the privacy and security by design framework into our processes. Our customers entrust the personal data they gave us, and we had an obligation to protect it as best we could. One of the things we did was ensure that our product people, even though they aren't daily privacy professionals, had the same training and if possible certifications. With regulators worldwide calling for tech professionals to factor data privacy into their products and services, we thought it was best that we extend the responsibility of privacy outward into other job requirements. The privacy department had its own certified experts, but at times it was necessary to give others the knowledge, certifications, and power to make the necessary changes to products and services based on privacy factors earlier in their development lifecycles to alleviate mistakes and changes earlier. At the end of it, we were able to certify engineers, product, marketing, sales, and HR people with the necessary tools to be successful in their day jobs, but also help protect Return Path and the data we've been entrusted with. The most interesting thing about working with others in helping them learn privacy is how it can change their career. One person managed a product line, but when an opening came up in the privacy department, they were able to apply and obtain the new position and continue and build a new career path. Today, that employee is now the Senior Director of Privacy at another company and just received another certification in privacy all from being a part of the overall privacy knowledge plan we put in place.

Now you understand how the app is going to collect data, how it is going to share that data, how the data is going to be moved around, where it's going to go. You'll understand this as you're building the architecture out. You're a much stronger company if you're building privacy in versus pushing the app out there, letting people use it and then realizing that you just collected data and illegally moved it to another country. You'll then have to delete all of that data because it was illegally obtained. And so privacy by design is literally starting privacy as a first thought in your development processes, asking yourself, "What do I need to think about when it comes to privacy?"

One of the easier ways to think about that is with email marketing. Marketers love to send emails, right? They just love to. So, if marketers just decided to create a template email and then take a database of email addresses and start sending the email to people, they are missing key steps that can land them in hot water. Do you know whether you have permission to send emails to all of these people, or have you confirmed somebody hasn't unsubscribed previously? Do you know the country of the email owner, and is it in the European Union or Canada where you have to have permission to email somebody?

I work with a lot of startups and founders as an investor and mentor, and some have lived their entire life giving their data to access something they want, never giving it a second thought. One company did not follow the privacy by design approach and they illegally collected data on 13-year-olds and younger, which is against the law without permission from the parents. The 1998 Children's Online Privacy Protection Act (COPPA) imposed requirements on operators of websites or online services directed to children under 13 years of age, and on operators of other websites or online services that have actual knowledge that they are collecting personal information online from a child under 13 years of age. They ended up paying a fine to the U.S. Federal Trade Commission, they had to delete all the data they collected and when they did, they didn't have any data to really run the business as a social app and they had to close the company. Everyone lost their job over the silly fact they did not think about privacy in their designs.

So, my advice for startups is to take data protection, privacy, and security as seriously as you do any other early-stage startup tasks. Embrace privacy and don't fear it. Leverage privacy as a competitive advantage. My ideas in Chapters 111–120 will provide you with the documents you need to protect and show how to build a privacy team that can add value to startups and scaleups.

Chapter 111

Legal Documents

When it comes to legal documents, you have to do a couple of things right to protect your company, and you have to be hyper-transparent with anybody who might be involved with your business. With customers (or even visitors to your website), you'll have to ask permission to collect data and if they provide it, you need to explain to them why they're giving you their data, why it's important, and what you're going to do with their data. You'll also have to tell them how you're going to protect their data. There is widespread concern from consumers about their data and plenty of high-profile data breaches (see Facebook example in Chapter 109) justifying those concerns.

Privacy Policy

The first document you should create is a privacy policy. This is a statement made by you as a company that says to a visitor or consumer what data the visitor or consumer will need to provide to receive service from your company. And then you'll tell them what you are collecting, how you'll use the data and, more importantly, how you're not going to use the data they provide, who you share it with, and how you protect it. Even though a customer or visitor will not "sign" the privacy document, they will need to check a box and/or click a submit button and it then becomes a contract between you and people coming to your website. Then, people who are supplying their personal data—even if it's an email—will trust that you do what you say you're going to do with their data.

If you breach that contract, then you're responsible for damages if and when you misuse or change the use of data on them. So, if you tell people in your privacy policy that you'll never share data with a third party but then six or 12 months down the road you change your mind and share it, you've violated your privacy policy. In this case you'd have to send an updated privacy policy and get permission from visitors and customers to share their data. It can be easy to say "Hey, I can violate my policy a little bit. It's MY policy, after all. Or I'll just change it." But I'd encourage you to think about your privacy policy as a contract between you and your users. Violate it, and you violate your contract. You wouldn't do that so quickly or freely with a client, would you?

At one time, privacy policies were a best practice but now they are regulated by laws. Today a company must have a privacy policy when it's collecting anything that is related to PII. That's the kicker, though, what is PII? It's a broad definition and it could mean a website visitor or customer's name, it could be an email address but it could also be a social security number or a passport number. A company could gather PII without thinking that they were by using a tracking technology to understand how many visitors went to their website. Since most computers' and browsers' beacon send out an identifier that's related to a person's computer and to the browser they use, and the tracking technology captures that, pretty soon there's PII being collected on your website through your analytics.

All these pieces of data are now identifiers and now you have to have a privacy policy regardless of whether you're using that data for anything. Your privacy policy can state, "We don't collect email addresses, we don't do this, we don't do that. This is who we are and if you visit our website you may get a cookie or a tracking, or we use Google Analytics. But we use our data to ensure that our website is working to its full potential."

A privacy policy could be as simple as that. It could be one page or 15 pages, depending on how much you have to say about the data. If you do have a long and complicated privacy policy, then you need to consider a "just-in-time" notice. The just-in-time notice digests the longer privacy policy into short, manageable segments that are provided to consumers and users of a product as they need them. Instead of delivering thousands of words to be read prior to adoption at a time when the user may or may not be familiar with them, a just-in-time privacy notice gives the user precisely what they need to know shortly before the decision needs to be made. This can be done by including the text in the appropriate section of the website, a pop-up information box, or a brief explanation with a link to a longer disclosure. Just-in-time privacy notices improve transparency and build consumer trust by providing individuals information about data collection, usage, and sharing at the time that it is needed by the individual to make an informed decision. It does not rely on obtaining consent through a disclosure months earlier, that the user may or may not have read and comprehended. But you have to have a privacy policy and it has to be on your website on day one, and it has to reflect exactly what data you collect and what you do with that data. Remember, if you say it, you need to do it. If you change your business practices, the privacy policy needs to reflect that.

Internal Privacy Policy

An internal privacy policy tells employees how they can use the data the company collects. It is another document that many startups neglect to create. For example, Marketing might collect email addresses to send out marketing emails, but does that mean that the Chief Technology Officer can go grab that same database and send out a bunch of emails about how great our technology is? If that's not what people signed up for, you don't have permission to use that data source in that way. The internal policy needs to be included with your onboarding and training to make sure employees understand acceptable and unacceptable use of data based on the purpose under which it was collected.

The internal policy also tells employees the things that you expect them to do and not to do with the data that's been entrusted to the company. Employees are responsible for protecting the data, whether that's through encryption, or telling people not to put data onto USB drives and take it home, or download massive databases from protected production systems onto their laptop and take it home. It's a very specific policy that says, "Do not do this, or do this when it comes to the data types that we collect." A lot of companies forget to do that. They tend to focus on policies externally, but don't think about training annually and reminding and putting their employees under a rule set or a policy that says, this is what you can and can't do.

The policy and the recurring training are really important; employees can do dumb things because they are unaware of the risks they are taking. But if you put that policy out, and every year when you make them go through their training and sign the privacy policy, it will help remind employees of their obligations with the company's data. Make sure you have an updated privacy policy and that employees are aware of what it is and what they should or should not do.

Terms of Service

Privacy Policies and Terms of Services (TOS) agreements are both, as the names imply, legally binding contracts. The main difference between these two types of agreements is this.

A Privacy Policy agreement exists to protect your clients. A Privacy Policy agreement is required by law if you collect or use any personal information from your users, e.g., email addresses, first and last names, etc. The purpose of this agreement is to inform users about your collection and use of personal data of users.

A TOS agreement exists to protect you, the company. This agreement sets forth terms, conditions, requirements, and clauses relating to the use of your website or mobile/desktop app, e.g., copyright protection, accounts termination in cases of abuses, and so on.

When it comes to the terms of service about 90% of it is a legal contract—you'll pay us "x" and we'll provide the service and warranties and you won't sue us and we won't sue you. But what many companies forget to put into the terms of service is something telling the user that they are responsible to stay within the realms of the law if they collect data. For example, when you rent a car, the car rental agency provides you with a contract and you agree that you will drive at the posted speed limit for whatever city, county, state or country that you're currently driving their vehicle. It is your responsibility to be aware of what the laws are and how to drive. And that Terms of Service needs to have a privacy flavor which tells companies that they are responsible for the privacy laws that affect them.

Another key element of the Terms of Service that startups and small companies often forget is to tell their visitors or their clients, "By the way, when you give us your data, if we ever sell our company, your data will be moved over to the acquiring company." The acquiring company will become the new owner of the data and the visitor's or client's contract would then be between them and the acquiring company. The acquiring company is now responsible for protecting the data and they will need to abide by the original Terms of Service and Privacy Policy. And if they want to do something different with their data, they would have to go back to that consumer and say, "We acquired this company, we were told that this is how they were using the data, we would like to make a modification to how we're going to use that data." And if any changes are going to be detrimental to them, give them the opportunity to opt out of sharing that data or being a client.

Too often, startups forget to include this language, and it negatively impacts a merger because the acquiring company would not have permission to use that data. And they would go and throw their arms up and ask, "Well, then what am I buying? If I'm buying a company, I want all this information and systems but I also need the data to go along with it. And if I can't transfer the data because you didn't tell people that it will transfer over, then your company has no value to me." It happens too often because startups forget to put that one simple line or paragraph into their documents at the very beginning. For example, Facebook's acquisition of WhatsApp raised certain privacy issues. WhatsApp had made certain promises to its users about what data it collected, maintained, and shared—protections that went beyond those promised to Facebook users using their app. The Federal Trade Commission (FTC) made it clear that regardless of the acquisition, WhatsApp must continue to honor its promises to consumers.

Chapter 112

The European Union

One of the biggest drivers of privacy policies is the European Union; some of its Member States have had privacy laws since the 1970s. It's at the forefront of privacy and security today. The European General Data Protection Regulation (GDPR) protects European Union citizens and also visitors by giving control over a person's data to that person. That's why, if you visit Europe, you'll see pop-ups asking for permission to do anything on the Internet. More important than individual control, for the startup company, is that European regulations dictate where data can be stored.

The European Union is concerned that the massive amounts of data belonging to their citizens will be stored in countries with inadequate privacy and security protections for those citizens. Europe has identified countries that have proper privacy and security laws and those that don't. Canada, for instance, is not part of the Union but they have adequate laws and regulations protecting their own citizens and visitors so that Canada is recognized by the Union as a country that can move and process data from European citizens, back to Canada. Canadian companies are able to process that data, send out newsletters, conduct data analytics—whatever they want, since citizens are protected under an over-arching federal framework.

Other countries, like the United States, may not be recognized by others as having adequate privacy and security standards. In the US we approach privacy with a series of sectoral privacy laws like HIPAA (health information), PCI (credit card information) and the Gramm-Leach-Bliley Act (financial information), but unlike Europe and Canada, we don't have umbrella or country-wide federal regulations. Since the US has some of the largest technology companies in the world (Microsoft, Amazon, Google, Apple) and their data centers are located in the US, in order for companies in the US to transfer data between the US and EU, they need to utilize one of the following methods: Binding Corporate Rules (typically used by very large enterprise companies) or through Standard Contractual Clauses (SCC) sometimes referred to as model clauses. Until July 16, 2020, when it was invalidated, companies could also use the US Privacy Shield program.

Currently, the more common approach is for companies to sign the SCC, a legal agreement stating how they would protect European citizens' data in regard to their laws.

Standard Contractual Clauses (SCC) state that US companies will protect European data to the same standards as European countries and allow US companies to move the data to the US and then process, analyze, or monetize it the way they want. The clauses contain contractual obligations on the data exporter and the data importer, and rights of the individuals whose personal data is transferred. Individuals can directly enforce those rights against the data importer and the data exporter. When you agree to the standard contractual clauses, you'll basically state that you'll use things like encryption to move the data, you'll encrypt it in a database in the US, you won't allow anybody access to the data by installing a firewall or intrusion detection system. They govern the transfer of personal data to countries that are not recognized as providing adequate protection measures for such personal data processing. You'll also agree that you have privacy policies internally and externally on how you're going to handle the data and you'll be able to very quickly state the technologies or processes that you will go through to protect that data and to meet the standards of GDPR in the European Union.

By signing the SCC, you're really saving yourself a lot of time and hassle because otherwise you'll have to contact each person in the European Union and ask them individually for permission to use their data and process their data in places like the US. So, when an application crashes on your computer and a bot asks whether or not you want to share that data with the company so they can better fix their products, if you have signed the SCC, you can do that in the background without having to ask permission every single time. A standard contractual clause allows you to say to your customers that any data you're moving is going to be protected to the same standards as the European Union.

Chapter 113

Data Mapping

When you create a data map, it's sort of like thinking about your daily technology life in a sense. If I were to ask you what you did on technology today, where you sent emails, what you tweeted, or what you posted on Facebook or LinkedIn, for example, you wouldn't be able to tell me 100%. But each of those interactions, at each of those websites you visited, left little data breadcrumbs related to your digital persona. Unfortunately, that happens in a lot of companies. There are so many things that are going on day-in and day-out, whether it's Marketing or Sales or HR. We find over time that we start stuffing data into databases without having a full understanding of where the data is, what it's used for and, more importantly, the privacy and security obligations for its use.

What we have today in most companies is similar to what people experience when they live in a home for a long time. When you first move into a house, it's very clean and pristine, but 10 or 20 years down the road, you realize that your closets are full of things you thought you needed 15 years ago but they have been sitting here collecting dust and not doing anything. Maybe I should do a spring cleaning and get rid of things?

When you do a data map, it's the same process. It's taking an inventory of what data you have in the company, where you're storing it, why you're storing it, what you're using it for, and how long you're going to keep that data there before you get rid of it. And what happens in privacy is that data can end up almost anywhere today in the cloud arena. You no longer put data on a hard drive or on a computer, or in a data center sitting in the middle of your building. Companies are now using cloud services like Google, Microsoft, or Salesforce. And in the course of doing their jobs, employees are creating spreadsheets, templates, databases—whatever—and storing them in Dropbox, or Google, or somewhere in the cloud. Well, where is that cloud? Who controls that cloud? What security is around that? What privacy issues are around that as well? When you start doing a data map, what you're doing is talking with each owner, each responsible party and asking, what data have you been collecting?, what do you do with it?, who has access to it?, where do you store it?, how do you protect it?

Your first questions should be about what data each functional area has been collecting and where the data is. Where does HR, for example, store all their data? HR may respond to me, well, some things are paper-based and they're sitting here next to my desk. But what about vendors you work with? HR will possibly have a vendor for payroll processing and benefits so that vendor has all sorts of personal information like a social security number, home address, and bank routing information. HR also has other systems like an employee's annual or quarterly reviews, their review of their boss, their boss's review of them. Pretty soon you begin to understand that every function—not just HR—uses multiple systems and all sorts of data and personal information is out there somewhere in the cloud. Where is that data being held? If vendors are global, is that data being held in North America? Is it in Canada, Europe, China, or is it in some other country where it shouldn't be?

As you fill out your data map, you'll find that your data is spread out across many places with varying degrees of protection. It's like driving from a house in the country to a major metropolitan area. As you drive, you'll go on bigger and bigger roads and the more you do that, the farther away you are from your starting point. That same thing is happening with data. Every time we do something with these systems, that data gets pushed out further and further and further away from the protection of your company, from the policies and procedures we trained people on.

So, data mapping literally is finding out what that pathway is to the goal. And you're going to finally figure out where all your data is, how it's being moved around, who has control over it, why they have it, how long they've had it, when they're going to be getting rid of it, so that you always know where data sits in case you have a data breach. If an employee asks what data you have on them, or a consumer asks what data you have because they have the right to know, under certain regulations, what data you have and where it is, you'll be able to answer those questions. And if you can't answer those questions, you can be in a lot of trouble.

Data mapping should begin as soon as you start your company and continue until an exit of some sort. The later you wait to get a full picture of your data, the more problems you'll have in creating the data map and the more risk you'll be putting on your company.

Chapter 114

Data Breach

As the Chief Privacy Officer there's nothing that will get your adrenaline going faster than hearing the word "breach" from someone in, or out of, your company. Breaches can also be the CEO's worst nightmare. These breaches of security and privacy can happen to even the most sophisticated companies out there. For example:

- In September 2017, Equifax announced a data breach that exposed the personal information of 147 million people. The company had to agree to a global settlement with the Federal Trade Commission, the Consumer Financial Protection Bureau, and 50 US states and territories. The settlement included up to $425 million to help people affected by the data breach. Several top-level executives also lost their jobs over this breach and the company took a big hit in the press as well.
- Marriott International announced in November 2018 that attackers had stolen data on approximately 500 million customers. The breach initially occurred on systems supporting Starwood hotel brands starting in 2014. The attackers remained in the system after Marriott acquired Starwood in 2016 and were not discovered until September 2018. The attackers were able to take some combination of contact information, passport number, Starwood Preferred Guest numbers, travel information, and other personal information. The credit card numbers and expiration dates of more than 100 million customers were believed to be

stolen, but Marriott is uncertain whether the attackers were able to decrypt the credit card numbers. The breach was eventually attributed to a Chinese intelligence group seeking to gather data on US citizens.

- Then, in 2015, a *Guardian* writer named Harry Davies reported that Cambridge Analytica had collected data on millions of American Facebook users without their permission, and used their likes to create personality profiles for the 2016 US election. Cambridge Analytica executives were called to Congress to answer for their actions over the 2016 election. Yet the conversation about privacy largely has moved on from the now-defunct firm, which shut down its offices. As Cambridge Analytica faded to the background, other important questions emerged, like how Facebook may have given special data deals to device makers, or why Google tracks people's location even after they've turned location tracking off. On a more frightening note, the number of phishing attacks and data breaches rose to new highs in 2014 with claims of 43% of all companies experiencing a data breach in 2014.
- North Korea's aggro about a movie resulted in a data breach from Sony where personal emails were released publicly.
- Home Depot announced a large data breach and warned its customers of the theft of their email addresses and warned them of future targeted phishing attacks.
- Target, who had their data breach at the end of 2013, discovered intruders accessed their systems via a spear phishing campaign against one of their vendors, and their sales decline in 2014 was blamed on lack of consumer trust after the data breach.
- Even Return Path had its own data breach in 2014 started by a phishing scam before I came to the company, but a breach that even changed how things were being done and changes I needed to inherit and continue to support. The hackers tricked an employee into clicking on credential-stealing phishing email and from there the hackers gained access to clients' administration contact data points. Then, a relatively small list of our clients' email addresses was taken from us, meaning those addresses were

> then targets of the phishing campaign that are intended to compromise those client systems. Two clients actually had their email systems compromised and spam sent from them since the users had the same username and password there that were used on our systems.

The most common way that people think about breach is that someone has hacked into your systems and stolen all the data from the company and they have all the sensitive personal information, like social security numbers, or worse, an employee is doing something illegal with your data. What's becoming more commonly understood is that a breach comes in so many other forms. Maybe somebody didn't steal the data but they've got access to the system and they looked around and now they know a little bit more about your company. Or maybe they looked around your systems and now they know more about your customers and their purchase history, some of it highly personal and confidential. The third way is a breach of trust, the violation of that trust, and perhaps a loss of confidence in the company. This breach of trust can be a lot more damaging to a company especially if you do something with data that people were not expecting. Like the Facebook example I mentioned earlier, who promised consumers that they would never give personal information to a third party, but then went back on their word and gave it to Cambridge Analytica, who did use that data.

Facebook had a breach of trust with their customers and it became a political nightmare, or perhaps a PR nightmare for them. And that's where a breach in trust really becomes damaging to a company, when people have been trusting a company for years and now it's all bad news.

Now, interestingly enough in that Facebook example, is the issue outside of the company or inside the company? Studies have found that 97% of IT leaders say insider breach risk is a significant concern. Not always, of course, because there are employees out to do something terrible, but it's usually through the fault or people just not thinking about things and moving too fast. A study had found that 41% of employees who had accidentally leaked data said they had done so because of a phishing email; 31% said they caused a breach by sending information to the wrong person, for example, by email.

I personally like to have a zero inbox at the end of every day, I hate clutter, my desk is clean, and I don't have stacks of paper or tasks undone at the end of the day. Some people can't work that way, and they're rushing around so much that it's always crazy. As a hacker if I wanted to get into someone's systems, the best thing to do is to confuse them and hack in during times of craziness. You'll see in some of these breaches that employees are simply making a mistake by clicking on an email or a phishing email not realizing that it's not really from this other company. Or, it's common for employees to get a phishing email purporting to be from the company's CEO asking to send over information, or asking you to transfer $10,000 to an account to get a deal done. Nobody wants to hold a project up. Nobody wants to upset the CEO. And unfortunately, what we would typically do is just to get the work done and to keep moving. We do exactly what we thought that person was asking us to do, when in reality it was a hacker who was posing as the CEO. And so a lot of breaches really come down to just human mistakes and also some breaches happen because of social engineering.

Social Engineering is the art of manipulating people so they give up confidential information. The types of information these criminals are seeking can vary, but when individuals are targeted, the criminals are usually trying to trick you into giving them your passwords or bank information, or access your computer to secretly install malicious software that will give them access to your passwords and bank information as well as giving them control over your computer. Criminals use social engineering tactics because it is usually easier to exploit your natural inclination to trust than it is to discover ways to hack your software. For example, it is much easier to fool someone into giving you their password than it is for you to try hacking their password (unless the password is really weak). Security is all about knowing who and what to trust. It is important to know when and when not to take a person at their word, and know that the person you are communicating with is who they say they are. The same is true of online interactions and website usage: when do you trust that the website you are using is legitimate or is safe to provide your information?

We want to satisfy everybody, we want to satisfy ourselves, and we want to get our work done as expeditiously as possible.

Especially now, during the pandemic, it's even more important for companies to realize that the weakest link is the employee because we went from being in our businesses or offices where it was all protected, locked doors, firewalls, intrusion detection systems, keeping each other accountable to now being in our homes where we don't have the firewalls that we used to have, we don't have to lock doors anymore. We have our children bothering us, because we have to teach them now remotely during a pandemic, we have a lot of other things that are going on.

On top of managing all these other things, we're also not thinking about privacy or security 100% of the time. And so unfortunately now what we're seeing is a lot more data breaches are occurring because hackers know that people's brains aren't 100% on their work. They're thinking about nine other things happening right now, and where the pandemic is today. So, the breaches can come in all different forms, it could be stealing data and taking it somewhere and it being used. It could be somebody getting access into the system to see that data, or even encrypt that data through malware, or it could be a misuse of trust or a breach of trust, because the company misused the data when they said they wouldn't ever do something like that.

So it's not *if* a breach occurs but *when* you'll have that data breach. It will happen. There is no such thing as a 100% secure system. However, there are many things the CPO can do to minimize the risk of a breach.

Chapter 115

Least Privileged Access

One of the best things you can do as CPO is to be very thoughtful about access. Who in your company has access to data and how can they use that data? In the privacy and security realm we use a process called "least privileged" to help make those decisions. What that basically means is that an employee should only have access to any systems or to any data for the purposes that they need it for. In the government we know that there are different levels of secrecy and some people have a low level of clearance while others have a very high level. That same mentality exists around controls of your systems so that I, as a CPO, should absolutely not have administrative access to the HR systems around payroll. There is no reason for me to have administrative access to the HR systems because I'm not in HR and why would I need access to that information?

Does the Vice President of HR need access to that information? Probably not, but an HR manager would need payroll information. Does the Chief Technology Officer need to have full blown access to every single database that the company has? Probably not. The CTO's job, from a policy procedure perspective, is to help run the entire department, but they're not a developer. They're not the person who will be accessing any of the consumer data to develop products. So when you look at the controls that you're putting into your products and your services, you should first look at it from the employees' perspective and ask which employees should have any control, any sort of viewpoint in any of the data.

As CPO, I don't need access to HR data but I would need access to HR's data map; similarly, the CTO doesn't necessarily need full access to the data map, so you may only give them access to the IT portion of the data map. A best practice is to start the least privileged process from the employee side and if you were to look at it that way and ask the question, do employees have too much access? You're likely to answer that they probably do. And hackers know that employees have way too much access, too. An example of this happened at Twitter when 50–60 Twitter accounts of famous people were hacked by a Twitter employee who had way too much access to the systems. Twitter found out that many employees had too much access to all the systems. In Twitter's case, the incursion seems to have involved phone calls and happened in multiple phases. Not all of the employees that were initially targeted had permissions to use account management tools, but the attackers used their credentials to access their internal systems and gain information about their processes. This knowledge then enabled them to target additional employees who did have access to their account support tools. The attackers exploited that human factor to gain access to Twitter's internal systems and the accounts of some of the world's most prominent figures. Twitter may not have been practicing the idea of least privileged controls. So, the first question to ask, as you're defining the least privileged list is, who really needs access to the data?

The second thing to look at is access to data from the consumer side. Yes, consumers have a right under most privacy laws, like the GDPR in Europe and CCPA in the US, to access the data you collect on them. You're required to give them access to review the data that you have on them, to edit, correct, or delete it. Some things can't be deleted because it's required to keep records of purchase agreements. So while you can't delete a record of the actual purchase, a consumer could delete the marketing data that got them to purchase your product or service.

A best practice today, especially on social media platforms, is to give people the ability to control their profiles and you would engineer or code for this as you're building the product. You'll want to allow consumers to change their email address, phone number, name, birthday or any other non-relevant or personal information. And obviously, as I have already addressed, a best practice is to be transparent with customers about what you're doing with their data and provide them with an opportunity to opt out of providing their data to others. By giving control back to the consumers, you literally are then respecting the bounds of the relationship, the same way that you respect the bounds of individuals in your lives.

What you're saying to consumers is that you respect them, they'll always have access to their data, they can see what you've collected, correct it, delete it. And, by the way, here's how you do that. The simplest form of control that you can give a consumer is to have an unsubscribe button on an email, so that's the lowest standard. But you can go far beyond that and provide consumers with more control, and you can be far more transparent with consumers, too.

A powerful way to make your least privileged systems impactful is to segment your systems and network access. What happens in a lot of companies is that everybody has a username and password to access a system and then employees are given another ID and password for a different system, and so on. And by the time it's all over, you realize you've got 30 different separate logins for different systems and with different passwords. That's fine, but what most people end up doing is using the same password for all the systems and when one of those systems gets breached, all the systems get breached. Also, it's a lot more work for employees to keep track of their passwords and to keep them secure, and you'll spend a lot of time creating new passwords for employees who forget their password so that they can access the system.

A way around that is to set up a single sign-on process that allows employees to have one password and one ID and add a two-step factor of authentication on that as well. The two-step factor usually comes in as a code prompt on the employee's phone, which is far more secure than a password by itself. When you're building a company, and you're building all of these different systems, whether they're Production, HR, Sales, or another function, you're telling people how you access systems to do their job, here's your ID and your password. And, by the way, the first time you're logged into all these systems, you're going to get two factor authentication, or maybe because the system you're about to log into is highly sensitive, because it contains your HR data, you will get two factor every single time that you visit, because we really want to make sure that it's you and that you're the only one looking at your private HR data.

And so when we talk about least privileged access, we're also talking about the segmentation and segmenting ourselves off by putting an additional factor authentication in between all these different systems.

Chapter 116

Employee ~~Training~~ Engagement

The strikethrough of "training" above in the chapter title is not a typo on my part, but a subtle note to the Chief Privacy Officer that the best way to protect your data is by engaging with employees, not merely telling them what to do, or training them on what to do. Or worse, punishing them when they do something wrong. What I've learned over time is that privacy and security are not sexy topics and they're basically like law enforcement. Nobody ever wants to get in trouble with law enforcement. They don't want to get pulled over for speeding or some other infraction and if they do, they're not happy about it. It's not fun for them.

So, what I've had to learn is to approach staff by explaining that privacy and security are not things that should be feared. They're not in place to make employees' jobs harder, they're an essential part of how we do business that will avoid problems as we scale. Just as privacy by design brings people into the fold at an earlier stage, so does talking to staff early to help bring them into the fold. Engaging with employees helps to get them thinking about their actions as it relates to privacy and security. The way to do that is to integrate the privacy team into what employees are working on and vice versa. You want to create a culture where you can have a two-way conversation between the privacy team and other employees.

One of the ways that you build that culture is to ask if you can attend weekly functional meetings. If the technology team is meeting, you can ask if you can listen in and to be able to give advice on when things need to happen in a certain way to be compliant with privacy laws or regulations, or even how to adhere to the policies that your company has in place. At the same time, you would want to invite the technology team to the privacy team meetings so that they would be able to understand what things the privacy team looks out for and what we have to be aware of when it comes to regulations and policies.

By being transparent and open with employees, they'll understand that the role of the privacy team is to protect the company and, if we don't, the company could be fined, it could be sued for things, or it could have a bad PR experience. In the worst case scenario, the company could fold and all of us would lose our jobs.

Another thing that I've started to do to help build that culture of engagement is to establish a Privacy Council, which is like a roundtable. But rather than use the council as an enforcement team, I use it to get people from each functional team to help find and understand the policies and regulations that impact their functional area. Ask people on the council to carve out 10–20% of their business as usual time for thinking about privacy. Meet once a month or once a quarter, depending on how big the company is. Understand and be proactive about any new regulations that are going to have an impact on the company.

To start our Privacy Council at Return Path, I went to each functional team—to Sales, Marketing, Production, IT, HR—and said that I'm somewhat deputizing you in this whole process. Your role is to keep an eye on what's happening and not tattle, but to report to us what's happening. As the CPO, you're also not only giving them a little bit of idea of what's happening, but you're making them feel good about themselves. The Privacy Council is particularly useful for a startup because what I've found is that by deputizing one person in each division, I then don't have to go out and hire five or six additional full-time headcounts in privacy. I can put some of that work onto the people who understand the systems like in Sales and Marketing and HR and work with them and teach them a little bit. In some cases, believe it or not, I can actually bring them into the privacy team for training and get them certified.

The Privacy Council works really well in a startup and as you scale and build your team, what I've found that works well is to designate one person to be the liaison between the privacy and each functional team. They are in charge of checking in with these groups on a regular basis, they become the one person who has all the information and all data, and they're also the person who is project managing. If we are working with Marketing trying to understand what they're going to be doing in the next month or two on email campaigns, or how they're going to buy new data and send out emails, then we can get a better handle on the privacy and security issues before they do that. The privacy team member assigned to Marketing would be doing the back and forth, they would be tracking all this information and what's happening and what's occurred in those meetings. And they can bring it back to the privacy team and ask for a decision, or more support, or additional meetings. So, the privacy team can be more proactive, earlier, and can help Marketing get through what they're going through.

By talking to staff, having regular attendance in their meetings, and creating a Privacy Council, we've managed to change the mindset of employees so that they don't fear privacy and security, they don't see us as enforcers, but instead see us as collaborators. By providing them time and decision making on privacy and security for their area, we find higher levels of engagement and respect. So now, when employees see privacy people coming down the hallway to talk, they're not hiding from us. They're coming to us with ideas and results and this makes them feel engaged.

Chapter 117

Building Your Privacy Team in a Startup

If you're an early-stage startup, you'll probably be the only person solely devoted to privacy and security and in that case, there are a couple of things you can do to ensure your company is paying attention to privacy issues. One way to expand your reach and influence within the company is to create a Privacy Council. As I mentioned in Chapter 116, members of the Privacy Council are people you choose, one from each functional area, who you'll work with to keep you up to speed on privacy issues in their domain. The Privacy Council will help you gain leverage and develop greater awareness about privacy and security without additional headcount.

The other thing that I would absolutely insist on, especially as a startup, is that the privacy team should be independent of all other functions in the company. The privacy team should not report to Product, or IT, or any other functional area. It should be its own department for the same reason that we (in the US) have separation of Church and State or separation between the legislative, judicial, and executive branches of government. You don't want someone within a functional area having influence over privacy because there will be times when the needs of Product, for example, differ from what's best for privacy. What really matters is that the CPO has sufficient top management buy-in to be able to influence compliance decisions. Therefore, as long as the CPO has that influence—even if they are one or two levels below someone who really has access to the CEO or Board—the requirement will be met in practice. Usually reporting to the Chief Operations Officer, the legal counsel, or in some cases the Chief Financial

Officer is a good place to start because they are "neutral" in the direction of product and use of PII.

What I suggest for startups is that you have a separate department that's in charge of understanding things like regulations, the impact those regulations would have on the company, and the decisions that need to be made. It is especially important to have this separation because your mandate as CPO is to put people and their data first, ahead of revenues, products, marketing, and other functional areas.

A third thing to do as a startup, when you're trying to keep privacy at the forefront, is to undertake an assessment where you list the benefits and impact of conducting business in various ways. The issues between privacy and the other functions can get heated, and one way to remain objective is to create a privacy impact assessment. The thing that the company is about to do in Product or Services is going to benefit somebody, and in that case, who is it and who gets the benefit? Especially when you're collecting large amounts of data and processing that data and maybe thinking of selling that to a third party, it's easy to figure out the company side of the equation. But what about the consumer who has trusted you with their data? What if you were sending a newsletter but sold that email list? That could have negative consequences for the consumer and, since their purpose was to access the newsletter, you selling their data could cause them harm.

And so, you take a look at all these scenarios and you run these sorts of tests, but you need to do it in a separate department that is independent from every other sort of decision within the company. Your primary responsibility as CPO is to look out for the best interest of those who have trusted the company with their data and at the same time, ensure that the company is compliant when it comes to a legal regulatory perspective. Over time, people in the company will begin to think about privacy and security when they make decisions, and you won't be backtracking or bolting on privacy solutions after the fact. It's similar to any new law that takes time for people to adjust to. When mandatory seatbelt laws by states started to be put into place in the 1970s, there were all sorts of people who didn't wear seatbelts, and they would get tickets for that. Today, most people—especially younger drivers who haven't learned any other way—use a seatbelt. It's a normal part of driving for people today.

So, when you're a startup, there are a couple of important things you can do to elevate privacy and security to the company and to ensure that the staff are thinking about privacy in their day-to-day tasks. Create a Privacy Council to educate people on privacy issues; create your privacy team as a separate function; and create an assessment that is transparent and objective to weigh the benefits and impact of decisions in light of regulatory and legal issues.

Chapter 118

Building Your Privacy Team as You Scaleup

If you follow my advice for startups, you ought to have widespread adoption of the privacy first perspective that's necessary to creating a great company. As you scale your business, you'll be able to bring additional people onboard and here's a list of possible roles to consider when you do that.

One role that you'll definitely need as you scale is a project manager, someone who understands how to build and track projects, timelines, and budgets within the context of privacy issues. Another role you could bring in is someone who is an expert in technology, a privacy engineer. Their job is to understand all of the ins and outs of coding, like scripts, how cloud technologies work, and how data may or may not be moved or touched within that technology. You could also hire a public policy person, somebody who understands the regulatory issues on state, federal, or even a country-wide basis. They would understand how to lobby or how to work with groups of other companies who are looking to either help push a law forward, kill a law, or help edit a law. The policy person would need to understand the regulatory process, and the legislative process. In some cases, you also might have someone who is an expert on security on the team as well. Privacy vs. security … isn't it the same thing? Not really. But they are kissing cousins. Data privacy is focused on the use and governance of personal data—things like putting policies in place to ensure that consumers' personal information is being collected, shared, and used in appropriate ways. Security focuses more on protecting data from malicious attacks and the exploitation of stolen data for profit. While security is necessary for protecting data, it's not sufficient for addressing privacy on its own. However, in privacy operations, we are seeing a merging of the security and privacy teams, in which the CPOs are leveraging the security team's expertise to implement and manage technology in order to simplify regulation adherence.

When you're scaling, you'll have all these different roles on your team. They'll be working on specific projects at the same time but each one will bring a different skill set and lens to the project.

Chapter 119
Certifications

One thing that's important in the privacy world is certifications, which not only keep your team members up to date on the latest changes and how those changes might impact your company, but also signal to others in the market that your company takes privacy issues seriously. Privacy is a very big topic, it's a global effort, and things are constantly changing around the world as countries create new laws and regulations. As you bring new people into the privacy department, one of the things that you would like to do is make sure that they have the necessary knowledge in privacy. There are multiple opportunities to help them gain that knowledge.

There are continuing education courses available through webinars and conferences that offer training on what's happening and when things are changing. Sometimes completion of these courses will result in a regulatory certification that comes from organizations like the International Association of Privacy Professionals (IAPP). There are also certifications for subject matter experts in EU laws like the General Data Protection Regulation (EU GDPR) and other member countries that have their own regulations on their books. For example, Canada has an umbrella framework for privacy regulation called Personal Information Protection and Electronic Documents Act (PIPEDA), but each province also has their own regulations, so you can be certified in Canadian law. The same is true for Asia and many other regions these days.

Certifications also exist for managers in privacy, for technology people in privacy, and they exist for subject matter experts, so regardless of a team member's role, there are avenues for them to build skills and keep current on regulations and changes that happen globally. To be honest, it's quite a feat to get these certifications, and it takes quite a bit of training to understand the nuances of technology and law and how they intersect and how they don't intersect. It takes a lot of work and knowledge to understand what's important and what's not important in those intersections. A person might spend months and months, sometimes even years, studying for certification tests, and many of the certifications need to be renewed annually. So, you should plan your budget and workload to allow for privacy team members to continually learn about privacy issues.

Chapter 120

Assessments

When it comes to security and privacy, I'd like to say that everybody who's in charge of privacy and security should know what they're doing to protect data and stay compliant with regulations and policies. The only problem with that is that we all get busy and as you scale, there are a large number of products and lots of things happening. Things do get missed, or dropped, or ignored, and that's just the way that we, as humans, operate. We are all fallible. One of the things to prevent that is to get an assessment done by a third party. In the security world these assessments come in all different forms. Some consultants will come in and perform an audit like a System and Organization Controls (SOC 2) audit, an ISO 27001, or even a ISO 27018, and you're basically paying them to come in and test whether you're doing what you say you are. In your terms of service and in your internal and external privacy policies you're making a statement to people about how you're securing their data. Well, do you do those things or do you just say them?

ISO 27001 is a top-down view of security that establishes the core controls and principles of a service organization's business model regarding data management. An SOC 2 report provides an assessment of the controls that help to support that business model. ISO 27001 involves more work, but it does more to protect organizations from information security threats. The main difference between SOC 2 and ISO 27001 is that SOC 2 is focused mostly on proving the security controls that protect customer data have been implemented, whereas ISO 27001 also wants you to prove you have an operational Information Security Management

System (ISMS) in place to manage your InfoSec program on an ongoing basis. ISO 27001 requires about 50–60% more time to complete than SOC 2. Typically, it takes approximately three to six months to complete a SOC 2 Type 1 certification from start to finish, depending on how long it takes you to implement all of the security controls, and another three to six months to achieve SOC 2 Type 2. ISO 27001 usually takes 12–18 months to complete, again likely due to the additional process and documentation required to install an operating ISMS. Once completed, SOC 2 Type 2 needs to be renewed annually. As for ISO 27001, most engagements include a three-year commitment where you have a point in time audit in year one, and renewals each year thereafter.

An assessment follows the idea of who's watching the watcher? Again, I'd love to be able to say that I'm doing everything right every single day. But I'm fallible, things don't always go the right way and things also change behind my back. The auditors will come in and ask questions of the staff of the company and while some of the questions are directed to the privacy team, others are directed to functional departments that control the data. They'll assess your privacy and security program, identify risks, and call out the appropriate remediation. For example, do you have a way of managing or maintaining data, like logins? While people might know that it's a good thing to change their password every 30 or 60 days, most people don't do that, so they check that. They also check whether you have things like two-step authentication, they check if you have monitoring technologies, firewalls, and intrusion detection systems. They check whether your internal policies are followed, like, can people put consumer data on a USB drive and take it home?

If you're answering these questions in the affirmative, great, but what you're really looking for are the holes or gaps you have in your processes, systems, and technologies. The auditors will come back with an assessment and say, here are the things that we found to be untrue, or here are the things that you should be doing that you're not doing today, or even, here are some problems we found that allowed us to inadvertently access your data. That's a problem! We've operated by thinking that people couldn't get to a certain system, which can be a huge security breach. Typically, the auditors will use white-hat hackers to "break" into your company, both physically and electronically. To be honest, about 100% of the time they will come back with some problems. It's nothing for you to fear, it's nothing for you to be concerned about, but it sure is helpful to see what you need to fix.

An audit can take several months or in some cases, years, but at the very end of all this you will have been given an opportunity to address any of the issues around security and privacy. So, whether it's your policies, controls, technology, physical space, electronic aspect of security, or something else, the auditors will help you make sure that all those are fixed. And at the end of it all, once you fixed all the problems, they hand you a certification which is validation that you're the best company you can be when it comes to privacy and security.

What can you do with a certificate besides proudly display it on your website? Certification is actually a competitive advantage because it's recognized globally and when you're dealing with clients who would like to understand your security, instead of taking weeks to prove to them your security measures, you can literally hand them the certification and start them off on the right foot to understand that you take security seriously. The other area where a certificate makes a big difference is during mergers and acquisitions, especially in the due diligence stage. Usually, people think of due diligence as looking at the finances, the sales numbers, and the legal aspects, but one of the things that's forgotten on a pretty regular basis is your privacy controls. Due diligence during the early stage of a merger or acquisition either advances the transaction or causes it to terminate. Well-executed due diligence is one of the key factors that leads to a successful closing and contributes to the overall positive impact of the transaction. Carving out a detailed cybersecurity and data privacy phase that is separate from a traditional information technology review is a critical component to thorough due diligence. By taking a deep dive into a target company's cybersecurity and data privacy history, policies, and practices, potential threats and liabilities can be uncovered and analyzed before the transaction continues. Is the company doing anything wrong that would be against the law when it comes to privacy and security? Is there an unknown risk the acquiring company may be taking on? Does the company have the right policies and procedures in place when it comes to privacy and security? How could an acquiring company know the answers to these questions?

A lack of due diligence in evaluating a company's cybersecurity controls and privacy requirements during the M&A process can result in a host of short-term and long-term problems. This may include subsequent data breaches or privacy complaints, loss of business revenues, higher cybersecurity premiums, underperforming stock value, and loss of consumer confidence. The FTC can hold an acquirer responsible for the bad data privacy practices of a company that it acquires. The target's response to due diligence requests should be used to negotiate appropriate pre-closing conditions, indemnities, and the ultimate transaction price. If you recall in my high profile breach examples in Chapter 114, Marriott, for instance, inherited a massive breach crisis in its 2016 acquisition of Starwood that went undetected at the time of the merger. In contrast, gaps in due diligence may provide an acquiring company a competitive advantage as in the example of Verizon's reduction of its purchase price of Yahoo by $350 million dollars after a significant data breach came to light before the acquisition was finalized. With security and data incidents so widespread and potentially damaging to acquiring companies' valuation and reputational health, a target's cybersecurity vulnerabilities and privacy risks should be as closely investigated as the financial documents within the M&A due diligence process.

Even at Return Path, we had to do many due diligence processes on companies we wanted to acquire. I recall one in which early on my group identified some systematic problems with their privacy legal documentation such as privacy policies and their data sharing disclosures or lack thereof. We did bring this up to their management team to see if we could get them to address it. But they didn't seem too concerned about the findings. In the end, the deal never came to fruition for other reasons. A short time later their company was acquired by someone else, but soon after that, the FTC brought a case on them for the exact issues we had pointed out in our due diligence process regarding their privacy policies and lack of third party sharing procedures and disclosures and they had to pay a big fine. This meant any escrow monies in holding for the stockholders were reduced to pay the legal fees and fines and less money went to the shareholders and employees in the end.

Failing to conduct adequate due diligence for cybersecurity and privacy risks during the M&A process can negatively impact the organization after the deal is closed. After all, no entity wants to have malware injected into its system that causes the purchasing entity to suffer a breach or system failure because of a failure to recognize a security risk prior to integrating the new company with its current systems.

Again, rather than responding, "trust us, we have great privacy and security controls," you can instead give them your assessment and certificate, proving that a third party tested your processes, systems, and technology and you're current in all of those. So, the assessments, the audits, not only help you internally to understand where you may be falling short in protecting data, but they can help speed up the M&A due diligence process.

One thing you don't want to do when it comes to audits and certifications is to set aside the results, forget about the issues, and not do anything about it for years. Even if you fix all the issues and you get a certificate, you can't stop there—technology, laws, and regulations change rapidly so you'll want to continue with assessments every couple of years. Besides that, certificates expire, so you'll have to continue with them. You can also continue to test yourself against that certification internally and ask, have we made any changes that would make us invalid or make us no longer true? Has anything new been added? Has anything at all been taken away? Once you have the initial certificate these are the questions auditors will ask, but you can ask (and answer) them yourself in the off-years when you're not doing a formal assessment.

I said earlier that audits come in many different forms and they also come at different price points. I've seen them as low as \$30,000 to \$70,000 for a small company, and higher than that for larger corporations. An assessment is definitely not a cheap thing to do. When it comes to startups, it's one of the hardest things for them to ever get done because of the cost, and unfortunately, it ends up on the sidelines. Generating revenues just to stay afloat and pay employees is what most startups spend their money on. An assessment also requires a team of people or a strong team of security and privacy professionals to be able to manage the project itself and if you don't have that talent, the cost of an assessment will increase significantly.

Some startups will buy an assessment kit for less than $10,000 and do it themselves, but it's not third-party certification so it doesn't carry the same weight, especially with clients and acquiring companies.

If you're a startup CPO, you should keep in mind the benefits of an assessment and certification and work with your executive team to plan and budget for that to happen.

Final Thoughts

Out of all the functional areas in a startup, privacy and security are the easiest ones for teams to ignore, to delay taking action on, or to take short-cuts. I have raised a number of reasons why you wouldn't want to do that, the primary one being that data is the new gold standard, the most valuable asset in your company. Ignoring these issues or being casual about data comes with enormous risks for all companies, not just startups.

But startups and scaleups have a competitive advantage over established companies because as a startup you can design your products, services, and systems in a calculated and efficient way by using "privacy by design" principles. That's a lot better than doing nothing or cobbling something together piecemeal as you go. As the Chief Privacy Officer you can also institute a number of measures internally and externally that can mitigate privacy issues. For example, thinking carefully about who within your company has access to what customer data and what they can do with that data will go a long way in preventing issues. Creating a data roadmap will help in this regard and is a good first step to developing robust data management and protection. Your terms of service and privacy policy can help customers trust you and build your brand.

Many people within a company rarely think about privacy and security and when they do, they often see it as an enforcement function, something to be avoided. I suggest instead that the CPO should engage with employees on a regular basis, by forming a Privacy Council, by assigning a privacy team member to a function, or by inviting others in the company to attend privacy meetings. I also think it's a best practice if the privacy team is somewhat independent of control from others as there will be inevitable conflicts of interests between Privacy and other functions. Also, it's a best practice to give the knowledge and power to all in your company to help protect data. People who produce data—through their functions and processes—should be held accountable for how they produce that data. This includes the quality, accuracy, and value of the data they produce. Basically, every person in your organization has a relationship to the data. Therefore, Everybody is a Data Steward and should help protect it at every step.

The privacy landscape is rapidly changing and can be very complex since each country, state, and even jurisdiction may have different regulations and laws. Rather than avoid or delay privacy and security, the great startups put measures in place on Day 1 and ensure that their company "gold" remains valuable as they scale.

Chapter 121

CEO-to-CEO Advice About the Privacy Role

Matt Blumberg

What comes before a full-fledged CPO? Most startups don't have a CPO and just rely on outside advice from external counsel or a privacy consultant. Some probably don't have any help in this area at all.

Signs It's Time to Hire Your First CPO

You know it's time to hire a Chief Privacy Officer when:

- You wake up in the middle of the night terrified that you're going to find your company on the front page of the newspaper or served a subpoena to testify before Congress about a data breach.
- You are spending too much of your own time trying to understand what PCI Compliance, or HIPAA, or GDPR means to your business.
- Your Board asks you what your data breach client communication plan is, and you don't have a great answer and aren't sure how to get to one.

When a Fractional Chief Privacy Officer Might Be Enough

A fractional Chief Privacy Officer may be the way to go for most startups forever—sometimes in the same person as a fractional Chief Information Security Officer. You probably can't get by without a full-time leader in this area if you are large (>$50m in revenue) and are sitting on a massive amount of consumer data, especially PII, financial information, or health information. But if that's not you, a fractional Chief Privacy Officer may be the way to go. While a fractional executive is similar to an outside lawyer or consultant, an executive has a company title for external credibility and the personal commitment to the organization to ensure compliance.

What Does Great Look Like in a Chief Privacy Officer?

Ideal startup Chief Privacy Officers do three things particularly well:

1. They create educated evangelists inside the company. Our privacy team at Return Path under Dennis's leadership had a lot of experience and industry certifications. And they also took the time to make sure others in the company, especially in the product management and engineering teams, also got some of that same training and those same certifications. By not making the privacy team a single point of knowledge or failure, Dennis was able to make privacy part of our product strategy and an offensive strategy as opposed to a mitigation or defensive function.
2. They also handle the basics of InfoSec. If you're actually a security-related company or a massive consumer or financial organization, you may need a dedicated Chief Information Security Officer. If you aren't, then a good Chief Privacy Officer should be able to handle a number of the functions that a CISO would otherwise handle, especially on the policy and communication front.
3. They communicate well, both internally and externally, and they help connect you to the relevant members of your community or ecosystem. When we had the sizable data breach on Thanksgiving Day about 10 years ago that Dennis referred to earlier in this section, our fractional Head of Privacy Tom Bartel was on the spot, writing internal emails and external blog posts that needed almost

no review. He was also instantly communicating with dozens of his counterparts at related companies so that the industry knew where we stood and what we were doing about the problem. It was like an instant activation of an emergency response system!

Signs Your CPO Isn't Scaling

Chief Privacy Officers who aren't scaling well past the startup stage are the ones who typically:

- Look at you sideways when you ask for a strategy or even a mitigation plan for a breach. While we like to talk about things like privacy by design and using data protection as an offensive strategic weapon, the reality is that Chief Privacy Officers need to have actionable plans in place at all times for the areas where they judge your company to be the most vulnerable. If you ask to see the plan or get briefed on it and you get back a blank stare, you know you have a reactive person on your hands for what needs to be a thoughtful proactive role.
- Would rather lecture you on GDPR than talk about why your data protection plan will win business. Privacy people can be geeky, legally oriented, policy-focused, and very technical. All that is well and good. But if your Chief Privacy Officer can't engage in strategy with you and other executives and understand the levers of your business and how their role can help further them, you may as well use an outside law firm instead of taking up a valuable seat at the table internally.

How I Engage with the CPO

A few ways I've typically spent the most time or gotten the most value out of Chief Privacy Officers over the years are:

- Getting them to deeply understand our business. Part of what I think we got right in this regard at Return Path was that we almost always made this a fractional role that was combined with other responsibilities—Tom Bartel, Dennis, and Margot Romary almost always did other senior jobs in operations or product as well. This is what most likely enabled us to play more offense with the function than defense.

- Role modeling training and compliance. Mention of the word compliance doesn't usually get anyone's juices flowing inside a company, but it's important for the company to live up to its obligations with customers or its own policies, which frequently involve a certain amount of employee training every year, not to mention policy compliance. I always try to be the "first done" on an online training course and make sure to follow related policies so that our Chief Privacy Officer has air cover ... and so that I can ask others to do the same with a clear conscience.
- During a crisis. I may interact with Privacy infrequently, but oftentimes when I do, it's because something has gone wrong, or we're worried about something going wrong. That's OK! As long as you can be there to support your Chief Privacy Officer on an emergency response basis and practice some level of servant leadership in a crisis ("How can I help here ... who do you need me to call?"), you're doing your best work in this department.

Part Ten

Operations

Jack Sinclair

Chief Operating Officer

It is impossible to nail down the definition of the startup COO. People in the COO role are often the "second-in-command"—but not always. All roles in a company have to "operate," so why would you need a Chief Operator? Of course, the person in this role will need to be an excellent operator, communicator, and builder, but their actual responsibilities can vary greatly from company to company. You will find many of the tasks and teams the COO will manage are already found elsewhere in this book, but for a lot of startups you need a COO, you need someone who has their pulse on all of the operational details, someone who can see the big picture, who can understand the vision for the future, and who can make sure that vision becomes a reality. Here are a few key areas that most COOs end up managing.

Types of COOs

There is an HBR article ("Second in Command, the Misunderstood Role of the Chief Operating Officer," Bennett and Miles, 2006) that details seven types of COOs and while that's helpful, it's also focused primarily on large enterprises. For many modern startups, I feel there are not seven types of COOs, but three types of COOs. Some COOs are market-facing and include the sales team in their purview, some are product-focused, and others only look inward managing groups like Finance, Legal, IT, and Sales Operations. Table P10.1 defines these three types of COOs.

A Few Common Areas for the COO

As I mentioned earlier, many of the areas for a COO are just pieces of other roles and as a startup matures, the COO role will change, diminish, or maybe even be replaced by functional leaders. But there are a few critical, business shaping functions an early stage COO should be managing.

Company Operating System

The company operating system for a startup is how the company is run, how it's managed, what meetings take place, what measurements and tracking are collected, what policies exist for travel, paid-time off, or other key tasks. Some of the key items include:

- What are the standing meetings, the meetings that are planned, scheduled, and required? Standing meetings include all-company meetings, executive team meetings, Board meetings, and team meetings (especially cross-functional ones). A COO should also include 1:1 meetings within the company operating system. And inevitably, they'll have to make sure there are not too many meetings!
- How does the company set quarterly and annual priorities and goals?
- How does the company track deliverables and commitments to drive accountability and alignment?
- How is information delivered to teams and the company as a whole? What frameworks like OKRs or Agile frameworks will the company use to help run parts (or the whole) company?
- How do things like travel, vacation, leaves, and internal company events work?
- What tools are used to track everything and how are they operated?

Managing the Marketing/Product/Sales Triangle

Early on in a startup, people are making decisions quickly and working on many areas at once, some for the first time. Typically,

Table P10.1 Types of COO.

Type of COO	Definition
Product Focused: Owns developing the product and/or services for the company	• Product-Market fit for business and needs to understand the competitive landscape, pricing, packaging, and other items that ensure the product is right for the market • Often leads the product and engineering teams in Product Development and is able to get into technical details. They will manage the technical stack decisions, managing costs, and performance • Has a clear understanding of data science and database architecture, cloud computing, and basic data regulations • Serves as the cross-functional coordinator for the product. Helps determine how it is sold, what features the product includes, what is the product roadmap, and communicates it to employees and outside stakeholders
Go-to-Market Focused: Owns revenue generation and operations for the company	• Is a key stakeholder in building the sales and marketing functions ensuring alignment between the two with clean data and processes • Leads the product-market fit discovery and packaging • Participates in and sometimes leads the early key sales and partnership • Develops pricing models ensuring sufficient margins considering customer acquisition and service costs • Establishes a service function that matches the business need in the short term that prioritizes timely and clear customer communication • Builds a scalable customer acquisition machine • Has a clear understanding of the role of brand development with a consistent brand message and color scheme • Establishes sales operation functions that include a scalable customer relationship management tool. The team should be able to use the system to provide a real-time look at the entire sales funnel • Builds teams and hires/mentors key functional leaders • Establishes key go-to-marketing metrics and reporting • Evolves team operating system that oversees effective cross-functional partnership with Product and Finance • As the company scales, develops a plan to move the functions out of the COO suite and into functional leaders • The role may end up disappearing if the COO moves to run a different function or moves out of the organization
Internal Operations Focused: Owns the inside operations of the business and ensures it is a well-run company, with the right information to the right people at the right time	• Establishes the inside processes and beginning of teams including Finance, Legal, IT/Facilities, and Sales Operations • Creates key accounting functions such as accounts payable, accounts receivable, and overall chart of accounts management. Starts to establish the key information systems and legal agreements used in the business. Ensures that there are clear sources of truth for all of the company data • Sometimes will manage the key human resources back-office operations including payroll, benefits, and state and federal compliance • Builds key management operating systems to include functional leaders in some of the teams, such as Finance and Accounting and mentors junior people in the organization • Creates interest to order and order to cash flowcharts to highlight key areas for improvement and reporting • Establishes sources of truth for all corporate data and a map of all key corporate systems • Creates or builds a company operating system to include all-company meetings, leadership meetings, off-sites, team meetings and expected 1:1 meetings. The goal in any company operating system is to ensure all people have the right information needed to make decisions • Often this COO will also build the sales operations function and system but will need to be sure to be closely aligned with the sales and marketing leaders

I am a fan of quick (and informed) decision making but the risk is that required communication is skipped, resulting in misaligned teams. This misalignment becomes especially painful when Marketing, Product, and Sales are not on the same page. The most common situation is where the sales team sets their sales targets without coordinating with the other teams. Then Sales becomes frustrated when Marketing does not deliver enough leads or the product is slow to market. Or, if Product does not hear from Sales about what is working or not working in market, they may have the wrong priorities. The COO will manage and coordinate between Marketing, Product, and Sales.

Metrics and Measurements

The COO often has the most experience in understanding what the key drivers are in the business, and they'll know what metrics demonstrate company health and what measures indicate early warning signs. For a startup, you want to ensure you are measuring the items that will show areas of constraints or poor resource decisions. For example, if you are doing well in creating leads but have no visibility to what happens to them or even if they are all followed up, you are missing a critical early part of growing the business. That's something that a COO would definitely provide to a company.

The COO should also ensure that the metrics the company uses actually help drive decisions. To go back to the sales leads example, if the metrics show that a lot of leads are not being responded to in the first week, it could drive a decision to hire more sales development reps. When the startup lacks a CFO or VP Finance, it often falls to the COO to ensure that the cash projections and gross margins are done in a timely manner and accurately to make sure the business stays a going concern. Lastly, the COO usually owns the corporate data. So, ensuring that there are single sources of truth, that the data is stored in a place so people can get to it to make decisions, and that it is delivered to the right people at the right time are hugely important.

While not every startup will need an actual COO, you can see from what I have outlined that what the COO provides is critical, especially as a company scales, has more products, more employees, and more customers. A good COO will smooth operations across functional teams, make sure that communication is timely, and that teams are aligned.

What I Look for in a Chief Operating Officer

Brad Feld

Foundry Group partner, Techstars co-founder

If your company is smaller than several hundred people, you are too small for a Chief Operating Officer. Instead, keep it simple and use a two-level executive hierarchy: CEO and VPs. The VPs are often creatively titled (SVP, EVP, CXO), but they are the direct reports to the CEO and have a meaningful span of control over the business.

Things get complicated as a company scales. Eventually, you have a leadership team structure with a CEO and many VPs (VP Engineering, VP Product, VP Finance, VP H&R, VP Sales, VP Marketing, VP Customer Care, VP Operations). Often, titles are inflated (CTO, CFO, Chief People Officer, CMO, COO, CRO, …) or artificially demarcated (VPs, SVPs, and EVPs). Titles shouldn't matter much in an early-stage company, other than helping people outside the company understand who does what.

As the company scales, a CEO should start focusing more time working "on the company" instead of "in the company." However, when the number of direct reports to the CEO expands, the overhead of managing the team increases. The CEO puts more systems in place to manage, where much of this activity is of the leaders reporting to them. Instead of getting leverage from leaders, more CEO time is consumed by them. The CEO is either exhausted by managing the company, ends up trapped in it rather than able to work on it, and spends less and less time with customers, partners, and investors. Decision making slows down because too many things route through the CEO. Now you need a COO.

Your first COO hire should complement the CEO, providing a skill set that is additive rather than duplicative. The CEO doesn't need a "mini-me," but preferably someone who can grab a chunk of the business and manage it. Ideally, the COO would take over "operations," but this can scale well beyond just a few functions. Many product-oriented CEOs want to maintain control over product while giving everything else (including Finance, Marketing, Sales, and Customer Support) to the COO. Other CEOs, especially ones who are sales-oriented, include products in the COO's purview.

If you are struggling to think about how to split up things with a COO, use the "Three Machines" metaphor. The three machines are (1) the Product machine, (2) the Customer machine, and (3) the Company machine. As CEO, you'll likely continue to manage the Company machine. Your COO should either operate the Product machine or the Customer machine.

Finally, look for a COO who is a better operator than you are. You want to learn from your COO, let them take over big chunks of the business, and free you up to move the business forward and into new areas, without worrying about big swaths of the day being taken up by activity. While this is true for any executive you bring on (e.g., you want your CRO to be better at selling than you are), this is particularly important for the COO.

Chapter 122

CEO-to-CEO Advice About the Operating Role

Matt Blumberg

Complexity or Crutch?

As we suggest in this Part, COOs come in many shapes and sizes. Some COOs are just accumulations of other functions, whether "inside" functions or "outside" ones, leading to a larger title of COO or even President—these people are really partners to the CEO and in many cases are essentially a co-CEO. Sometimes a COO is really just a Head of Operations. But if your business has a COO who is managing all functional departments, or all except Finance and Legal, as often happens, that's a good point to reflect on the underlying reasons why—you might discover something important about your business.

Does the COO role exist due to the complexity in the business? Some businesses are so complex that even a strong CEO can't effectively manage the whole of the business. A COO might be incredibly helpful or a better solution might be to think about ways of simplifying the business by streamlining the organization structure or product set, or if the things this kind of COO does could potentially be done by a less expensive resource like a Chief of Staff.

Is the COO a crutch for a new or weak or ultra-focused CEO? Some CEOs are great leaders in some respects but awful in others. If you're a CEO who, to pick one example, is a great product visionary but completely uninterested in mastering the rest of the craft of being a CEO, you may need a strong executive reporting to you to keep the trains running on time and the company from falling apart. Although sometimes this is a positive thing—good, seasoned operators like this kind of COO can "mentor up" an inexperienced founder, for example—every time I hear a CEO say "What I really need is a good COO," all I can think is "Are you sure you want the CEO job? If so, what about trying a good coach or mentor first?"

Understanding which of the three scenarios that Jack discusses your business has is the key to understanding why you have an additional expensive executive around the table and whether or not that role is necessary or is masking some larger problem.

Part Eleven

The Future of Executive Work

Chapter 123

The Future of Executive Work

Matt Blumberg

While the ideas in this book are focused on scaling up your company, it's not the case that every company will be able to scale up with their existing talent. Even after you read the book and try to implement the ideas, you may still find that you're not ready to bring senior people in, you're not able to level up your existing leaders, or you may just not have enough people and not be able to afford the talent you need. What are your options if you're stuck in the middle? We've got an answer for you here.

One of the things we're most passionate about is the emergence of a new role that aligns perfectly with the dynamic talent needs that startups have: the fractional executive. Bringing in seasoned, experienced, executives on a fractional basis to build a function is a great alternative to the traditional executive hiring model that lets startups scale more elegantly. In fact, we're so excited about the possibility of helping executives and the startups and scaleups that need them that we created a new company, Bolster, that matches executives (members) with companies (clients).

What we see in today's business world is increasingly shorter tenures for senior executives—people with deep experience operationally and strategically—and enormous difficulties for startups and scaleups to identify, vet, and hire people who can help them get to that next level.

Welcome to the Executive Gig Economy

The gig economy that started with roles like artists, drivers, and web designers is quickly expanding to include executive-level roles. There are a few trends in today's workplace that are driving this expansion. Startups and scaleups have more flexible, remote friendly work environments and are looking for creative, less expensive ways of accelerating growth. Executives have shorter average job tenure and are more often displaced or between roles, and they are also interested in the flexibility that gig work can give them.

In a study conducted by MavenLink/Research Now, "The White Collar Gig Economy," 47% of companies state they are looking to hire contractors to fill management and senior executive roles, including C-suite contractors. At the same time, 63% of full-time executives would switch to become a contractor, given the opportunity. These trends will be accelerated by the current economic downturn and recovery, as some companies have fewer resources, and more executives are displaced.

While executive coaches and consultants can be considered "gig executives," the gig economy reaches well beyond coaches and consultants, and we're now seeing a new up-and-coming type of executive gig role: the fractional role. The expectation for fractional executives is that they'll be involved in a long-term relationship with a company on a part-time basis. For example, they might work a limited number of days per week or a certain number of hours per week. The fractional executive is responsible for all functional areas same as a full-time executive in that same role. The company may be too small to need (or afford) their level of expertise on a full-time basis, but needs more than just an advisor or project consultant.

You may be thinking, isn't a fractional role just like a consultant, or independent contractor, or interim? That's a good question and we'll answer it here. Consultants and independent contractors normally have a well-defined and limited role within a company. They don't have any decision-making authority (hence the title "consulting"), they are unlikely to be able to direct any team members within a company, and the scope of their work is typically highly specialized. Once the project is finished, the consultant walks out the door and leaves the implementation to you.

An interim role is often open-ended and the person filling that role is trying to demonstrate that they are the next "right" person to fill the role full-time. Interim roles usually emerge suddenly and unless you have a deep bench, you will have to scramble to fill the role. Interim roles are common in professional sports where coaches get fired or leave because of "family circumstances," and often an assistant coach or staff member fills the role until the end of the season. The motivating factor in having an interim is as a stop-gap measure, to keep the trains running on time, to make sure that the organization keeps plugging along. One of the biggest hurdles in getting value from an interim is their credibility with existing employees. Since interim roles are typically filled from an internal pool of candidates and it normally happens quickly, many interims struggle with authority and actually getting employees to follow their lead. This is compounded by the fact that an interim role is temporary, so employees have many reasons to wait for a full-time replacement.

A fractional executive role is different from consultants, independent contractors, and interims on a number of levels. Fractionals have far greater operational experience than consultants and are normally given wide discretionary powers of authority to make changes. Unlike (some) interims, fractionals show up with a long list of achievements with other companies and in other verticals so they have credibility along with experience. Most significantly, fractional leaders can provide more than operational and strategic expertise and the good ones put a lot of effort into coaching and developing their teams, setting them up for long-term success.

But don't take our word for it: we've assembled several fractional executives and asked them to write up their ideas on this newly emergent role (Chapters 124–132).

Chapter 124

Fractional Chief Financial Officer

John McCarthy

The Need

Every company needs a Chief Financial Officer. Some companies need them two hours a week; others need them 60 hours a week. But every company needs somebody to help the CEO and the Board understand the finances, and they need somebody to make sure the financials are organized. You can call it a Chief Financial Officer or an acting CFO or Head of Finance, but that person is really needed at pretty much any company that has taken venture capital (VC) funding or has aspirations for taking VC funding.

Because once you've taken VC money, you probably need to tighten up your act. Some companies are very good at it—they're keeping good books, they track metrics and KPIs on a regular basis, and they close the books quite quickly. Others, not so much. It's important for founders and early-stage startups to recognize that what's made you successful so far has been your product or your customer acquisition strategy, but it has probably not been your finances. And founders need to realize that the day is going to come when they'll need to have their finances in good shape. If you've been able to successfully raise your Series A without having your books in order, you pretty much got lucky, but you'll soon realize that won't be able to raise your Series B until you're able to quickly turn around financials to get through due diligence. You'll have to be able to answer the basic unit economics of your business.

If you don't have your financials in order, then that's the time to bring somebody in. And whether it's a fractional person or a full-time person depends on how complex your business is and how quickly you're growing. The benefit of a full-time person is definitely share of mindset but what you'll find with a person who takes on a fractional role is someone with a broader experience base and they can usually get to the answers a little bit quicker than people who are seeing the problem for the first time. You are also only paying for a fraction of their time, so they are cheaper! It's a tradeoff, of course, but getting a feel for the experience that the fractional person has in your industry or understanding how they can take care of issues that you're now facing is an important thing to get a handle on.

The Structure

Unlike a full-time role—50+ hours a week, for example, as fractional CFO, you'll have a lot more flexibility on days and times that you work. Your first question to answer is, how much time does the company need you? They might not know so I usually keep it open-ended. For the open-ended situation I can be with the company anywhere from six months to five years, depending on their situation. Sometimes a company brings me in for a defined time-period. These are the types of situations where someone goes on maternity or paternity leave, or where the company is transitioning from one Head of Finance to another and they need somebody to bridge the gap. In those situations, the timeframe tends to be defined.

Some people set up their schedule and will say, "I'm going to work for you on Monday and Thursday." From my experience that doesn't work very well because the CEO or the Board wants an answer—maybe not immediately, but as close to immediate as possible and if it's Monday and you tell them you'll get back to them on Thursday, that's not really acceptable. For simple issues that might be OK, but it's not a great strategy for you, the company, or the CEO. If, for example, you work all day Mondays and Thursdays for Company A, expect to check emails periodically on the other days, and prioritize responding to the most important requests.

Another thing you'll want to be very clear on as a fractional CFO is decision making. In my mind, to be able to make good decisions, I have to be highly involved at the executive level. It will obviously change as they get more comfortable with you, but I want to be viewed as a member of the leadership team. I want to be in all the regular weekly or staff meetings, I want to be introduced as John, the company CFO, not John, he's going to be working with us on a temporary basis. I find that I'm much more effective that way.

Another issue to address is how you want to structure compensation. For a fractional role, compensation is usually more weighted toward 100% cash at the start with an option to add equity once both sides are comfortable. From the fractional executive's point of view, you don't really know what the equity is worth until you really get in there and understand the company and the opportunities. Once you've done your homework, you can have a more thoughtful conversation with the CEO and the Board as to what the proper cash and equity split could be. But going in with a single simple formula for what an equity split should be for all companies isn't the right approach. It just doesn't work that way, and it's definitely not a cookie cutter conversation. Whether you charge hourly, weekly, monthly, or project-based is something that is case-specific but the only advice I can give is that you'll need to be flexible.

The Role

This probably goes without saying, but what's important to me is I need to be 100% committed to the company. I need to commit to responding to requests on a daily basis, and I go into a new company assuming that I'm working for the company five to seven days a week, just like I would for a full-time role. Some people get lured into a fractional role by saying, "Oh, it's only a day or two a week." But that one or two days a week, from my experience, usually gets spread out over five to seven days during the week. So, you're putting in your 16–20 hours per week, but it's over seven days a week. The actual work can ebb and flow based on each client, what they have going on during the year or quarter, but no matter what, I'm always 100% committed.

One of the keys to doing the fractional role well is balancing across clients. I usually work for multiple clients—sometimes as many as six at a time, but I'm more comfortable in the three or four range. That means instead of having one CEO or Board to keep happy, you have multiple people looking for quick answers and turnaround time. You need to know each one's personality, you need to become comfortable with their work style, how they want information presented, and how quickly they want information turned around. You also have to realize that a system that's worked with one CEO may or may not work with a different CEO. You may have a CEO who is very quantitative and then you may have one who's not and you've got to find a way to quickly figure out that CEO's learning and management style and adjust. And there's always the chemistry factor. I have turned down roles because I could tell from the conversation that I just would not have chemistry with a CEO—the way they were approaching questions and the way that they were handling the conversation led me to decide that there's not a good fit for me.

Being able to turn down positions where there's not that chemistry is important, even if all the other criteria like industry, structure, compensation, and tasks are lined up. If the chemistry isn't there, then you may find it very difficult to have a big impact, so keep that in mind.

The Skills

If you're thinking of becoming a fractional CFO, or hiring one, there are a couple of key skills you'll need, like flexibility, strong time management skills, and organizational skills. One basic example, you'll need to be very good at password management because with the seven or eight tools you're using in Finance at a company multiplied by the three or four companies you're working with, you will have a lot of passwords. And the same is true with emails. I've got a very good setup and it's taken me a while to get comfortable with it, but each company is represented on a toolbar and it's fairly well organized. Strong organizational skills are important, which is true of any executive position, and even more critical when you are juggling more than one position. You can't afford to spend time, or waste someone else's time, because you're not organized.

The ability to truly multitask and to frequently context switch is also really important. You really need to be able to, on a dime, take a phone call or a Zoom call from the CEO right after you've gotten off the phone with another CEO. You have to be able to go from zero to 60 in two seconds, with the second CEO. There's a lot of juggling of time, a lot of being able to multitask and truly multitask, not just pay lip service to it. You're frequently context switching, you're going from literally looking at a company that's running out of cash next week to a company where their biggest problem is that they can't hire people fast enough. You've really got to be comfortable in switching contexts and making sure that you're not mistakenly telling the CEO that is hiring as fast as they can that they're running out of cash next week. You have to make sure that you're keeping the two contexts completely separate in your mind.

One of the biggest assets that a fractional CFO brings is broad experience. This essentially defines the fractional executive. You'll bring your experience to a company and one great benefit is that you have the ability to counsel and to act as a coach, which is not necessarily something that a junior person would be able to bring to a company. That's one of the things I really enjoy about what I do. I'll help counsel CEOs on what the numbers are really saying, what the vibe in the company is saying, or counsel them about the future of where the company is going. I can put that into a context based on all the things that I've seen—the successes, the failures, and all the stuff in between. Having the ability to sit down one-to-one with the CEO and give them my honest assessment of what the numbers are saying (and what the numbers are saying in context for the company) and provide some guidance is one of the best aspects of a fractional role. When, for example, the Board says x, what are they really talking about? What are they really thinking about?

In bringing in a fractional CFO, you're paying a little bit more than you would if you hired somebody who had a controllership or a director or VP finance role. But you gain extensive experience from a fractional CFO that can help you as a CEO, help your Board, and help the next full-time CFO you bring into the company.

The Exit

The fractional role is going to end, no question about that. But how it ends can make a difference for you and the company. I've been in a couple of situations where my exit was due to an acquisition. When that happens, I am basically turning the keys over to the finance team with a larger company that's acquiring us. A more common situation is bringing on staff from the very beginning with the understanding that I'm not going to be around as a CFO forever. My role there is finding a way to hire people who could take over the role when the needs of the company exceed my availability. So, I can help them on the recruiting side to find the right person to take over the full-time responsibilities.

Chapter 125

Fractional Chief People Officer

Courtney Graeber

I've been working in human resources and operations for over 20 years and I feel like all the positions that I've been in have been positions of transformation, where a company would bring me in at the beginning of something that needed to be changed. Maybe it was with an early-stage startup with no systems or processes in place, or a company that needed to pivot, or an organization that really wanted to do some kind of transformation of the business. What I've learned is that whether you're in a startup or in a larger organization that's looking for change, a lot of the characteristics and qualities are similar.

One thing I realized about myself is that what really got me excited was working with startups. I enjoy the infancy stages of a company and everything that goes along with it, like the wacky space, the weird furniture, and the people doing everything even though they don't know what they're doing. I'm excited by the messiness, the passion of the founders, and I just want to get in there and make it all work.

Fractional HR Tasks

I recently wrapped up an assignment with a startup that provides an example of what a fractional executive does and the value they can bring. I got a call from someone who I used to work with who started a company and said, "We're in big trouble and we need you to help us out." Originally, they just wanted an HR presence that would help employees if there were any employee relations issues. But within one day of me being with them I took over the entire role of running an HR department, and I ended up completely revamping HR operations. At the time, HR was being handled by the CEO and when they realized I could do all of the HR, they gave it to me to run. This allowed the CEO to focus on the business, not on HR. In the beginning, it was a lot of cleanup from an HR operations perspective. My goals were to make all aspects operationally sound, fully utilize all the technology that was there, ensure the data was accurate and everything that was in place was compliant, and quickly identify elements that were missing.

From a culture perspective, if a company can't nail the basics, you lose credibility. Nobody wants to talk about having free lunch on Wednesdays, for example, before they have the basics of their employment contract honored. New hires need to be onboarded correctly and benefits and paychecks need to be in effect when originally communicated. When I work with a startup or any company, I always dig into this first and I refer to nailing the basics as part of Phase 1 of an HR, People, or Talent organization. When your employees can rely on you that the promises the company made are being delivered, then you can talk about building a culture together.

I've found, with HR-related work, that once you're involved with a company it can be like a Pandora's box because if one thing needs adjusting, so do a lot of other related components and they're all connected. I'm an all-in type of person so I don't mind doing the work to get it right.

That said, in a fractional HR role, it's important to understand what the priorities are. And sometimes after a discussion between the company and fractional HR person, the priorities change. But you want to discover that in the early stages of the assignment so everyone is aligned. Your time and resources are limited, which is why the role is fractional to begin with. Additionally, as a fractional HR person, you'll need to be flexible with what you do. While most executives are used to working at a strategic and higher level of thinking, some of the fractional roles are very hands-on and require execution. Once priorities are clear and the fractional HR person is comfortable with the skills and experience needed, a fractional role is a great short-term solution to address priorities of the leadership team.

In any job, whether it's a full-time or part-time role, you want to make sure you are aligned with the leadership team and you believe in the company's vision. These relationships and the passions for the vision will be a good foundation for success. In a fractional role you don't have a lot of time to "get onboard," you need to figure things out quickly and make an immediate, high impact.

Startup or Established?

If you have been a Chief Talent Officer or Chief Human Resources Officer, then you're used to working at a very high level, but if you're working in a fractional role at a startup, you'll likely be doing a lot of execution with an eye on strategy. Typically, resources are limited, so you may be a single-person team or have someone within the organization (usually an admin) help you execute on priority or high impact tasks. As discussed earlier, a fractional HR role in a startup is usually nailing the basics, which involves a lot of execution, hands-on, roll up your sleeves kind of work. You may have to dig into the current HR operation structure, which could include partnering with a third-party HR vendor like Trinet. But you also need to keep an eye on strategy and in every action you'll need to think about how this will impact the company in six months or a year. Companies hire a fractional executive because they are looking for someone who does have that strategic mind and understands the big picture.

If an established company is looking for a fractional head of HR, then the situation is a little different. The fractional HR person should always be thinking about agreeing on the priorities up front but also you'll have to answer the question, why am I here? You must understand the relationships of the leadership team and the existing HR team. You need to specifically learn about the HR team capabilities, including what everyone does and how the department manages the HR function for the company. Understanding relationships and roles are key things to figure out quickly from the beginning. Also, it's important to understand the reputation of the existing HR team and their perception internally. If your priorities are to build the HR department and make it more effective, that means you still need to dig in but you have people around you who can actually help to execute. Another key element of this role is to make sure you assess the HR talent and understand what their strengths and weaknesses are in being an effective team. And then put a plan in place to make sure you have the right talent and experience to meet the priorities of the department.

So, not all fractional HR work is the same and the stage of the company has a big impact on the tasks that you'll do.

Transitions

Once you fulfill the priorities of the role you were hired for, a conversation needs to happen about the future. Sometimes it makes sense for the fractional HR executive to stay with the company. Other times a company will realize they don't need someone at this level yet and a solid mid-level person can handle the work for the next year or two.

One of the key reasons you'd want to use a fractional person is either to make high impact quickly for your population with someone who has a lot of experience with building and growing an HR department OR to allow a company to have access to someone with this experience at a more manageable cost. The fractional role allows the company to understand the impact of the function with limited commitment and cost. For the fractional HR executive, the benefit of being fractional is that it allows them to do what they love in a part-time capacity.

In my last role as a fractional HR executive, once it was clear they wanted me to manage the entire HR function, I created a roadmap. The first phase was all about nailing the basics and ensuring the HR operations were sound and sustainable. Phase 2 was about understanding the company's growth plan. How many people do they plan on hiring in the next 3, 6, 12 months? It also involves understanding the company more deeply and learning about the existing people, departments, job descriptions, performance review plans, compensation strategy, and other key characteristics. Phase 2 is really about understanding the company's expected growth and making sure you have the right talent to drive that. Phase 3 is discussing the future of the fractional HR role. I begin the conversations about what the leadership team needs what they can afford to sustain the work that has been done. In my role, we realized that they really didn't need someone at my level. They had a more limited budget for HR, and the plan I put in place was solid enough to hand over to a seasoned mid-level person who could handle the execution and build on the strategy put in place.

Qualities of a Fractional HR Person

The qualities of a successful fractional HR executive are as much about skill as they are about mindset. In a lot of cases, you need to be comfortable in self-managing and building something from nothing. Get jazzed and excited about things being messy and the desire to fix it. You need to have the comfort level and skills to actually know how to fix things yourself or manage someone to do the work. As a fractional person, in a startup environment or in an established organization, you have to have that all-in mentality like it's a full-time role, be curious, and keep digging for what needs to be done, while maintaining the company priorities for hiring you. I feel all of these components absolutely need to be part of your personality, to be successful in this role.

Chapter 126

Fractional Chief Marketing Officer

Scott Kabat

I've worked in every stage of a company from co-founding a two-person Y Combinator education technology company up to being acquired by Cisco and working in their enormous machine. I'm a firm believer that every company will need a fully functioning marketing team, and it's important to figure out, when you factor in hiring time and budgets, what the right prioritization of the full-time roles are. And then you need to think about where you can supplement those people with senior guidance or boots on the ground so they can get more done. The fractional CMO can be an extremely valuable option for some companies, especially startups and scaleups.

Fractional Opportunities for CMOs

There's a big need for fractional CMOs and it's largely driven by the fast-paced changing nature of marketing. My experience, and one that I think a lot of CMOs are experiencing—especially in late-stage startups—is that invariably you eventually hit a point where the strategy changes. The needs of the organization change and you realize that you're not right for the company anymore. And for a lot of CMOs, their tenure with a company is 6–18 months and then they're back in the job hunting market. I didn't want to cycle in and out all the time and the fractional role allows me to be in market rather than in between jobs. I started doing fractional

CMO work almost exclusively with startups that had raised some capital, launched a product or were about to, but didn't know a lot about marketing, and frankly, weren't at the point where they needed to hire an expensive full-time CMO. I found that there was a real opportunity not only for startups but also for later-stage companies going through big strategic transitions with a change in business model, or a private equity firm investing in an old line business to digitally transform it.

My experience in fractional roles comes in two different flavors, and I always seem to have a composite of each. There's the kind of fractional CMO where you're running the marketing team on a limited basis, which I generally do no more than two days a week with one company. And I spread that time allocation through the week. And then there's the advisory CMO which is generally when the company is already executing and my job is to weigh in and give guidance. In the advisory-type role, we have a weekly rhythm of checking in but operationally I am less involved. It's worth noting that especially in marketing there are different approaches to being a fractional CMO. Some people have a standard assessment and playbook and they run you through their analysis, usually with a particular emphasis on performance marketing and the client side of it. I generally operate more as a full stack and I'll say to a company, "Let's look across the mix at both the client side and the softer side and figure out what's right for that company." Different people approach it differently.

In some respects, your role as a fractional is the same as if you were full-time. You are still leading from the front on strategy and working with key stakeholders, with the executive team and Board, and within the marketing team to put a plan and strategy together. You're still looking ahead to what's coming and creating a roadmap, which is actually really important for startups. It's common in a startup to get so into the weeds that you're not looking at where the step function changes need to happen. So those tasks—creating a strategy and a roadmap, are the same in either a fractional or full-time role.

The biggest difference is really figuring out the line between your role as a fractional and the business. You have to be able to back off and understand that this is their thing. I will give advice and drive solutions and guide them but at the end of the day I'm not a part of the team and I have to back off at certain points. I generally try to work as an embedded member of the team so I err more on being integral rather than being hands off. But, because you're not really a full-fledged member of the team, that's different. And then, obviously, I will get involved in a lot of the people and the organizational dynamics but I'm not writing performance reviews, I'm not managing the operational back end of leading a team as an executive.

I believe there is a lot of opportunity for fractional CMOs, but it's not the same as full-time and it might not be for everyone, especially if you need to be highly involved and control the entire marketing team.

Qualities of a Fractional CMO

It's critical for the fractional CMO to see the big picture at all times. Where I'm catching startups is when they've raised a round of funding, they just sold a dream, and there's this temptation to want to do it all. And part of where I try to add some perspective is, "Hey, guys, let's look ahead to what the next big milestone is, maybe it's the next funding round or the next product launch that's a little further down the road. Let's develop a prioritized plan to get there." So, a key skill is to maintain a degree of objectivity and to view the bigger picture through a strategic lens.

Another key skill is being able to context shift because I'm generally working with multiple companies, sometimes with five at a time. And during the course of any given day, I'll have a meeting for one client, a call for another and back and forth. I find it stimulating and I've noticed that the pattern recognition has sharpened my toolset as a marketer. But it can be dizzying at times. There is another model where some people work full-time four days a week, but I have steered clear of that. I generally work no more than one and a half or two days a week with one company.

The other skill that is critical I mentioned earlier, and that's the ability to be able to walk the line between speaking with conviction and adapting within the existing organism of the culture. No one is hiring you to be a fractional executive to just keep the trains running on time. But at the same time, it's not the sort of role that works well if you're just a bull in a china shop. I don't think there's any way I could be successful at this if I hadn't had deep experience as an operator and dealing with all the messiness of making the sausage because you have to appreciate that. And things go sideways with these businesses all the time, so I spend a lot of time reassuring them like, "Hey don't worry about this, I've seen it before and it's OK. We're here to help."

One thing I really liked about being an operator that I've tried to hold on to as a fractional is the chance to mentor and advise people. Because particularly in startups, this work is very personal, and I enjoy being able to guide startups through the journey. That's not really a skill or a quality you need to be a successful fractional CMO, but the bigger point is to find what you like doing and bring that to every situation.

Fractional Opportunities for Companies

When I talk to CEOs, they'll say to me, "Look, I have to decide if I'm going to do this fractional thing or hire someone." I don't see that as an either-or scenario because what I generally see, particularly when it comes to hiring a Head of Marketing, is that there are 1,000 different flavors of marketers these days and if you don't know exactly what you need, what the musts are, and what the wants are, and how it fits into where the business could go, can you really hire someone? People might not last long or the process of hiring might take much longer than you think.

A fractional CMO can help you through that process and help you think about the job profile, help you vet candidates, help to onboard and advise those new hires so that it shrinks their time to impact. In a world where the CMO's tenure can be very short, you can't really afford to have someone come in and get a slow start. So, one role of a fractional, one benefit to the high growth company, is that this person can get involved upstream to assist with finding, vetting, and onboarding a full-time CMO. The fractional CMO can help them through when they start and then eventually recede into the background and support those hires with operational risks. It's rarely as much of an either-or thing as people think.

Chapter 127

Fractional Chief Revenue Officer

B.J. Bushur

I've always been in B2B sales and in technology companies. I started my career as a sales rep at a company that got bought by PricewaterhouseCoopers where I learned both sales and management skills. I got the amazing opportunity to go to Netscape right when the first browser was launched in 1994. Netscape didn't have sales reps, just tech support, so I went there as the first sales rep starting as inside sales and then took over the inside sales organization. And over the five and a half years I was there, I got to see how the rocket ship took off when we went public. By then we were selling enterprise solutions, and I ran a 100-person inside sales organization with quota bearing reps, sales development reps, and account managers who would renew and upsell customers.

After we were bought by AOL, I decided it was time to follow Marc Andreessen and Ben Horowitz over to what was then called Loudcloud. We were a managed services organization (in 2000) selling the whole web operation, people, and technology stack to SasS companies. That's when I really learned lots of the ins and outs of changing the go-to-market models, when I'd had my first baby, and when I decided I loved working with startups. Ever since then I've been working with B2B technology companies to set up or strengthen their sales foundation for scale. I work with them to surround their sales organization with the processes, tools, systems, compensation, enablement, and metrics to scale a fast growth startup.

Becoming a Fractional Executive

Becoming a fractional executive is part skills and part mindset. By that I mean, one of the things that holds talented, credible, impactful people back from taking on a fractional role is the fear of the unknown. Can I do this? Will I make enough money? How do I get clients? Will the work be steady or will I constantly be searching for new opportunities? There are a lot of big unknowns in the beginning, and those fears can prevent you from doing something you love. What helped me to overcome these fears and doubts was working with a leadership coach to peel back those unknowns. Could my family afford it? Could I get enough business? What would the barriers be? How difficult would it be to work with different companies? The coach I worked with helped me uncover those unknowns, and I researched them so I could make that informed decision on whether a fractional role would work or not.

I also worked with a financial planner because so much of the risk that you take is actually not about being capable of doing the work, but about whether or not you can survive financially. Working with a financial planner really helped me and my family figure out if it was feasible. We had to understand our health insurance options because we were both self-employed. We had to figure out our childcare situation and costs. So, a first step was just researching the unknowns and asking, can we live with that? Because your income looks variable at first, when you're a fractional leader, I had to know if I would be able to weather those unknowns and I had to trust that even if the worst-case scenario happens, I could live with that.

The other thing that's important to know, if you're thinking about the fractional role, is what you're good at and what you love. They're not always the same. I needed to understand, what are the things I enjoy doing and I'm good at? What are the necessary evils of running sales that maybe I'm not good at, or bore me, or frustrate me? For me, it was systems. I knew how to use Salesforce.com (back then it was a newer technology) and I was good at it. But it bores me. I didn't want to get deep into the technology, and I'd much rather work on solution selling, coaching sales reps on deals, hiring great talent, and enabling sales reps. So, my first year was a lot of experimenting on what are the projects and the situations at the client that excite me, energize me, that I'm good at, and that I *want* to do, versus going down the rabbit hole of things I don't really enjoy doing.

One situation really drove this point home for me. A client hired me just to recruit their inside sales function which I happily accepted because I thought I would have a lot of interaction with the sales management team. But I was siloed off just with HR and never really engaged with the sales side of the house; I just helped screen candidates. And I realized when I was pigeonholed into a small piece of sales management that it frustrated me; I wanted a bigger picture, I wanted to be working hand-in-hand with the sales leader and help with bigger strategy. I learned to say "no" to those things, and now I partner with people who are really good at those other things. Surrounding yourself with the rest of what needs to be successful for your C-function is a key to being successful as a fractional executive.

Finding Fractional Roles

Once you're established, it's relatively easy to find fractional roles. All of my work comes through referrals, from past clients in my network, from clients or CEOs who have moved, and from VCs primarily. But what about starting out? Before Bolster, there wasn't really a network that brought together companies and fractional executives so you had to leverage your own network. I started by conducting free webinars and (before COVID) I'd do in-person breakfasts or lunches on a topic that's really important to CEOs. I did seminars on common misconceptions about high velocity selling or sales and marketing collaboration—topics that I knew a lot about and that some CEOs would be interested in. That's one way to do it.

Another way to get traction when you're starting out is to network with the most networked people in your network. For example, I had two bosses who retired, and one of them really retired and didn't network while the other one was out there with VCs networking all the time and connecting people with great talent/portfolio's needs. So, I spent a lot of time with the second one, not the first one. So, figure out who those people are who are networking and make sure they know and trust you so that you can meet the people who will want to hire you.

Cautionary Advice

One thing I learned very early on in doing this is to trust your gut because when you start out, you're in a new situation with a lot of unknowns and you don't know the lay of the land; you haven't figured out yet what could be a problem. During my first year as a fractional executive, I got introduced to some teams where they wanted me to change their go-to-market strategy. I got a call from one leader who asked if I could listen in on a strategy session with the whole C-level team. I responded with, What would be your goal of my participation? The executive countered with, Just come for half the day, not the whole day. My gut reaction and what I asked was, Why would I want to do that? Why do you need me there? And what would be the outcome of that? The executive couldn't tell me.

I called my former boss whom I trusted, immensely, and who was a very good friend of mine as well as one of my mentors and heroes. I told him that I'm new to this whole fractional thing and should I be expected to just go for a half day not understanding what's going on? He said, "No, there's some management dysfunction there." But then I questioned myself. Should I have just gone? I didn't have anything else going on. It ended up not panning out and the company was dysfunctional. So just because you're new to being a fractional leader, still trust your gut and if a company can't articulate exactly why they need you and what they want you to be doing for them, it's probably not going to work out. If I'm in that situation now, if I realize that it's not a cultural fit or they know they need help but can't pin it down, I just say I'm too busy to take on the role.

When a company can't describe the deliverables that they want, or what skills they need from you, or what outcomes they want to happen—if they can't describe those, then they're probably not ready for any fractional executive. If a company wants a bunch of free strategy sessions, then trust your gut that if they're not willing to pay for even some small assessment that's probably an indication that they're not valuing your skill set or they don't know what they need enough. Getting a sense of what are the macro goals that they're trying to hit, understanding their current challenges, and making sure that your skill set, your experience, and what you want to do are a match for you in that situation are the keys to being a successful fractional executive.

One other important thing for a CRO: you definitely need to understand whether the company you'll be working with has product market fit because if they don't have product market fit for whatever they're asking you to sell, you won't be able to make an impact. I had a project where the CEO thought they had product market fit, and the CEO was passionate about it. It was a younger CEO, a first-time CEO, a technology CEO, and I should have paid more attention to what others were telling me about where they were in their product lifecycle instead of taking the CEO's word for it. I'm sure it differs by domain for other C-level roles like Finance or HR, but you need to make sure that you can actually impact the company. You need to make sure that some of the fundamentals are in place, for whatever role you're doing. You may have to triangulate with the other managers to be able to answer the question, what needs to be in place for you to be able to have an impact?

I'm passionate about fractional roles because I think people can have a much better life if they choose this route and they can make a truly impactful contribution to a company. You just have to peel back those unknowns and go for it!

Chapter 128

Fractional Chief Revenue Officer

Sherri Sklar

One of the things that I like about working in a fractional role is that elements from your entire career can come into play, enabling you to make a large impact in a relatively short period of time. Many emerging tech companies often get to product market fit, only to stall when trying to grow and scale. The talent that got the company to that point may or may not be enough to take the company to the next level. From a revenue perspective, there are timely decisions that need to be made operationally—in your go-to-market strategy and in your execution—that can make or break your next growth cycle. It takes a seasoned senior executive who has lived the various parts of the revenue function under many different previous stress tests, to know which strategy and which move is the least risky and most likely to succeed.

I started selling IBM mainframes right out of college and when I left IBM I was running a $40 million business. When I went to the startup world everything went out the window. In the new world of tech startups, I started at the basement level to learn the business from the bottom up. I began again as a sales rep and from there to district manager, regional manager, VP Sales, and then went over to the marketing side to become a VP Marketing, and finally became a VP of Business Development. Over the course of my career, I had the opportunity to wear every kind of revenue hat in large organizations and in startup organizations. As a fractional executive, that provides a lot of credibility. When I walk in the door, everybody knows that I've been in their shoes at one point or another in my career; it creates instant trust knowing that I have carried a bag, that I know what I'm talking about, and that I can speak as well from a Sales perspective as I can from a Marketing perspective as I can from a Customer Success perspective.

Companies that I work with bring me in to help run their organizations, and they typically are a Series A or Series B company who may not have a revenue leader or the right kind of leadership in place. These companies are at a point where they need somebody who understands how to achieve growth and scale. They need a fractional CRO to take the revenue functions off of the CEO's plate so the CEO can focus on other critical actions to help the company grow, such as fundraising, product strategy, development, people, culture, and driving the vision for the company. The fractional CRO can enable the CEO to move forward, confident that the right decisions are also being made on the revenue side of the house.

I love this work so much because I get a chance to implement my twin passions for helping companies to grow and helping companies become customer-centric so they can truly make a difference in their customers' lives. I usually come in right before a CEO realizes they need somebody permanently. So, it's almost like they could try it out with me and then sometimes they realize, "Oh, my gosh, I need somebody full-time." So, my role is to get them structured and set up for really great results and help pave the way for a full-time CRO.

Qualities of a Fractional CRO

One key to being successful as a fractional CRO is to be comfortable in your own skin. You have to believe in yourself and be a great leader in the face of many people who may question your every move, which can happen for a variety of reasons. You might be the only one in the company at this point who knows the best way forward, and you need to remember it's your experience and your know-how that got you here. Believe in yourself and stand up for what you believe in. Help others in your new companies understand the path they must take, even if it's unfamiliar territory to them.

It's also important to know how to be a leader as a fractional because you may not have full authority that everybody who works for you thinks or expects you to have. Many times I have been brought in as the CRO but did not have full authority to price deals, make hiring decisions, or manage certain personnel issues without having the CEO or some other executive approve. Constraints on your authority can be a consequence of nothing other than being in a position of authority but technically not being a full-time employee. Sometimes, unintentionally, that can send the wrong message—that you are not trusted by your CEO. Other times it can be incredibly frustrating to your people, because without having full authority, processes can get longer rather than shorter. One way to counteract this situation is by having a proactive discussion with your CEO up front before your contract starts. Ask them if there is a way around some of these legal technicalities so that you can gain full authority. If you can't get full authority, maybe you can set processes up that serve to streamline rather than frustrate people.

Another quality that a fractional executive needs is to be comfortable with ambiguity and uncertainty. How long will you be with the company? A lot of times you don't know—it could be three months, six months, a year, or even longer. Or it could be at some trigger event. I worked with a company to help them in a fundraising round. When they successfully got it, the PE firm that took them over wanted to put their own person in. You have to be comfortable with the fact that you might do a wonderful job, and they may still not want you to stay as the full-time CRO. On the other hand, you also should expect that if you do a wonderful job, you may very well be asked to stay on as the full-time CXO.

What to Expect as a Fractional CRO

You should expect that in a startup or scaleup company that the day you show up you'll be faced with a myriad of issues that have to be addressed. A fractional CRO needs to be aware that there's going to be a ton of processes and urgent situations that need addressing and need fixing. Fractional CROs need to expect that many parts of the organization they are walking into are broken.

This is not a "red flag" about whether this situation is a good opportunity for you or not. You're there precisely because many things may be broken and because you are the best at fixing and optimizing. The fractional CRO will relish these challenges—this is how you know you are in the right place at the right time.

If you have to move on, it can be difficult. If you are there for a limited or certain period of time only, it can be hard to move on because you end up loving the people with whom you work—whether it's the senior executive team or the people that work for you. It's just like any kind of job where you get into a new relationship with people and a company—you form relationships, you have achieved a certain set of results, you're proud of your work, and it can be hard to move on from that.

Another challenge about the job is that as a fractional executive, it's often hard because the expectation is to achieve the full result within a fraction of the time. Many times you end up putting in 100–150% of your time, simply because the workload is so hefty, you will feel like you have to address the issues head-on and immediately. Often the issues you are being asked to address can't wait for you to be available during the part-time hours that were pre-arranged. So, you have to be careful to set expectations going in, that in the event that the workload is much higher than anticipated, you can reset expectations and renegotiate your contract based on what has shown itself to be more realistic.

The Good Part

The good part of being a fractional executive is that you get a chance to do what you love and really make a difference to a CEO who really needs you. And that is exciting. You're bringing your expertise that nobody else in the company has, you're helping that CEO and the Board get results that they never could have gotten without you—all in a very compressed amount of time. You get the chance to prove yourself, to make a difference for somebody. And in the fractional world, you can help many companies in this way.

Chapter 129

Fractional Chief Business Development Officer

Jon Guttenberg

I have spent most of my career working in and around media companies. After 15 years inside of media companies, I've spent the past 20 years outside of them advising them on developing and executing growth strategies. My work involves (1) helping companies better define their customer, product, and positioning; (2) developing and executing against that; and (3) supporting their efforts to grow by other means (e.g., M&A, fundraising, partnerships).

In a fractional role, people need to get comfortable with you, they need to understand what you can do for them, and they have to get a sense of your style and approach to see if it's compatible with theirs. Trust is essential. Engagements oftentimes require several introductory meetings to help people get comfortable and during those meetings you will be giving "free advice." I don't worry all that much about giving some advice up front since ultimately you are looking to build a long-term relationship. If you could answer all of someone's questions in an hour and you didn't get paid for that single hour, that's fine. I look at this for the long term. If you can answer someone's questions in an hour or two, you are better off having that person come back to you in the future when their problems are greater or refer you to other clients since they now have a better understanding of what you do (although I will say there are people that even though I didn't plan on charging them, insisted on doing so).

But one caution: you need to be careful that someone is not using you for endless free advice. Fortunately, most people respect that and understand that your business revolves around getting paid for your time. It is usually fairly clear when someone does not have a sense of boundaries around advice and when that happens, I point it out. More often than not, those situations have turned into paid engagements.

Alignment

It's important to make sure that incentives are aligned between you and your client. You should always be paid for your time—other than the free introductory meeting mentioned above. Compensation can also include a success-based milestone. Try to avoid situations where you are *only* compensated on success, since even though it may not seem like it, when you are only compensated for success, your interests and the client's are not in true alignment. It is essential that your client have "skin in the game" since they need to understand that your key asset is your time and they should pay you for it. They also need to understand that your success is usually dependent upon their performance. Your reliance upon their performance creates risk for you, even with very established companies. If they drop the ball, you don't get paid, even if you did everything right. You can find yourself left with a significant investment in hours that will go uncompensated.

Additionally, what is initially defined as "success" may not be what your client ultimately wants and a success-based compensation structure can find you at odds with each other. For example, if you are compensated on closing a deal for them and they determine, at the eleventh hour they don't want to do it, your agendas are no longer aligned. A success-only deal would mean that you spent time and effort at their request and on their behalf which went uncompensated. That is why it is important that you get paid for your efforts, so you are not lobbying them for a deal that turns out not to make sense for them, but still does for you, since that is your only form of compensation. That does not mean there cannot be some success-based compensation, it just means that there should be a mix so that all of the risk is not borne by you.

Many fractional executives and clients will lean toward a retainer instead of an hourly rate since a company desires cost containment while the fractional executive often wants revenue security. But unless these retainers have a tie to hours invested (with overage if hours are exceeded), this structure can be problematic and risky. In retainer relationships without a tie to hours, I have seen people endlessly debating with their client about what is "in-scope" and what is "out of scope." In most engagements I have done, I have found that many projects evolve and it is hard to know, up front, exactly where they will go. Charging for hours does not force you to have to restructure or rewrite that deal multiple times as the work evolves; it holds throughout. For the client who wants to manage expenses, I simply provide an estimate upon the "known" scope of work, provide them a range, and inform them that I will alert them if I feel that something has changed over the course of the assignment to possibly have them go over that estimate. This allows them, at that point in time, to determine if it is worth investing further based upon whatever may have changed. For example, I was hired by one client to help them build a strategy on an initial engagement of 45 days. That project evolved into also helping them execute that strategy and the project ultimately finished seven years later without ever having to change the structure of the original agreement.

Objectives

It is important up front to clarify and define the objectives for any engagement. However, as you define the problem, it is also important to understand that often the problem your client thinks they have is not the real problem. Sometimes you can spot that up front, but usually it will take getting into the engagement to understand what the real problem is, and it's important to share that need for a discovery process up front. Clients use fractional people to either fill a gap or solve a problem that they and their team can't solve on their own. Part of the issue is that people in the company are often too close to the problem and only see it from a narrow perspective. The reason they need a fractional person is to bring in another set of eyes with different experience and insights, oftentimes from completely different industries. A fractional executive is also divorced from protecting "turf" or a position in the company. For the company, this new set of insights and objectives is often invaluable.

Quite often I find the work that I do involves pattern matching, which is why I often find myself saying "I've seen this story before and I know how it's going to end" and use that experience to inform the work I do for the client. I've been fortunate to have a wide range of experience, playing roles in Operations, Strategy, Marketing, Market Research, Product Development, Finance, Business Development, M&A, and fundraising. That broad experience allows me to look at problems and possible solutions from multiple perspectives and to suggest a wider set of solutions. For example, I was asked by a client to develop a plan to build out the company's sales and marketing efforts and in doing the research it became clear that the client had a more significant issue they needed to resolve before they could expand their business. In talking to current customers, it became clear that they were not using my client's product after they bought it. The client had the data but missed it. So before expanding their customer base, it was imperative for my client to solidify their existing base since current customers would be relied upon as references to win new customers. What started off as a sales and marketing engagement became a product engagement. That's why it is more important to tie compensation to hours, since the problem isn't always what it seems. That's also why the first goal of a project is to develop a hypothesis of what the problem might be and an approach to validate that hypothesis. The goal is not to define everything around that hypothesis as your full scope of work, since things aren't always what you (or the company) think they are initially.

Trust

Trust is always important in any relationship, but even more so in a fractional role since you're typically involved with a company for a finite period of time. It's important for people to trust you and to trust that what they share in confidence will stay in confidence. If you have high levels of trust, people in your client's organization will share more with you and typically most of the information that you need to be successful on a given project resides within the company. Oftentimes some of the best ideas are ones that come from people who have had their ideas shot down multiple times by their senior managers. Additionally, in getting people within the company to trust and open up to you, they will point you to a treasure trove of data that is essential for your success on a given project, but nobody thought it was necessary for you to do your job. Bottom line, the more people trust you, the better chance that they will share more. All of the sharing is a direct result of people trusting you and trusting that your goal and intentions are really aligned with theirs. They're trusting that you're solely focused on how to make their company a better company, which typically makes them more successful.

By way of example, years ago, I was asked to help a company that had built multiple new lines of business over the past few years to rationalize which of those different businesses they should remain in and which ones they should exit. In talking through the issues with the General Manager of one of those businesses we both figured out that this business was not viable. And you might say, "Wow, that was really a bad career move from this person." But the reality was that the manager was beating his head against a brick wall trying to solve a problem that really was not a solvable problem because in this case the product they built really had only four potential customers. Even if they won 100% of the market, the operating unit still wouldn't be profitable. So now the question was, OK, so we can't make this work, what can we do with it? That's a much better opportunity for the General Manager and the company because now they're not trying to solve an unsolvable problem and instead can focus on what to do with what they have already built and how to make that viable or valuable in a completely different way.

Passion

One other element that I have found key to my success has been passion. Oftentimes, even though you are hired to do something to make your clients more successful, you find yourself "fighting" with your client to defend a recommendation. It is oftentimes not just presenting and justifying your recommendation but showing the real passion behind why they need to do it. Many companies are stuck, and it takes more than just a simple recommendation and a set of supporting data to get them to make a change. Rather, it takes you passionately getting behind the idea and selling it to your client.

If you can get alignment with a company, come to a mutual understanding on objectives, develop deep levels of trust with everyone at a company, and show passion and conviction for your recommendations, fractional work is a huge benefit for you and clients.

Chapter 130

Fractional Chief Customer Officer

Amy Mustoe

I have been in customer-facing roles my entire career, leading teams in customer support and product to customer experience and success. In my fractional roles I was working as a principal consultant for The Success League. I love building up and building out customer-obsessed teams for startups and growth companies.

One thing I've learned as a fractional executive, is that you need to move fast and make the right moves. There is typically no time to do experiments. Your honeymoon period is condensed, and you need to get the lay of the land and show your value quickly. It does help to have done this a few times and have experienced successes and failures with certain models. There is no one-size-fits-all solution, so you have to be able to look at the puzzle pieces and figure out where the customer success team needs to go and make swift decisions. You won't have all the time in the world to sit down and interview people and ask lots of questions. You have to hit the ground running and have a pretty good track record and instincts about how things could work. Although it has a VP or a CCO title, this interim role may not have any management or director level support. You are the executive, the director, and the manager. You don't just attend meetings and give orders. You are the strategic arm and the doer and trainer.

For the customer organization there are a lot different ways that you can work with Sales, Product, and Engineering and when you get dropped into this role people want to know, who should own renewals? Do all customers have a CSM? Do we need a digital engagement program for our smaller customers? What should the comp plan look like? For the CCO fractional role, the company expects you to have an opinion and a way of moving forward. A company will typically give me the support to implement whatever processes and programs I feel will move the needle forward. Typically, there is not a lot of questioning because what they really want is somebody to come in and solve their problem so they don't have to deal with running such a critical team while looking to bring in a full-time leader.

Things to Figure Out Quickly

What is the temperature of the team? If you go into a situation where everyone's unhappy, upset, and wants to leave, that's a different type of thing, because now you have to spend a lot of time doing team building and a lot of time building back trust. In this situation you have to be eyes wide open that half the role is going to be just retaining the existing team. I was in a role where there was a lot of turmoil in the company and a lot of turnover at the leadership level in the prior two years. People on the team needed stability and I was just fractional, so I had to do more than introduce scalable processes and measurable metrics in supporting the customer success organization. I had to build trust in the team that I would find them a great leader and wouldn't settle for less!

Another thing to figure out quickly is knowing how big the problem is. Are you being brought in with the sole purpose of hiring a full-time VP of success? Or are you also fixing bad decisions from the past and socializing the customer culture? Is there a churn problem? Doing this discovery quickly can help set the right expectations with management on what you can affect during your tenure. If there are truly things that can't be fixed or are outside your influence, you should share those up front. That doesn't mean if the CSMs say the product is always broken that you go with that. There are many ways to discuss bugs, pricing, or renewals that not all CSMs have training on. I do help with that. The Success League has great content on the hardest CSM skills for new and tenured CSMs.

Define success and share that with your management team. Success could be very straightforward, like the company wants you to replace yourself and they want you to hire your replacement.

Or it could be that in addition to finding your replacement, at the same time they would like the renewal rate to go up. Some companies just want you to be able to keep the wheels on the bus and keep the status quo and just make sure the team's doing what they need to be doing, putting some process in place. But if they really don't want to skip a beat, you need to come in as a fully functioning executive and you are in this role until someone else is hired. You should expect to be participating fully in that executive role, attending executive meetings and possibly Board meetings, and working with the other functional leaders. It's important that you can build cross-functional relationships quickly and build credibility if you need buy-in from other departments. From my perspective, I need to be able to know what success looks like for the company and what it looks like for me to do my best job.

Should You Pursue a Fractional Role?

A fractional role is not the same as an independent contractor or a consultant because you're part of the organization, you interact with the executive team, and you have some independence in making decisions, some authority. I've found that people at a particular stage of their career are suitable for a fractional role, that they would like it and thrive in the fractional environment. If you're at a place where you don't want to make the commitment to the day-in, day-out of a long-term engagement (4+ years average for VPs), or maybe you don't want to deal with company politics, a fractional role could work. It's possible, sometimes, that as a fractional you can work a three-day work week for the same pay as a five-day, so that might work for you. Also, if you're at a place where you feel like you've got a lot of skills and you can come in and make change quickly and get people on the right track, that's the reason to be fractional. In the end, it's a hard job and you're not doing anything easier than if you were in a VP role. I treat my roles as if I was holding this title long-term, not as a contract position where I may not put my total self out there. There's a lot of commitment that comes along with that, and you don't get to reap a lot of the fringe benefits such as equity, vacation, and healthcare. For the right person, the benefits of a fractional role can give you a more manageable work week, freedom from the long haul commitment, more time with your family, a chance to grow and learn from being a part of various executive teams, while putting a few new things in your toolbox each time.

The biggest reason to choose to go fractional is to be at a place where you really don't want to make that long-term commitment. If you are giving it your all and doing all the stuff it can seem, like, why I am doing this when I can just get a full-time role with all those benefits? There are times when you're in a fractional role where you can stand back a little bit easier and not let the issues overwhelm you. There are times when it's not going well but as a fractional, it's a company problem. Another benefit of being fractional, besides not internalizing all the issues, is that you get instant credibility. When you're fractional you come in because you have this skill set and the expectations are high, but you have the ability to come in and make some really big changes quickly. You can help a company really get things back on track and then they're on to their next thing, their next challenge.

And you're also on to your next thing, as a fractional. You learn a lot in one company over a six-month period and then you go do it again in a different company. You learn a whole other way of doing things, of dealing with issues, and of interacting with different people versus being in the same company for three, four, or five years and only know a few ways of hands-on success.

Why Companies Need Fractional Executives

One of the main reasons why a company needs a fractional executive in the CCO slot is because they don't want to lose ground on customer satisfaction, onboarding success, renewals, or expansion while the team is without a leader. When you lose or replace that leadership at the top level on the CS side and you take three to six months (average) to find that right person, you can lose so much ground, especially if you don't have a senior member of the team that could drop into that role. Probably the biggest reason to hire fractional on the CS side is because there's a lot to lose it if you just put everything on ice until the next leader comes in, especially if there are churn issues, employee engagement problems, or process improvements needed to move the team forward.

I see the fractional role as a win-win for executives who have skills and a passion to make a deep impact quickly, and I see it as a win for companies that need skills, experience, and wisdom and don't have the time to lose ground.

Chapter 131

Fractional Chief Product/ Technology Officer

Drew Dillon

Getting Started in a Fractional Product Role

A lot of people in Product have a random resume where it's a little bit of sales, a little bit of engineering, a little bit of design, and then all of sudden you're in Product because if you average all those together you've basically defined "product." When I first spoke with CEOs I was often the first person that the CEO had ever met who did a product fractional job. They really had no idea what I did or how they could work with me; I was just a person that their VC told them they needed to talk to. For product and technology roles there are two hurdles to overcome because you have to sell your role and then sell yourself on top of it. The term "fractional" changed all that and takes a lot of guesswork off the table. Maybe a company doesn't need me five days a week, maybe just a couple of days a week. I'll come in, we'll work together, and maybe it will be a full-time gig, or maybe we'll get to this good spot, and I'll help them find a full-time leader.

I would suggest anyone considering a fractional role to not leave their full-time job until they at least have a couple of people interested in hiring them. Making fractional work sustainable is less about the first round of leads and it's really about the next round. So, planning for that period when that first set of contracts ends you need to be able to answer the question, how are you going to get that next set of folks? What's your unique value? And how do you brand what you do in a way that's really easy for a company to hire you?

I had a friend who was doing engineering and CTO work and he suggested starting by writing a blog post and say exactly what I am going to do. Then, when people ask, "What do you do?" I sent them the link to a fairly tactical blog post with a menu of items to choose from. That worked great for me, and later I followed up with a blog post on scaling product strategy which highlighted how I help companies at various points in their lifecycle. When you're first getting into this, it's nearly impossible to describe exactly what your benefit is going to be for a company and so writing the blog post forced me to be concrete and explicit. It also forced me to outline some broad challenges that startups and scaleups face, like every company at some point faces a gap in product leadership. That's something that resonates with lots of founders so from there I needed to help people see that I could not only fill that gap but help them accelerate through it.

What Does a Fractional Product Person Do?

There's probably a lot of ways to work with companies but I usually work in three areas. The first one is like a Think Tank where they will give me a lot of information, and I'll go conduct some research and come back with results, data, and my thoughts on their questions. The second way I work with companies is as a coach to the executive, and I will participate in their executive meetings and facilitate longer-term planning. And a third way is to help build a product team and help interview and recruit a full-time replacement. The third way is actually a true fractional role, and I lead the function as though I were a full-time employee and help hire my full-time replacement.

One thing that a fractional executive does, or should do, is to be honest about what they see. I charge by the project so I'm not obsessed with the hours—I don't need a company to keep me on for 12 months. So, I'm going to give them the best answer I possibly can to help their business because I feel that's ultimately what they're hiring me for. Even if it's tough advice, they're going to get that advice and they're going to hear the truth and hear what I think. Some people tell a client what they want to hear or they want to raise issues or problems so they can stick around and bill more hours. My perspective as a fractional is just the opposite: I give you the straight truth and my goal is to bring my best to the company so that they can move forward.

Advice for Companies

If you're considering a person to come in and help your company, one of the key differences between a fractional role and a consulting or advising role is that you really need the fractional person to own the function. While a lot of people might have the skills, a fractional person is someone who will make a much bigger impact in your company because they'll work closely with your team, technology, and markets. One way to screen for that is to look for curiosity. How curious is a person about the problem? Are they really getting in there and learning about your challenges? Are they really going to get their hands dirty and help you with the problem? Or do they give you solutions quickly, like they have a solution and they're looking for problems?

One thing I find in working with companies, especially startups, is that they can't pin down exactly what they need so they should think about precisely what they're looking for. If it has something to do with product and a company can't pin it down, they should at least be able to come up with some broad categories, like it's a strategic project, or it's writing specs, or it's someone to come in and focus the product team, or we need someone to help the executive and management teams. I worked with a company once where they said, "We don't know what we don't know. We don't know whether we need you to execute. We don't know whether we need you to tell us what we're doing. But what we want to know is what having somebody like you around will do for us. We need to know what having an executive product voice is within our company so we can know whether to hire a person." In that case, I helped them calibrate what level of help they needed. So, even if a company doesn't know what they need, a fractional person can still be helpful.

Another thing, and this might be specific to product or technology, is that we operate in a space that is unique to the core DNA of a company. I initially thought that there wouldn't be a need for a fractional executive because the founders or someone close to the business wouldn't see a need for help in this area, or that they wouldn't be open to an outsider coming in and understanding their core technology.

I haven't found that to be a problem and instead I get all sorts of input from founders as to how they feel about that. Some of them will come back with, "Well, we put a lot of effort into training people and we don't want to put that money into you because as a fractional you'll leave." That's a genuine concern but my usual response is that if a company feels like they need to put a lot of money into training me, then I'm not very good at my job.

The other thing that I wondered about was whether or not the existing team would connect with me, especially whether engineers would work with a product person. Would they see me as a threat or as a person who wanted to help them? From what I've seen, as long as the morale of the company isn't shot, as long as there isn't broad mistrust in the company, then people are going to give you the benefit of the doubt that you're being brought in as an expert. Obviously, your experience matters here because it gives you credibility in the game, but after the initial introductions, it's all about how you operate and work with the team. My first discussions with people are around the challenges they're seeing and as they describe them I can say, "Oh, I wrote something on that." And because I've written those blog posts and put my ideas out there, people know that I understand and care about their concerns. They will often say, "Yes, he really does get engineering, or data science, or he really does have a philosophy on product."

Final Thought

One other thing I haven't mentioned yet is that this is just a lot of fun. I was interviewing some candidates for a VP of Product for a company and at the end of the conversation I asked them if they had any questions and their main question was, "How do you get to do what you do?" That pretty much sums it up: interesting work, great people, new challenges, the ability to help people, and tons of fun.

Chapter 132

Fractional Chief Privacy Officer

Teresa Troester-Falk

Although I am an attorney by background, I started my career nearly 20 years ago as a "privacy compliance associate" at DoubleClick, really, at the very beginning of what has become a high-growth profession. Since then, I've practiced for large global corporations, been a chief global privacy strategist for a leading privacy software company and have worked with leading regulators on privacy and compliance across the world. I began working as a fractional executive about a year ago because the demand for expertise far outweighs the supply of experienced privacy professionals and the risks to companies in not getting privacy and compliance right can be disastrous.

One of the things I've noticed is that when it comes to compliance, I've seen too many companies, especially startup and fast-growth companies, create a check-list of do's and don'ts around privacy that inevitably end up on the shelf, or privacy software not being used or not optimized. In the early stages of a company's growth it's usually the IT team that has responsibility for privacy, but the privacy law landscape is constantly changing and the skills needed to keep up are beyond most IT teams. Finding a full-time privacy officer is out of the reach of a lot of companies since privacy and compliance professionals are in short supply and it's a seller's market.

A fractional privacy officer makes a lot of sense for many companies because that person can help ensure that you build privacy into your enterprise, product, and brand. This helps you go beyond strict compliance with the law, and helps you build trust with customers and gain a competitive edge. As you scale you can consider a full-time privacy officer but when you're starting out that's tough to do. The problem is, the privacy laws are not going away, they're not becoming simpler, and the consequences for violating laws are becoming more severe.

Qualities of a Fractional Privacy Officer

There's no one path to becoming an expert in privacy—some people have a law degree like me, while others have deep experience as practitioners and more of a technology background. But this role requires someone who has the ability to influence and guide leadership to focus and align on creating an infrastructure for privacy compliance, otherwise you're constantly putting out fires.

While a good fractional privacy officer will need to be persuasive, you'll also have to "roll your sleeves up" and become involved in creating, supporting, and executing the strategy. This is not the type of role where you can just "provide insights" or ideas, or even provide benchmarks or best practices. Fractional privacy officers need to be involved in the operations and execution as well.

Another quality that a fractional privacy officer ought to bring to their role is the ability to understand the business goals, the risk profile (including tolerance), and the resources. You won't have a blank check to create privacy and compliance protocols as you see fit and you'll more likely have to be scrappy and creative to get things done. Even if you have a big budget, you can still waste a lot of business resources if you don't clearly understand the company's objectives and risk, their desired business outcomes, and their timeline. So, a good fractional privacy officer will be able to balance privacy compliance requirements and goals along with the resources, business outcomes, and risk profile. Usually there is not a one-size-fits-all approach that will be effective and each company's issues and ways of interacting with customers will dictate how you'll solve the company's problems.

Of course, it goes without saying that as a fractional privacy officer you will need to hold yourself to the highest ethical standards. Trust is driving the next generation of growth and innovation, and you will be guiding the company not only to comply with privacy laws but also to create trust with their customers and other stakeholders—like consumers and regulators. Of course, you wouldn't choose the role of privacy officer if you didn't have a deep commitment to privacy, responsible use of personal data, and a belief in compliance and upholding regulations.

What a Fractional Privacy Officer Does

What role you play within a company will obviously differ from company to company, and it is highly dependent on the type of data the company processes. It is a constantly evolving space with new regulations and laws coming into force regularly. There are a few general areas that a fractional privacy officer would be involved in, regardless of the specifics of a company.

For example, a fractional privacy officer would need to have a deep understanding of the overall privacy initiatives and be able to manage the reporting of those initiatives. You'll also have to maintain and update existing data inventories and, probably, create new ones, since new regulations are constantly being implemented.

There is a whole body of work involving policies and procedures that a fractional will be responsible for. In some cases, you'll be at a company that doesn't have any policies or procedures so you'll have to start from scratch and not only create these documents, but make sure that they are understood and implemented. That could mean training sessions with employees or with a subset of employees. In other instances, companies will have these documents but you'll still need to understand whether they are current, whether they are compliant with new laws and regulations, and whether they are actually implemented at the company. Just because policies and procedures are in place doesn't mean that they are supported, implemented, or up-to-date.

A good fractional privacy officer will be up to speed on privacy laws like GDPR and CCPA and will have deep knowledge of pending laws, such as new US state laws. While some of these regulations and laws might not apply to a company today, they could come into play as new products and services are rolled out. This means that the privacy officer has to have a good working relationship with Product and other innovative teams because you'll need to make sure that privacy has been considered before a product is launched.

There are other tasks that may or may not be expected of a fractional privacy officer, like answering customer privacy questions or managing consumer rights requests. You may also set up and carry out ongoing privacy awareness and training or help to build a culture of privacy within the organization. Sometimes, as a short-term fractional, the best you can do is to provide training since cultural change often takes years to affect. But you can put in place the foundations of a privacy-aware culture and educate other leaders on the need to continue along that path.

Finally, a common role for a fractional privacy officer is to ensure that the company has a robust vendor management process to mitigate potential risk from third party use of data. Vendor processes are critical for many companies and as a company scales and gets more complex, developing these processes along with the checks and balances can be a monumental task.

Advice for Companies

Similar to other consequential business decisions the degree to which a company needs a full-time or fractional privacy officer needs to be understood within the context of your objectives. Are you comfortable with minimum compliance? Is it only a particular law or regulation that applies to your business? Or is privacy going to be a cornerstone of your brand, your culture, and your interaction with customers? The answers to those questions will help get you started on prioritizing what would be best for your company.

One mistake that I have seen, unfortunately, is that companies oftentimes under-invest in privacy compliance, and they do this by "promoting" an existing employee to privacy officer to handle these responsibilities. Although this person will know the ins and outs of your company, they won't necessarily have a working knowledge of privacy compliance and they may not have a deep understanding of the issues and regulations that apply to your situation. Your risk in promoting a person with little or no privacy compliance expertise is significant, and it's likely that you will miss important privacy-related milestones for your company. Privacy and compliance are complex topics and experienced privacy professionals understand the best practices, terminology, and use cases across industries. A fractional privacy officer can help mitigate risk in the short term and also provide guidance on helping you find a full-time leader for the role when you need it.

The final bit of advice is that data privacy is not a set-it-and-forget-it task. Similar to other parts of your business, the regulations and requirements for data privacy are constantly changing, they can be specific to verticals, they can be relevant at the state, federal, and country level, and they can make a big impact on your business. A fractional privacy officer can quickly bring you up to speed and put into place robust policies and procedures that will allow you to have a proactive, rather than reactive, approach to privacy.

Conclusion

One of the challenges in writing a "Field Guide" is that it could lack depth, it could just barely scratch the surface of the ingredients needed to scale all of the functions in a company. We wondered if it would be a mile wide and an inch deep. There's nothing wrong with providing quick, sound-bite-sized advice and learning, but that type of information is more useful for very early-stage startups, where there is a founder or two and they have an idea but not much else.

For the early-stage startup the first part of the epigraph we opened this book with makes total sense:

If you want to go fast, go alone.

The early stages of creating a company are about speed, about making mistakes, about changing direction, about creating the minimum viable product as quickly as possible, about learning if you have any customers and figuring out whether they'd buy anything you've created. Early-stage tasks involve a lot of trial and error and experimentation.

But after that early stage, then what? That's where the second part of the epigraph comes into play:

If you want to go far, go together.

That's why we wrote this book, to help the entrepreneurs who have something of value but can't scale. It's written for the team that wants to go far and needs to figure out how to manage all the complexity and uncertainty to do that. Maybe what's holding your startup or scaleup back is specific functional advice from people who have been there before.

Our experience is that many startups and scaleups struggle not because of a lack of effort, but because they're learning about themselves and their business as they go. They're "flying the airplane as they're building it." As a startup or scaleup you're usually seeing something for the first time, or maybe you don't know what you're seeing, or maybe you don't know what to look for in the first place. But without a clear-cut sense of what's going on, or what to look for, you end up making mistakes, taking the wrong path, making the wrong hire, or trying a myriad of other "solutions." But in the end, you're still no closer to scaling than you were before.

We've been there and each of the contributors to this book have been there—they've been staring at a problem they've never seen before and neither they nor anyone else at the company has the experience or understanding to step in and help. That's where *Startup CXO* is most valuable to a startup or scaleup because now you can consult the specific function when you're seeing a problem for the first time and now you're provided with ideas on what worked and what didn't work in solving those problems. And not just any ideas, or "thought leader" ideas, but ideas from people with deep experience in startups and scaleups.

It would be impossible in this conclusion to provide some overarching takeaways for the reader because each functional area is different, has different challenges, and was written by a different contributor. What we do hope you take away from the book, though, is a greater understanding of all the functional areas and a renewed sense of purpose to apply the lessons here to your own situation.

After all, you …

Haven't come this far, only to come this far.

Epilogue

Pete Birkeland

In the fall of 2019, the manuscript for *Startup CEO* Second Edition came across my desk and I was supposed to edit it and work with the author, Matt Blumberg, on revising it and adding some new material. I had never heard of Matt Blumberg and hadn't read *Startup CEO,* but I gave it a quick read and came back with the thoughts, "this is pretty comprehensive" and "at over 370 pages and 50+ chapters, is anyone really going to slog through this?" I put it on hold until 2020.

I returned to Matt's manuscript in January, 2020, and read it multiple times and I changed my mind about the book. It *was* comprehensive (and the new material on selling a company—successfully—added important information), but it wasn't a slog to read. It was well-written and engaging. And I realized that there were so many insightful ideas in the book that the audience who would benefit from reading *Startup CEO* is much greater than entrepreneurs, founders, and people in the startup community. Anybody—from the youth soccer coach to the pastor to the law enforcement captain—could easily read *Startup CEO* and learn how to be better as a leader or as a team member.

I sent an unsolicited email to Matt sometime in the winter of 2020: "I would have loved to have worked with someone like you during my career." I meant it, and that's not because I've worked with capricious, ego-centric people (I've had my share of those, too), but because there's an authenticity that resonates from the pages, there's a sense that Matt deeply cares about helping people who are facing challenges in their business. What I didn't realize until I re-read *Startup CEO,* is that the book really speaks to organizational life: How can people make any organization they're involved in better? What do they need to do to turn things around, to move forward, to make an impact, to realize their vision?

I was delighted when Matt asked me to work with him on *Startup CXO,* and I was happy to meet and work with the other collaborators. The content is different but my thoughts are the same: I would have loved to have worked with any one of the contributors at some point in my career. That wish actually came true for me through writing *Startup CXO,* but I hope people who read this book can emulate the authors and raise the level of engagement and decision making in their own organization.

References

Bennett, N. and Miles, S. (2006). Second in Command, the Misunderstood Role of the Chief Operating Officer, *Harvard Business Review*, 84(5), 70–78.

Blank, Steve (2013). *The Four Steps to the Epiphany*, 5th ed. New York: K & S Ranch.

Blumberg, Matt (2020). *Startup CEO: A Field Guide to Scaling Up Your Business*. 2nd ed. New York: Wiley.

Cagan, Marty (2018). *INSPIRED*. Hoboken, NJ: Wiley.

Catmull, Ed (2014). *Creativity, Inc.* London: Transworld.

Christensen, Clayton M. (2016). *The Innovator's Dilemma*. Boston: Harvard Business Review Press.

Feld, Brad and Jason Mendelson (2019). *Venture Deals: Be Smarter Than Your Lawyer and Venture Capitalist*, 4th ed. New York: Wiley.

Lencioni, Patrick (2002). *The Five Dysfunctions of a Team*. San Francisco: Jossey-Bass.

Lencioni, Patrick (2004). *Death by Meeting: A Leadership Fable about Solving the Most Painful Problem in Business*. San Francisco: Jossey-Bass.

Lencioni, Patrick (2012). *The Advantage: Why Organizational Health Trumps Everything Else in Business*. San Francisco: Jossey-Bass.

Maurya, Ash (2012). *Running Lean*. Boston: O'Reilly.

Moore, Geoffrey (2011). *Escape Velocity*. New York: Harper Business.

Patton, Jeff (2014). *User Story Mapping*. Boston: O'Reilly.

Ries, Eric (2011). *The Lean Startup*. New York: Penguin.

Acknowledgments

I'm uncomfortable with being one of multiple authors in a collaborative effort and to be the only person writing the acknowledgments section of *Startup CXO* because this isn't MY book, but the TEAM's book. It was a truly collaborative and collective effort.

The list of people to thank for their role in the development of this book is long and has to start with my current and former colleagues who were the primary contributors to the book: Jack Sinclair, Cathy Hawley, Shawn Nussbaum, Ken Takahashi, Nick Badgett, Holly Enneking, Anita Absey, George Bilbrey, Dennis Dayman, and Dave Wilby. I could never have pretended to write this book on my own, and although I conceived of it, wrote a chunk of it, and helped edit it, it's a truly collaborative work by all of us. Pete Birkeland, who edited the second edition of *Startup CEO,* was a tireless collaborator across all sections of the book to make sure it came out well-written, balanced, and consistent.

I would also like to acknowledge the rest of the Bolster team that made this possible—our founding board members Scott Dorsey and Melody Dippold, and our two other co-founders Jen Goldman and Andrea Ponchione. It wasn't easy writing a 180,000-word book in the midst of the hectic first six months of a new startup, not to mention a pandemic, but with everyone's support, we got it done!

Although the professional lives of the contributors are now primarily at Bolster, most of us worked together for many years at Return Path, and all of us would like to thank our Board and shareholders, particularly Fred Wilson, Greg Sands, Scott Weiss, Scott Petry, Jeff Epstein, and Brad Feld (more on Brad in a minute) for giving us the opportunity to learn on the job as we scaled ourselves and scaled the business over the better part of two decades. That experience is what led us to be able to write *Startup CXO.* We'd also like to thank all 1,300 colleagues from Return Path over the years who challenged, inspired, and taught us things every day—including a number of key executives who could easily have been primary contributors to this book (you know who you are!). Although he was not a Return Path or Bolster team member, Marc Maltz from Hoola Hoop Consulting, my long-time partner as an executive coach, has shaped the thinking of me and of a number of the contributors to this effort.

Startup CXO is part of the Startup Revolution series that was created by my long-time board member and friend Brad Feld. Brad's advice on all things business, personal, and writing has been invaluable for over 20 years and whether attributed or not, many of the ideas in this book are the result of many thoughtful conversations with him. I would also like to thank the team at Wiley (Bill Falloon and Purvi Patel) and at Techstars, including Carlos Almonte, for their help and support as editors, publishers, and marketers.

This book had a very large number of people who contributed their insights to the final form including sidebars by Rob Krolik and Jeff Epstein, Guy Turner, Greg Sands, Scott Petry, Brad Feld, Dave Wilby, and Scott Dorsey. We are also grateful for the contributions to our fractional section from John McCarthy, B.J. Bushur, Teresa Troester-Falk, Scott Kabat, Amy Mustoe, Sherri Sklar, Drew Dillon, Jon Guttenberg, and Courtney Graeber.

We received a number of thoughtful comments on specific functional areas from Rick Buck, Caroline Pearl, Diana Caleroni, Jen Goldman, Mike Mutone, Debby Meredith, and Chad Shinsato. Brad Feld and Scott Dorsey did a final read-through of the entire book (not a small feat!) and provided helpful suggestions to the final work.

I want to end by thanking my family for their unwavering support as I embarked on a second book while launching a second startup—a combination that I can't exactly endorse as being sane or smart. *Startup CXO* was a collaborative effort but I would have been the anchor holding us back if it weren't for my wife of over two decades, Mariquita. My thanks start and end with her. An executive coach for startup CEOs, Mariquita has been intimately involved in all my professional projects, providing advice, encouragement, and support to whatever I'm doing.

Matt Blumberg

About the Authors

Matt Blumberg. Matt has spent his entire career creating startups, scaling them, and sharing best practices of what works and what doesn't work for other CEOs and team members in the entrepreneurial community. He is the author of *Startup CEO: A Field Guide to Scaling Up Your Business* (Wiley, 2020), a highly influential book embraced by entrepreneurs, CEOs, founders, and board of directors in the entrepreneurial ecosystem that was an outgrowth of his blog, StartupCEO.com. In 1999, he founded Return Path, an innovative email marketing company, and helped grow it to $100m in revenues and to a successful exit in a strategic sale to Validity in 2019. Along with a number of his colleagues from Return Path, Matt started Bolster in 2020, a company focused on helping startups and scaleups grow, develop, and scale their leadership teams and boards. Bolster's unique combination of heavily curated executive talent, tools, and services provides a new way for CEOs to assess and benchmark their executive teams and boards and provides a new avenue for on-demand executives to work with startups and scaleups.

Before Return Path, Matt led Marketing, Product Management, and the Internet Group for MovieFone, Inc. (later acquired by AOL). Prior to that, he served as an associate with private equity firm General Atlantic Partners and was a consultant with Mercer Management Consulting. He also co-founded and chairs the board of Path Forward, a non-profit created and spun out of Return Path. Path Forward's mission is to empower people to restart their careers after time spent focused on caregiving by working with companies offering mid-career internships. Path Forward gives women and men a path to a professional career, while giving companies access to a diverse, untapped talent force. Matt earned his A.B. from Princeton University.

Peter M. Birkeland. Pete is a sociologist, author, and collaborative writer. He has spent his career spanning research and business roles, as Research Director at Accenture and research associate to Gary Hamel, to CEO of a steel fabrication company, to Senior Editor at Techstars. Pete is the author of *Franchising Dreams: The Lure of Entrepreneurship* (University of Chicago Press) and *Legal Cannabis: The Great Social Experiment* (forthcoming, 2021). As a collaborative writer Pete has worked with a number of leading business and entrepreneurship authors including Matt Blumberg (*Startup CEO* and *Startup CXO*), Brad Feld and David Cohen (*Do More Faster*), Will Herman and Rajat Bhargava (*Startup Playbook*), Claudia Reuter (*Yes, You Can Do This*), and others. Pete received his Master's and PhD in economic sociology from the University of Chicago, focusing on organizational behavior and quantitative methods.

Contributors

Anita Absey. Anita began her career on Wall Street in equity research, risk arbitrage, and corporate finance with Credit Suisse and Deutsche Bank, but for the past 20 years has had wide-ranging roles in sales. Anita has been in sales leadership positions in startups and scaleups including Abacus Direct, DoubleClick, Return Path, Voxy, and LRN. Anita views her success through the lens of revenue growth and customer satisfaction and her greatest satisfaction comes from watching people on her teams thrive in their careers. She earned a BA in economics at Fordham University.

Nick Badgett. Nick is an experienced marketing leader with more than 15 years of experience working in both startup and enterprise environments across the B2B SaaS, Telecom, and financial services industries. His approach to marketing has been shaped by experience in a variety of go-to-market roles including sales, sales support, and marketing. Before joining Bolster, Nick held marketing leadership positions with Emplify, Return Path, Salesforce, and ExactTarget. Nick holds a bachelor's degree in marketing and a master's degree in information & communication sciences, both from Ball State University.

George Bilbrey. George is an experienced technology executive with over 20 years of experience running Product Management, Analytics, Service, and Customer Success organizations.

Most recently he was co-founder, President, and Chief Customer Officer at Return Path and currently is the CEO at Signpost. George started his career at Mercer Management Consulting (now Oliver Wyman). George earned an BA from Duke University and an MBA from the University of North Carolina.

Dennis Dayman. Dennis has over 25 years of experience in security, privacy, and governance issues. He has worked with startups, scaleups, and global organizations to protect and improve data for consumers and companies. He is actively involved in Internet and digital communication regulations, privacy/security policies, and anti-spam legislation laws for state and federal governments. Dennis was appointed by the Department of Homeland Security to the Data Privacy and Integrity Advisory Committee (DPIAC) to provide input on policy, operational, technical, and other privacy-related matters. He is a graduate of Stephen F. Austin State University with a degree in criminal justice.

Holly Enneking. Holly is an experienced marketing leader who believes in the power of storytelling and finding innovative ways to communicate a brand's value through cross-channel experiences. Holly is currently the Vice President of Marketing & Alliances at Lev, and thrives on amplifying the Lev brand and helping fellow marketers understand how a consultancy like Lev can help them maximize their investment in Salesforce. Prior to joining Lev, Holly began her career in video and digital production before being a senior leader at Return Path and overseeing brand, digital, customer, and partner marketing. She earned a BA in communication from DePauw University.

Cathy Hawley. Cathy is an experienced fractional and full-time Chief People Officer passionate about building highly effective leadership teams and creating inclusive and people-focused workplaces. She has more than 20 years of experience in a variety of industries developing global People and Culture teams. At Return Path, she spearheaded the "return to work" program, spun it out to a non-profit, and is on the Board of Path Forward, a non-profit that helps companies create return-to-work programs for caregivers. Cathy earned a BS in business administration from the University of Colorado and a master's in European HR and Industrial Relations from Keele University.

Shawn Nussbaum. Shawn is a technical leader and consummate problem solver with over 25 years of experience developing software and leading high-performing teams. Shawn spent the last decade at Return Path where he led Product and Engineering and was responsible for setting the vision and culture of the organization, extracting meaning from email data, and applying technology as a force-multiplier to solve interesting and difficult problems for customers. Prior to Return Path, he helped launch two Internet companies and was an advisor and technical director for a consortium of insurance companies. He believes that empathy and usefulness are the keys to remarkable products.

Jack Sinclair. Jack is an experienced Chief Financial Officer and Chief Operating Officer with over 20 years in startups and scaleups, including Return Path, Stack Overflow, and Bolster. Jack began his career at Stern Stewart as an analyst but quickly figured out he enjoyed startups and scaleups and has since co-founded Return Path and Bolster. Jack earned a BS in finance from the Wharton School.

Ken Takahashi. Ken has spent more than two decades in the B2B SaaS space helping companies and industries grow. His career has taken him from product management to pre-sales consulting and industry relations, and is now squarely centered on business and corporate development. Ken has significant international business experience, launching offices across Europe, Latin America, and Australia. He serves on industry boards and committees and is a mentor and advisor to technology startups around product-market fit, geographic expansion, and exit strategy and execution. He earned his BS in business from the Leonard N. Stern School of Business at New York University.

Connect with author
Matt Blumberg, read his blog, learn about current trends and best practices at **StartupCEO.com**

Access more books on entrepreneurship
and connect with authors Brad Feld, David Cohen, Jason Mendelson, Matt Blumberg, Amy Batchelor, Mahendra Ramsinghani, and Sean Wise at **StartupRev.com**

Also look for *Startup Boards, Second Edition,* by Brad Feld, Mahendra Ramsinghani, and Matt Blumberg published in the Fall of 2021 by Wiley.

Index

1Password, usage, 93
8seconds, usage, 333
30-60-90-day, goals/plan, 439–440, 450–451
90-day reverse review process, 150

Access, considerations, 515
Accountability, empowerment, 455
Account development representative (ADR), 226
Accounting, 45–48, 52, 88–89, 99
Accounting Standards Committee (ASC) 606, accounting perspective, 46–47
Account management (strategic account management) role, 364
Accounts Payable (AP), 52, 53, 94, 97, 100
Accounts Receivable (AR), 52, 53, 97, 100
Acquisition, 468, 473–474
Activation (customer journey stage), 369, 376
Activation phase metrics, 376
Activity reporting, outcome reporting (contrast), 246
Actual *vs.* Budget (AvB) process, 68
Adventure game, creation, 403
Advertising, 184, 199–201, 230–232
Agencies, 230–232
Agile, 432, 455, 463–464
Airtable, usage, 430
Always be optimizing (ABO), 200
Analytics, solutions, 184
Annual planning framework, adoption/modification, 124–125
Annual pre-paid/liabilities, accrual (accounting perspective), 46
Applicant tracking system (ATS), 164
Architecture Council, creation (benefits), 421, 426
Artifacts, usage, 423
Asana, usage, 163
Assessments, 529
Assets, product marketing creation, 219–220
Assets under management (AUM) fee, absence, 96
Associate Software Engineer, role/responsibility, 448
Attitude metrics, usage, 374–378
Audition, 435–437, 443
Audits, 419, 531, 533–534
Average price (financial statement component), 31

Balance sheet (financial statement component), 27–28
Best action, usage, 196–197
Best case scenarios, assessment, 280
Bias, removal, 445
Big Data, importance, 497
Billing software, usage, 43
Binding Corporate Rules, 506
Board Books, 81, 122
Board of directors, 36, 77–78, 81–82, 482
Bolster!, 102, 109, 179f, 182, 312
Bootcamps, 441, 450
Boundary spanner, 318
Brainstorming, 194
Brands, 178–180, 191–195, 202–203, 212
Budget (CFO involvement), 67–68
Budget agreement, 389
Budgeting, CFO involvement, 67
Bug fixes/troubleshooting, 389, 418, 484–485
Build-Measure-Learn loop, 429, 430
Business Development (BD), 330, 348, 350
Business development representative (BDR), 226
Businesses, 20, 21t, 24, 98, 349, 412, 458, 480, 486, 584
Business-to-business (B2B) issues, 199, 288, 296, 303–304, 571
Business-to-consumer (B2C) issues, 200–201, 244, 246
Buy-side (M&As), 85–86, 341

Calendars, 92, 95
Calendly, usage, 95
Call Analytics, 361
Call To Action (CTA), 195–197, 200
Cambridge Analytica, 491, 511
Campaigns, execution, 237–240
Candidates, 437–438
Capability, absence, 458
Capitalization Expenditure (CapEx), 100
Capitalization (cap) table, 36, 97
Careers, 132, 187, 435, 446–448
Carta, usage, 94–95

Cash issues, 27–28, 49–51, 57
Centralized dashboard, KPIs (inclusion), 239f
Certifications, 527–528, 531
Channels, 60, 237–240, 289–290, 330336
Chief Business Development (CBD) lead, 319
Chief Business Development Officer (CBDO), 306–309, 311, 314, 336, 349–351, 580
Chief Customer Officer (CCO), 233, 354–357, 360, 361, 387, 391, 396–398
Chief Executive Officer (CEO), relationship, 323
Chief Financial Officer (CFO), 16, 19, 41, 64–66, 71, 98–100, 150, 168, 234–235, 556
Chief Human Resources Officer, 564
Chief Marketing Officer (CMO), 174, 243–247, 282, 567–570
Chief Operating Officer (COO), 542–549, 544t
Chief Operations Officer, reporting, 522–523
Chief People Officer, 102, 110, 167–171, 562
Chief Privacy Officer (CPO), 488–492, 523–524, 536–539, 594
Chief Product Officer, 402, 409–411, 482–486, 590
Chief Revenue Officer (CRO), 250, 284–285, 300–304, 388
Chief Technology Officer (CTO), 402, 406, 483–486, 502, 590
Children's Online Privacy Protection Act (COPPA), 498–499
Churn, service organization control, 357
Churn Zero, usage, 381
Clients, 89, 218, 345, 368–378, 380, 388–389, 581–582
Cloud services, usage, 508
Code troubleshooting, 484–485
Commission rate, 263
Communication, 83, 91–93, 128, 188–189, 212, 537–538
Companies, 68–69, 83, 106–107, 490, 522, 543–545, 563, 570, 592–593
Company culture, 186, 213
Company sale, 475–479
Compensation, 152–153, 252–265, 391–392, 484, 581–582
Competitive intelligence, usage, 183
Consumers, respect, 517
Content, 197, 212–217
Context, 457–458, 569
Contracts, 500–501
Contributor/manager, balance, 451
Controller, hiring, 52
Conversion rates, 185
Core business process, development, 304
Corporate areas, due diligence critical request, 88
Corporate Development, 325, 345, 348
Corporate documentation, 37, 38
Corporate information technology (IT), CFO/CEO management/partnership, 69
Corporate law, specialization, 492
Corporate social responsibility, CEO/founder perspective, 107
Corporate strategists, impact, 329
Corporate systems, CFO/CEO management/partnership, 69
Corporation organization, guidance, 55–56
Cost of goods sold (COGS), usage/importance, 47–48
Costs (financial statement component), 27
COVID-19 pandemic, 91, 163, 204–205, 437, 489, 573
Cross-channel campaign, 238
Cross-company launches, 463
Cross-departmental collaboration, marketing role, 218
Cross-functional planning sessions, quality, 467
Cross-sells, 392–393
Culture, 103–106, 111–112, 186, 213, 241, 271, 424, 435, 519–520
Current business status (pitch deck section), 24
Customer acquisition cost (CAC), 30, 62–63, 97, 185
Customer journey, 369–371, 388–389
Customer lifetime value (LTV), 30. 62–63
Customer Management System (CMS), solutions, 184
Customer Resource Management (CRM), 37, 72, 75, 95, 184, 210, 381–382
Customers, 30–31, 97, 177, 277, 351, 354, 360, 368–374, 380, 394, 399, 428, 485
Customer Service (CS) issues, 355–356, 361–365, 370, 379–384, 391–393, 396, 397
Customer Service Manager (CSM), 363–364, 366–367, 383, 586
Cybersecurity, 531–533

Data, 210–211, 244, 432–433, 476, 480, 494–495, 498, 507–509, 533
Data breach, 510–514
Data privacy/security, 88, 497–498, 531–532
Debt after financing, consideration, 50
Decision making, CEO/founder perspective, 106
Defections, prevention (service team focus), 358
Demand, content driver, 212–213
Developing contributor, subtractor role, 454
Development culture, shift, 424, 426

Development team, product leadership team alignment, 421
Digital agencies, usage, 230–232
Digital marketing, 195–203
Digital technologies, impact, 490
Direct sales channels, usage, 60
Disability leaves, complexity, 161
Distribution channels, expansion, 288
Diversity, Equity, and Inclusion (DE&I), 111–112, 127, 131, 145
Diversity, impact, 107, 442–443
Do Business As (DBA) filings, 88
DoubleClick, 119, 315, 337, 594
Dropbox, files (storage), 508
Due diligence, 37, 88–90, 468, 469, 480, 531–533

Early-stage acquisitions, 474
Early-stage startups, 499, 599
Earnings before interest, taxes, and depreciation (EBITD) (financial statement component), 27
Earnings before interest, taxes, depreciation, amortization (EBITDA), 34
Ecosystem, 317, 342, 350
Email Genius Academy (Return Path internship), 443–444
Emails, 92, 197–198, 498, 511–512
Emotional intelligence, 128
Emplify, usage, 136
Employees, 38, 81–83, 136, 140–141, 150, 156–162, 189–190, 372, 453–454, 491–492, 519–520
Employee Value Proposition (EVP), 147–148, 152–155, 160
Encryption, 476
End-to-end client experience, 387–389
End-to-end security, 496
Engineering, issues, 413, 448, 482, 484, 513
Engineering Leadership Playbook, 451
Engineers, issues, 448, 453, 484, 496
Entrepreneurs, market entry, 496
Equity, 80–84, 94–95
Ethnographies, creation, 373
European Commission, data protection level, 495
European Union (EU), 489, 505
Events, 204–211
ExactTarget, 221, 320
Exchange, usage, 92
Executive Business Reviews, design, 370
Executive Committee (EC), GTM subcommittee, 314
Executive gig economy, 553–555
Executive sponsorships, usage, 372
Executive team, impact, 317, 397
Executive work, future, 575
Expense management, Expensify (usage), 94
Expensify, usage, 94
Extended parental leave (employee benefit), 159
External counsel, usage, 40
External influence, 318, 323
External representatives, compensation, 74
External sales organizations, impact, 330–331
External trust, development, 327

Facilities, CFO/CEO management/partnership, 69–70
Fairness opinion, sell-side consideration, 90
Federal Trade Commission (FTC), 504, 510, 532
Feedback, 128, 147–148, 234, 282, 299, 373, 380
File storage/management, Google Drive Enterprise (usage), 92–93
Financial Accounting Standards Board (FASB), 46
Financial documents/analysis, importance, 38–39
Financial models, 87–88, 107
Financial Planning and Analysis (FP&A) reports, completion, 44
Financial plan/planner, 27, 572
Financials (pitch deck section), 24
Financial statement, year-to-date (YTD) income statement (inclusion), 27
Financial systems, 75, 94
Firewall, usage, 506, 514
Fixed assets, capitalization, 45
Fixed costs, understanding, 472
Fixed scope, fixed date (contrast), 464
Flag planting, 339–340
Flow of funds analysis, sell-side consideration, 89
Forecasting, challenge, 266
Forrester (analyst firm), 220–221
Fractional approach, reason, 589
Fractional CBDO, 348–349, 580–585
Fractional CCO, 397, 586–588
Fractional CFO, 97–98, 168, 556–560
Fractional Chief People Officer, 562
Fractional Chief Product Officer, 482–483, 590
Fractional CMO, 567–570
Fractional CPO, 536–537, 594–598
Fractional CRO, 301, 571–579
Fractional CTO, 483, 590
Fractional executives, 555, 572–573, 578–579, 582, 589
Fractional HR issues, 562–566
Fractional product, 590–592
Fractional role, 554, 568, 573–574, 588–589, 592–593
Friction, approach, 460
Front-line manager, placement/promotion, 450–452
Full-time manager, role, 451
Functional areas, push/pull, 459–460

Fundraising, 22–24
Funnel team, increase, 434, 442

Gated content, ungated content (contrast), 216–217
General accountant, hiring, 52–53
General Data Protection Regulation (GDPR), 197, 480, 489, 495, 505, 516, 527, 538, 596–597
Generally accepted accounting principles (GAAP), 16, 45, 47
Geographic expansion, 291–292, 338, 339
Glassdoor, usage, 147, 187
Global Account Management role, creation, 270
Global expansion, impact, 240–241
Good-enough solutions (product development culture), 414–415, 495
Good Standing certificates, usage, 88
Google Analytics, usage, 501
Google Drive Enterprise, usage, 92–93
Google Search, usefulness, 146
Google Sheets, 432
Google Translate, usage, 240
Gopher internship work, 444
Go-To-Market (GTM), 246, 314–315, 463–464
Grants, tax implications, 80
Gross margin (financial statement component), 27
Gross Retention Rate (GRR), 31

Hackathons, benefits, 426
Head of Partnerships, collaboration, 233
Health Insurance Portability and Accountability Act (HIPAA), 490, 505–506, 536
Hierarchy, CEO/founder perspective, 106
Hippocratic Oath, 489
Hiring, issues, 149–150, 254, 256, 258, 435, 452–454
Historical data, usage, 206
Homogeneous teams, collaboration ease, 442
HubSpot, usage, 93, 95, 380, 432
Human resources information system (HRIS), 20, 95, 164
Human resources (HR) issues, 70, 88, 155, 165–167, 491–492, 508, 516, 562–564

Ideal customer profile (ICP), 388–390
Impedance mismatch, 463–466
Inbound channels, 332–334
Inbound sales representative (ISR), 226
Inbound teams, marketing resource leverage, 227
Incentive compensation, 262–265
Incentive stock option (ISO), 83
Inclusive culture, building, 103–104
Incremental development, operational cost component, 418
Indeed, usage, 147
Indemnification, sell-side consideration, 90
Indirect sales channels, 60, 349
Industry analysts, leveraging, 220–221
Industry-specific content, library (creation), 216
Influence, 314, 318, 323, 325, 329, 330
Information Security Management System (ISMS), 530
Information technology (IT) issues, 69, 161
In-house accounting team, building, 52
Initial data team, building, 432–433
Initial public offerings (IPOs), 16
Input metrics, usage, 374–378
Inside sales function, recruitment, 573
Instagram, issues, 198, 201, 490
Insurance issues, 96, 160
Intellectual property (IP), 38, 56, 471–472, 475–476
Intercompany documentation, completion, 56
Interest to order (high impact area), 59, 75–76
Interim role, 554–555
Internal communications, usage, 83, 188–189
Internal influence, 314, 323
Internal privacy policy, 502–503
Internal promotion, bias, 148
Internal referrals, usage, 146
Internal representatives, compensation, 74
Internal systems, 91, 96
International Accounting Standards Board (IASB), 46–47
International Association of Privacy Professionals (IAPP), 527
International data privacy agreements, 480
International funding, consideration, 49–50
International operations, 55
Internship, 443–444
Interviewing, 434, 435–438, 443
Intrusion detection system, 506, 514
Investment bankers, due diligence consideration, 87
Investor ecosystem research, 32
Investor negotiation, transaction items (consideration), 35–36

Job-specific training, 150
Just-in-time privacy notice, 502

Key performance indicators (KPIs), 16, 30, 68, 230, 238, 239f, 389, 423, 458, 556
Kickoff meetings, design, 370
Knowledge management, 364, 397
Knowledge team, help center maintenance, 355–356

Lead development representative (LDR), 226
Leadership, 120–121, 234–235, 439, 453
Leadership Development, 115, 127, 228–229, 450
Leaders, issues, 449, 484
Lean Canvas, 427, 429
Learning, applications, 430–431, 451, 486
Least privileged access, 515
Least privileged systems, impact, 517
Leave of absence practices, complexity, 161
Legacy systems, usage, 420
Legal agreements, due diligence critical request, 88
Legal department, CFO involvement, 63–64
Legal documents, 162, 492, 500
Legal risks, 471–472
Letter of intent (LOI), usage, 88
Licenses, tracking, 475–476
Lifetime value (LTV), 30, 62
Limited partners (LPs), money (raising), 32
LinkedIn, 146, 201, 283, 507
Liquidation preference (transaction item), 35
Local partners, finding, 55
Lone Wolf, salesperson success myth, 260, 274

Major expense categorization, accounting perspective, 47–48
Malicious software, installation, 513
Marketing, 174–177, 184–192, 214–228, 236–237
  alignment, 282
  CFO partnership, 62–63
  content, delivery/creation, 197, 215–216
  digital marketing, responsibilities, 195–203
  function, analysis, 191
  initiatives ROI, CFO involvement, 63
  optimization, importance, 200
  partnering/collaboration, 233
  privacy team member, assigning, 521
  reach, broadening, 240–241
  regulations, knowledge, 197
  role, CEO-to-CEO advice, 243
  sales, combination, 207
Marketing/product/sales triangle, management, 543, 545
Marketing qualified leads (MQLs), 185, 210
Markets, issues, 182–183, 286, 291, 338, 344
Measurement, marketing operations component, 223, 224
Mental math, usage, 28, 100
Mentoring relationships, 450
Merger model, 85–86
Mergers and acquisitions (M&A), 85–91, 312–313, 326–327, 341–345, 348–350, 532, 580
Metrics, 184–185, 374–378, 556
Minimum viable product (MVP), 182, 409, 429–431
Monte Carlo analysis, usage, 73

National Center for Women in Information Technology (NCWIT), 112
Net Promoter Score (NPS), 364, 374, 378–379, 396–397
Net retention rate (NRR) (net revenue rate), 30–31
New debt/equity, approval (transaction item), 36
Nondisclosure agreement (NDA), usage, 63, 88
No-negotiation policy, 153
Nonqualified option (NQO), 83
Non-qualified stock options, incentive stock options (contrast), 80
Non-structured formats, usage, 346

Objectives and Key Results (OKRs), frameworks, 543
Office management, 162–163
Office manager, promotion, 114
Omni-channel approach, 201–202, 237–238
Onboarding, 149–151, 306, 375, 393, 397, 434, 439–440, 502
Onboarding Development, 115
Onboarding team, 149–150, 362–363
One-on-ones, 452
Ongoing maintenance (operational cost component), 418
Open vacation (employee benefit), 156–157
Operating manual, discussion, 439
Operating margin (financial statement component), 27
Operating role, CEO-to-CEO advice, 549
Operating systems, 118, 122–123, 127, 279, 325
Operational accounting, 42
Operational costs, components, 418–419
Opportunity, pitch deck section, 24–25
Options, exercise (process/timing), 80
Order to cash (high impact area), 75–76
Organization for Economic Cooperation and Development, 488
Organizations, CCO partnering, 387
Outbound channels, 330–332
Outbound teams, marketing resource leverage, 227
Outcomes, 374–375, 425
Out-of-the-box thinking, 486
Outside advisors, CBDO over-reliance, 350
Outside agreements, importance, 39
Outsourcing, CEO/founder perspective, 107
Over-engineering style (product development culture), 413

Packaging, issues, 294
Paid time off (PTO), 128, 156

Paradigm shift, challenge, 485–486
Parental leave, complexity, 161
Participating preferred (transaction item), 35
Partnerships, 335, 337, 580
Path Forward, 444–445, 486
Payment Card Industry (PCI), 505, 536
Pay-per-click (PPC), 199, 200
Payrolls, 95, 160–161
People, 102–110, 161–163, 167, 254, 382–384, 435, 446, 489
People Business Partner (PBP), 115, 128
People Operations, 115, 149, 154
People Pact employee handbook, 162
Performance Aspire (annual RP event), 139, 272
Performance, issues, 130, 220–221, 440
Personal data issues, 495, 497
Personal identifiable information (PII), 492, 523, 537
Personal Information Protection and Electronic Documents Act (PIPEDA), 527
Personal privacy rights, idea (understanding), 489–490
Pipeline, challenge, 266
Pitch deck, 24
Post-event, 210
Post-termination exercise plan (PTEP), 81
PowerPoint, usage, 251–252, 278, 282, 300–301
Pre-event, 207–209
Preferred investors, 409a waivers, 83
President's Club, 139, 272
PricewaterhouseCoopers, 571
Pricing/packaging, CFO involvement, 62
Principal Software Engineer, role/responsibility, 448
Privacy Council, 520–521, 524, 534
Privacy, issues, 476, 488–492, 494–496, 500–504, 523–526, 534–536
Problem interviews, 182, 428
Problem-solving audition, importance, 435–436
Processes, impact, 68, 98, 144, 223–224, 276, 382–383, 477
Product, 392–393, 418–421, 425–430, 477–478
  client-facing product collateral, creation, 218
  configuration/connection, onboarding team assistance, 355
  creation, 483
  due diligence critical request, 89
  efficacy, 470
  gaps, service (impact), 358
  improvement, RP commitment, 497
  launches, challenges, 464–465
  management, engineering (push/pull), 406
  managers, strength/ownership, 460–462
  market assessment/alignment, 286
  naming, consistency, 58
  offerings/innovation, delivery, 218
  organization focus, 388–389
  pitch deck section, 24
  pricing/packaging, issues, 294
  roadmap, 486
  role, CEO-to-CEO advice, 482
  storyboarding, StoriesOnBoard (usage), 93
Product Council, creation, 465–466
Product Development, 405, 407t–408t, 412–415, 414f, 420, 425, 456, 460, 481
  organization, leading/unification, 405, 459–462, 480–481
  team, 423–426, 439
Product marketing, 218, 219–220
Product teams, 389–390, 486
Professional Employment Organization (PEO), 55, 155–156, 160, 167
Professional services, impact, 362
Programmatic story-telling sessions, 150
Projected growth, discussion, 479
Project management, 92, 184
Promotion, factors, 447–448
Proportional approach (product development culture), 415
Proportional engineering investment, 416
Pro-rata investing rights (transaction item), 35–36
Public relations (PR), 214–215, 230–232

Quality assurance (QA) teams, 457–459
Quality of earnings analysis, 89
Quarterly business review (QBR), 383

Real estate, CFO/CEO management/partnership, 69–70
Real estate leases, 88
Recruitment, 142, 147, 186–188
Regret score, measurement, 376
Remote culture, embracing, 106
Renewal (customer journey stage), 369, 377–378
Reorganization, problems, 140
Reporting, 184, 223
Responsible-accountable-consulted-informed (RACI), 13–14, 461
Return on investment (ROI), 97, 169, 171, 185, 197–198, 210–211, 223, 479
Return Path (RP), 19–20, 102–104, 370
  acquisitions, 287, 313, 333
  annual planning framework, adoption/modification, 124
  benefits, 156–160
  Board meeting, 9–10
  brand, appreciation/evaluation, 192, 194
  changes, 121
  channels, 60, 335–336
  CMO promotions, 247

core business process, development, 304
data, protection/breach, 498, 511–512
DE&I focus, 112
due diligence processes, 532–533
Expectations, 132, 145
hiring model, adoption, 258
historical data, usage, 206
initiation, 91
internal promotion bias, 148
internship, 443–444
leadership development bootcamps, 450
marketing operations function development, 185
office manager promotion, 114
omni-channel approach, 238
operating system, change, 122–123
organizational interface, 463
organizational structure, changes, 120
People Pact employee handbook, 162
performance, 131
Performance Aspire (annual event), 139, 272
Privacy Council, initiation, 520–521
problem-solving audition, 435–436
Product Development team values, 425–426
re-entry program, 444
reports generation, 363–364
Request a Demo button, usage, 195–196
Returnship/Return to Work programs, 180
return-to-work program, creation, 486
rewards system, 138
sale, 319–320, 333
scaling problems, 460
senior leader programs, 129
service roles, 133
team diversity, 442
technical debt, 416–417, 462
values, 108–109
volunteers, usage, 117
Return Path (RP) product/sales, 2, 58, 253, 392–393, 420, 480, 497
Returnship, 444–445
Return-to-work program, creation, 146, 486
Revenue attribution, determination, 56–57
Revenue controller, hiring, 53
Revenue makeup (financial statement component), 27
Revenue recognition, 46, 56, 138
Risk assessment, 472–474
Role-specific learning/development, 134
Roll-up tables, Data Pipeline team creation, 432

Sabbatical program (employee benefit), 157–158
Safe case scenarios, assessment, 280
Sales, 266
commission, accounting perspective, 47
demand, generation, 181
development, 226–230
fake situation, creation, 255
machine, construction, 253
marketing, combination, 207
math, usage, 257–258
occurrence, 59–61
organization, scaling/onboarding, 268, 388
process, lessons, 468
process/methodology, scaling, 276
representatives, engagement, 485
role, CEO-to-CEO advice, 300
taxes, 43–44, 47
Sales accepted leads (SALs), 185
Sales development representative (SDR), 226, 228–230, 269, 283–284
Salesforce, 355, 380, 432, 508
Salespeople, 257, 260, 270, 274, 299, 465
Sales team, 218, 221–222, 262, 332–333, 370–371
Search Engine Optimization (SEO), solutions, 184
Secondary equity transactions, 82–84
Security issues, 476–477, 495–496, 513, 525
Segment analysis (financial statement component), 31
Self-hosted events, 205–206, 208–209
Self-starters, role, 455–456
Seller representations, sell-side consideration, 90
Sell-side (M&As), 86–90, 344
Senior leader programs, co-creation, 129
Senior personnel, hiring process, 143–144
Senior Software Engineer, role/responsibility, 448
Senior Team (pitch deck section), 24
Senior technical leadership roles, filling (difficulty), 450
Series A/Series B rounds/company, 34, 577
Service issues, 61–62, 89, 218, 221–222, 358, 364–365, 373
Service organization, 357–359, 362, 387
Shareholders, impact, 77, 90
Signpost, 363, 371, 393
Single sign-on process, setup, 517–518
Slack, usage, 91–92, 163, 188
Small and medium-sized business (SMB) sales, 268–269
Social media, usage, 198–199, 516–517
Social strategy, marketing responsibility, 198–199
Society for Human Resource Management (SHRM), usage, 161
Soft skills, training, 149, 383
Software-as-a-Service (SaaS), 32, 138, 196, 296, 355, 362, 430, 571
Software Engineer, role/responsibility, 448
Solution interviews, 182
Special projects, labeling, 315–316
Speed/quality, CEO/founder valuation, 107
Spreadsheets, CFO review, 98

Staffing, absence, 458
Standard Contractual Clauses (SCC), 506
Startups, impact, 71, 319, 360, 427, 432–433, 494, 499, 522
Stock compensation, 46, 80–81
Storytelling, 479
Strategic finance, 58
Strategic planning, marketing operations component, 223, 224–225
Strengthsfinder assessment, 126
Success League, The (consultant role), 586
Support tickets, summary (publication), 373
System and Organization Controls (SOC 2), 529–530
Systems, 68, 93, 98, 164, 426, 477, 517–518

Talent Acquisition/Recruiting, 115
Talent, issues, 130, 421
Teams, issues, 420–421, 455–476
  building, 113, 385
  Chief People Officer management/leadership problems, 169–170
  development, 124
  diversity, 434, 442
  growing, 236–237
  hiring, timing, 366
  players, impact, 274
  roles, 115t
  scaling, culture (usage), 271
  special treatment, avoidance, 272
  structure/prioritization, decisions, 61
Technical audition, 443
Technical debt, 389, 415–424, 431, 460–462, 470–471, 483, 484
Technical roles, 449–450
Technical strategy, 416
Technical updates, operational cost component, 418–419
Technology, issues, 183–184, 223–224, 380–382, 421, 436, 468–469, 473–474
Tender equity transactions, 82–84
Terms of Service (TOS) agreements, 503–504
Thinking sessions, usage, 246
Third-party bookkeeper, usage, 42
Third-party events, 205
Third-party systems, usage, 83
Thought leadership, 234–235
Tooling, efficiency, 421
Top-to-bottom alignment, 332
Total Addressable Market (TAM), 24–26, 286
Tracking systems metric, usage, 477–478
Training, internal policy (inclusion), 502
Transaction items, consideration, 35–36
Transaction specialist, hiring, 53–54
Transparency, impact, 106, 502
Travel and expense (T&E) software, usage, 43
Treasury management, 49
Trello, usage, 92, 93, 429
Two-factor authentication, 517

Unconscious biases, interruption, 128
Under-engineering style (product development culture), 414
Unit economics, 30
Upsell, 376–377, 392–393
Use phase metrics, 376–377
User consent policies, 476
User experience (UX), 373, 447
User interfaces (UI), 428–429
U.S. Privacy Act, 488
US Privacy Shield program, 506

Valuation, importance, 34
Values, 105, 108–109, 127, 439
Variable compensation, focus, 391–392
Variable costs, understanding, 472
Vendor agreements, 88
Venture-backed companies, post-termination exercise period, 80–81
Venture capitalists (VCs), 82, 556, 573–574
Video-conferencing areas, setup, 163
Voxy, 288–289, 291–292

Waterfall methodologies, Agile methodologies (mismatch), 463–464
Web analytics, Databox (usage), 93
Websites, marketing responsibility, 195–197
Well-being day (employee benefit), 160
Whiteboard, usage, 251, 253, 278, 300–301
Women, re-entry programs, 444
Workforce, reductions, 140–141
Working capital adjustment, sell-side consideration, 90–91
Work-life balance, CEO/founder perspective, 107

Yahoo purchase price, Verizon reduction, 532
Year-to-date (YTD) income statement, inclusion, 27
Yesterday work, 423
Yield, chasing difficulty (consideration), 49

Zoom, 92–93, 95, 163, 304, 560